ADVANCED GEOTECHNICAL ANALYSES

*Developments in Soil Mechanics
and Foundation Engineering – 4*

Developments in Soil Mechanics and
Foundation Engineering:

ADVANCED GEOTECHNICAL ANALYSES

Developments in Soil Mechanics and Foundation Engineering – 4

Edited by

P.K. BANERJEE

Department of Civil Engineering,
State University of New York at Buffalo,
New York, USA
and

R. BUTTERFIELD

Department of Civil Engineering,
University of Southampton, UK

CRC Press
Taylor & Francis Group
Boca Raton London New York

CRC Press is an imprint of the
Taylor & Francis Group, an **informa** business

A TAYLOR & FRANCIS BOOK

CRC Press
Taylor & Francis Group
6000 Broken Sound Parkway NW, Suite 300
Boca Raton, FL 33487-2742

First issued in paperback 2019

ISBN-13: 978-1-85166-623-2 (hbk)
ISBN-13: 978-0-367-86458-3 (pbk)

British Library Cataloguing in Publication Data

Developments in soil mechanics and foundation engineering.
  (Developments series)
  1. Soil mechanics
  I. Series
  624.1

Library of Congress Cataloging-in-Publication Data

LC card number: 86-657601

Photoset by Interprint Ltd, Malta.

Visit the Taylor & Francis Web site at
http://www.taylorandfrancis.com

and the CRC Press Web site at
http://www.crcpress.com

PREFACE

Geotechnical engineers have to deal with complex geometrical configurations as well as enormously difficult materials which exhibit, strongly, a path-dependent mechanical behavior. In addition, geological deposits display extensive inhomogeneities which are often difficult to define quantitatively. As a result most geotechnical engineering design problems require significant use of the engineer's imagination, creativity, judgment, common sense and experience. To many geotechnical engineers therefore the role of any advanced analysis, particularly advanced computer based analyses, remains undefined. The editors have therefore invited some outstanding engineers who are engaged not only in developing advanced level geotechnical analyses, but are also in consulting practice to write various chapters of this book. These chapters show that a careful blend of engineering judgment and advanced principles of engineering mechanics may be used to resolve many complex geotechnical engineering problems. It is hoped that these may inspire geotechnical engineering practice to make more extensive use of them in the future.

Because of the difficulties associated with complex geometries and material behavior it is not surprising that the advanced analyses described in this book make extensive use of modern digital computers. Simplified hand calculations, although they have the attraction of being very good teaching tools, are rarely able to quantitatively reproduce the complete physical characteristics of the problem.

Chapter 1 deals with the complex interactions between fluid and solid skeletons for both static and dynamic loading. The governing equations for the solid and fluid constituents have been set out in a general manner and a nonlinear transient finite element formulation for the problem developed. A centrifuge model test of a dike is then simulated by the analysis, and the success of the developed analysis was demonstrated by the ability of the analytical model to reproduce the physical observations in the centrifuge model.

Because the method of construction used has often significant influence on the mechanical behavior of geotechnical engineering structures, the next three chapters describe modifications to the finite element formulation to take account of construction-induced events. Chapter 2 investigates the effects of compaction-induced stresses on the behavior of retaining walls and buried flexible culverts. Chapter 3 describes the use of an anisotropic soil model in the simulation of excavations. Such unloading problems in saturated clay, which result in a transient safety factor diminishing with time, can create dangers on many construction sites. Chapter 4 shows the finite element analysis of either cantilever, propped or anchored retaining walls in which effects of construction methods (excavated or backfilled) are considered. The chapter concludes with the application of the analysis to a complex embedded retaining wall.

The development and the use of the finite element method in analysing a number of penetration problems involving sampling tubes and piles are described in Chapter 5. Such analyses can not only provide quantitative information on the changes in soil state caused by the penetration but also enable one to extract characteristic soil parameters from the simulation of actual field tests such as cone penetration and pressure-meter tests.

A simplified analysis of the dynamic pile driving problem by a combination of one-dimensional wave equations for the pile with an approximate plan-wave propagation model for the soil has been considered in Chapter 6. Such analyses which include soil inertia effects represent a considerable improvement on conventional analysis of pile driving problems. In many situations the soil properties do not vary in one or two spatial directions and it is then possible to use the finite layer approach described in Chapter 7. By using an analytical representation of the field in the horizontal direction, such semi-analytical methods became very efficient.

The final two chapters describe both the development and use of explicit finite difference methods for analysing a wide range of geotechnical engineering problems involving both continuous as well as discontinuous jointed media. Such distinct element methods may prove to have much wider applications in the future.

The editors hope that these demonstrations of advanced analysis applied to geotechnical engineering problems might encourage engineers to consider incorporating them in their strategies. Perhaps equally important, such analyses might enable them to extrapolate more effectively experience gained from one geotechnical site to another.

P.K. BANERJEE
R. BUTTERFIELD

CONTENTS

LIST OF CONTRIBUTORS

P.K. BANERJEE
Department of Civil Engineering, State University of New York at Buffalo, NY 14260, USA

J.R. BOOKER
School of Civil and Mining Engineering, University of Sydney, Sydney, NSW, 2006, Australia

A.H.C. CHAN
Department of Civil Engineering, University of Glasgow, Glasgow G12 8QQ, UK

M.B. CHOPRA
Department of Civil Engineering, State University of New York at Buffalo, NY 14260, USA

G.F. DARGUSH
Department of Civil Engineering, State University of New York at Buffalo, NY 14260, USA

J.M. DUNCAN
Department of Civil Engineering, Virginia Polytechnic Institute and State University, Blacksburg, Virginia 24061, USA

R.M. HARKNESS
Department of Civil Engineering, University of Southampton, Highfield, Southampton SO9 5NH, UK

A. KUMBHOJKAR
Department of Civil and Architectural Engineering, University of Miami, Coral Gables, Florida, USA

N.C. LAST
Geomechanics Section, Reservoir Appraisal Branch, BP International Ltd, Sunbury Research Centre, Chertsey Road, Sunbury-on-Thames, Middlesex TW16 7LN, UK

C.Y. OU
Department of Construction Engineering and Technology, National Taiwan Institute of Technology, 43 Keelung Road, Sec. 4, Taipei, Taiwan

M. Pastor
Laboratoire de Geotecnica, Cedex 28014, Madrid, Spain

D.M. Potts
Department of Civil Engineering, Imperial College of Science, Technology and Medicine, University of London, Imperial College Road, London SW7 2BU, UK

M.F. Randolph
Department of Civil and Environmental Engineering, The University of Western Australia, Nedlands, Western Australia 6009

R.B. Seed
Department of Civil Engineering, University of California at Berkeley, Berkeley, California 94720, USA

J.C. Small
School of Civil Mining Engineering, University of Sydney, Sydney, NSW, 2006, Australia

Y.M. Xie
Department of Civil Engineering, University College of Swansea, University of Wales, Singleton Park, Swansea SA2 8PP, UK

N.B. Yousif
Department of Civil Engineering, University of Basra, Basra, Iraq

O.C. Zienkiewicz
Department of Civil Engineering, University College of Swansea, University of Wales, Singleton Park, Swansea SA2 8PP, UK

Chapter 1

COMPUTATIONAL APPROACHES TO THE DYNAMICS AND STATICS OF SATURATED AND UNSATURATED SOILS*

O.C. Zienkiewicz,[a] M. Pastor,[b] A.H.C. Chan[c] and Y.M. Xie[a]

[a]Department of Civil Engineering, University College of Swansea, University of Wales, Singleton Park, Swansea SA2 8PP, UK
[b]Laboratorio de Geotecnia, Cedex 28014, Madrid, Spain
[c]Department of Civil Engineering, University of Glasgow, Glasgow G12 8QQ, UK

ABSTRACT

The behaviour of all geomaterials, and in particular of soils, is governed by their interaction with the fluid (water) and gas (air) present in the pore structure. The mechanical model of this interaction when combined with suitable constitutive description of the porous medium and with efficient, discrete, computation procedures, allows most transient and static problems involving deformations to be solved. This chapter describes the basic procedures and the development of a general purpose computer program (SWANDYNE-X) for static and dynamic analyses of saturated and semi-saturated soils. The results of the computations are validated by comparison with experiments. An approximate reconstruction of the failure of the Lower San Fernando Dam during the 1971 earthquake is presented.

*Much of the text and figures have been published in Zienkiewicz *et al.* (1990a, 1990b) and are printed here by permission of the Royal Society.

1 INTRODUCTION

The mechanical behaviour of saturated geomaterials in general, and of soils in particular, is governed largely by the interaction of their solid skeleton with the fluid, generally water, present in the pore structure. This interaction is particularly strong in dynamic problems and may lead to a catastrophic softening of the material known as liquefaction which frequently occurs under earthquake loading. Figure 1 illustrates a typical

FIG. 1.   Liquefaction of soil and collapse of buildings at Niigata, Japan, 1964.

problem encountered in an earthquake. Although this example is rather dramatic, the interaction is present in more mundane, quasi static situations typical of the foundation behaviour of most engineering structures and a quantitative prediction of the phenomena resulting in permanent deformation (or unacceptably high pore pressure increases) is a necessity if safe behaviour of such structures is to be guaranteed. In addition, of course, the phenomena are of interest to geophysicists (and geographers) studying the behaviour of surface deposits.

The two phase behaviour just described allows the solution of many problems of practical interest, but is not adequate in others where semi-saturated conditions exist. In particular, if negative fluid pressures develop, dissolved air is released from the fluid or simply enters into the mixture via the boundaries and thus both air and water fill the voids. Indeed it is this semi-saturated state that permits the negative pressures to be maintained through the mechanism of capillary forces. Such negative pressures provide a certain amount of 'cohesion' in otherwise cohesionless, granular matter and are necessary to account for realistic behaviour of only partly saturated embankments under dynamic forces.

The saturated behaviour is fundamental and, though understood in principle for some considerable time, can only be predicted quantitatively by elaborate numerical computations, which fortunately today is possible due to the developments of powerful computers. It is the aim of this chapter to present a full account of the development of such numerical procedures and to extend such formulations to problems of semi-saturated behaviour with a simplifying assumption concerning the air flow. The results of the computations are validated by comparison with model experiments. Such validation is of course essential to convince the sceptics and indeed to show that all stages of the mathematical modelling are possible today. It is necessary to generate a predictive capacity which in general, due to the scale of the phenomena, cannot be accurately tested in the laboratory.

The full modelling involves several stages each introducing some degree of approximation. These are

(a) establishment of a mathematical framework adequately describing the phenomena,
(b) establishment of numerical (discrete) approximation procedures,
(c) establishment of constitutive models for the behaviour of the components.

To each of these stages we shall devote a section of this chapter indicating the current 'state of the art' and presenting the, generally novel, procedures described in this chapter.

The formulation will be given in full dynamic context, which presents the most difficult situation. However, such phenomena as slow consolidation or even purely static behaviour will be immediately available from the solution as special cases.

The procedure presented here forms the essential stepping stone for formulations of multiphase behaviour. Indeed, the extension of such a procedure to three phase behaviour has been given in Li *et al.* (1990) and Zienkiewicz *et al.* (1991).

2   THE MATHEMATICAL FRAMEWORK FOR THE BEHAVIOUR OF SATURATED POROUS MEDIA

The essence of the mathematical theory governing the behaviour of saturated porous media with a single fluid phase was established first by Biot (1941, 1955, 1956) for linear elastic materials by a straightforward physical approach. Later the 'theory of mixtures' using more complex arguments confirmed the essential correctness of his findings (Green and Adkin, 1960; Green, 1969; Bowen, 1976).

The Biot theory was extended to deal with non-linear material behaviour and large strain effects by Zienkiewicz (1982) and Zienkiewicz and Shiomi (1984), and the basis, with the derivation of the essential equations governing the phenomena, is summarized below.

The mechanics of all geomaterials and indeed of other porous media is conveniently described on a macroscopic scale assuming the size of solid grains and pores to be small compared with the dimensions considered. This allows averaged variables to be used and we list below the most essential ones:

Total stress σ_{ij}: This is defined by considering in the usual manner the resultant forces acting on unit sections of the solid fluid ensemble. This definition will be applied to the current, deformed state and the Cauchy stress is written as σ_{ij} using the usual indicial notation for Cartesian axes x_i, x_j. A positive sign for tension is assumed.

Solid matrix displacement u_i: This defines the mean displacement of particles forming the solid matrix in the co-ordinate direction x_i.

Pore pressure p: This characterises the mean stress in the fluid phase. Of course deviatoric stresses exist in the fluid on a microscale but this overall effect can be represented by viscous drag forces exerted by the fluid on the solid phase which will be accounted for by the usual Darcy (1856) expression later. Pressure is defined as positive in the compressive sense.

Mean fluid velocity relative to the solid phase w_i: This is conveniently measured as the ratio of the fluid flow over the gross, deformed, cross-sectional area. It is important to note that the average relative velocity of the fluid particles is in fact w_i/n where n is the porosity of the solid.

With the above definitions we can proceed to individual basic relationships which will govern the problem analysis.

2.1 The Concept of Effective Stress

It is intuitively clear and also observable in experiments that when a sample of a solid porous medium is subjected to a uniform (external and internal) fluid pressure increase, only very small deformation occurs and this is due only to the elastic compression of the solid phase. In soil mechanics this leads to the concept of *effective stress* introduced by Terzaghi (1943) and Skempton (1961). This stress is defined as

$$\sigma'_{ij} = \sigma_{ij} + \delta_{ij}p \tag{1}$$

and σ'_{ij} is deemed to be responsible for all major deformations and rupture of the 'soil skeleton'. To account for the slight (volumetric) strain changes, a modification can be introduced to the above definition as shown by Biot and Willis (1957) and Zienkiewicz (1982), that is

$$\sigma''_{ij} = \sigma_{ij} + \alpha\delta_{ij}p \tag{2}$$

where

$$\alpha \approx 1 - K_T/K_S \leqslant 1 \tag{3}$$

and K_T and K_S are the bulk moduli of the porous medium and solid grains, respectively. While for soils the ratio of the two deformabilities is such that $\alpha \Rightarrow 1$ and the original effective stress definition is useful, for rocks and concrete, a coefficient α as low as 0·5 has been recorded and therefore the second definition is preferable (Zienkiewicz, 1982; Zienkiewicz and Shiomi, 1984).

In what follows we shall use the second effective stress definition and write all the constitutive laws with respect to this incrementally. We will present such laws as

$$d\sigma''_{ij} = D_{ijkl}\left[d\varepsilon_{kl} - d\varepsilon^\circ_{kl}\right] + \sigma''_{ik}\,d\omega_{kj} + \sigma''_{jk}\,d\omega_{ki} \tag{4}$$

where the last two terms account for the Zaremba – Jaumann rotational stress changes (negligible generally in small displacement computation) and D_{ijkl} is a tangential matrix defined by suitable state variables and the direction of the increment.

The incremental strain ($d\varepsilon_{ij}$) and rotation ($d\omega_{kl}$) components are defined in the usual way from the solid phase displacement as

$$d\varepsilon_{ij} = \tfrac{1}{2}[du_{i,j} + du_{j,i}]$$

$$d\omega_{ij} = \tfrac{1}{2}[du_{i,j} - du_{j,i}] \tag{5}$$

and ε_{ij}° refers to strains caused by external actions such as temperature changes, creep, etc.

2.2 Equilibrium and Continuity Relationships

We start by writing the total momentum equilibrium equations for the solid/fluid ensemble. For a unit volume we can write these in terms of the total stress as

$$\sigma_{ij,j} - \rho\ddot{u}_i - \underline{\rho_f[\dot{w}_i + w_j w_{i,j}]} + \rho b_i = 0 \tag{6}$$

where $\dot{w}_i \equiv d(w_i)/dt$ etc.

In the above we have assumed that the coordinate system moves with the solid phase (material coordinates) and hence convective acceleration in terms of relative velocity applies only to the fluid phase. The density ρ_f is that of the fluid and ρ is the density of the mixture written as

$$\rho = n\rho_f + (1-n)\rho_s \tag{7}$$

where ρ_s is the density of the solid phase; b_i is the body force per unit mass.

In eqn (6) we have underlined terms which in an approximate theory can be conveniently omitted and we shall pursue this notation in the present section.

The second governing equation is that defining the momentum equilibrium for the fluid alone. Again for a unit control volume (assumed attached to the solid phase and moving with it) we can write

$$-p_{,i} - R_i - \rho_f\ddot{u}_i - \underline{\rho_f[\dot{w}_i + w_j w_{i,j}]/n} + \rho_f b_i = 0 \tag{8}$$

In the above, R_i represents the viscous drag force which, assuming the validity of the Darcy seepage law (Darcy, 1856), is given by

$$k_{ij} R_j = w_i \tag{9}$$

where k_{ij} defines the, generally anisotropic, permeability coefficients. For isotropy these are conveniently replaced by a single k value.

We should note that in general the permeability may be a function of strain (via the changes of porosity) and of external temperature, that is

$$k_{ij} = k_{ij}[(\varepsilon_{ij} - \varepsilon_{ij}^{\circ}), T] \tag{10}$$

The final equation complementing above is one of flow conservation for the fluid phase. This can be written as

$$\dot{p}/Q + \alpha\dot{\varepsilon}_{ii} + w_{i,i} + \underline{(\dot{\rho}_f/\rho)w_i} + \dot{s}_0 = 0 \tag{11}$$

In the above Q represents the combined compressibility of the fluid and solid phases which can be related to the bulk moduli of each component as (Biot and Willis, 1957; Zienkiewicz, 1982).

$$1/Q = n/K_f + (\alpha - n)/K_s \qquad (12)$$

The last term $\dot{s}_0$ represents the rate of volume changes of the fluid such as may be caused by thermal changes, etc.

The above equations, valid in the saturated domain of the problem, govern both static and dynamic behaviour phenomena. When the constitutive parameters are defined these equations can be solved, as shown later, by a suitable numerical scheme providing appropriate boundary and initial conditions are correctly imposed.

The initial condition will generally specify the full field of u_i, w_i and p and the boundary conditions must define

(i)   the values of u_i or the corresponding total traction component and
(ii) the values of pressure p or corresponding rate of flow w_n (in the normal direction to the boundary).

The solution of the full equations in which u_i, w_i and p remain as variables is expensive and only necessary when high frequency phenomena are dealt with. For the majority of geomechanical problems we can omit the terms underlined in the various equations and arrive at a reduced system which leads to major computational economies. This simplified system is described in the next section.

2.3   The Simplified Governing Equations
The three governing equations (6), (8) and (11) together with the ancillary definitions (2), (4), (5) and (9) present a well-defined problem whose numerical solution has been discussed by Zienkiewicz and Shiomi (1984). When acceleration frequencies are low as in the case in earthquake motion, the underlined terms in eqns (6), (8) and (11) involving the relative acceleration of the fluid are not important and can be omitted as shown by Zienkiewicz and Bettess (1982) and Zienkiewicz et al. (1980b).

The omission of the underlined terms allows w_i to be eliminated from the equation system retaining only u_i and p as primary variables. In what follows we use a compromise writing r_i for underlined terms in eqn (6) and l_i for those from eqn (8) and omitting the underlined term from eqn (11) which is always insignificant. This allows an iterative correction (or at last an assessment of error) at any stage of the subsequent numerical computation.

With this simplification we can write the governing equation system as

$$\sigma_{ij,j} - \rho \ddot{u}_i + \rho b_i - r_i = 0 \tag{13}$$

and, by using eqns (8) and (9) to eliminate w_i from eqn (11),

$$\alpha \dot{\varepsilon}_{ii} - (k_{ij}p_{,j})_{,i} + \dot{p}/Q + [k_{ij}(\rho_f b_j)]_{,i} + s_0 - [k_{ij}(\rho_f \ddot{u}_j + l)]_{,i} = 0 \tag{14}$$

The above system with the ancillary definitions linking the stresses and strains to displacements u_i (i.e. eqns (1)–(5)) and the boundary conditions can be discretized and solved numerically using only two sets of primary variables u_i and p.

3 THE MATHEMATICAL FRAMEWORK FOR THE BEHAVIOUR OF SEMI-SATURATED POROUS MEDIA

In the semi-saturated state the voids of the skeleton are filled partly with water (or other fluid) and partly with air (or other gas). Denoting the respective degrees of saturation by S_w and S_a we observe that

$$S_w + S_a = 1 \tag{15}$$

If the pressures in the water and air are p_w and p_a respectively then

$$p_a - p_w = p_c \tag{16}$$

where p_c denotes a 'capillary pressure' difference. This effect of capillary forces is clearly only dependent on the intermaterial surface tensions and the geometry of the surfaces and hence on the saturation. Under isothermal conditions, for a given granular material and specific void ratio we can therefore assume that a unique function defines

$$p_c = p_c(S_w) \quad \text{or} \quad S_w = S_w(p_c) \tag{17}$$

With the above assumption we disregard a slight hysteretic path dependence when different pressures may develop at the same saturation depending on whether a decrease or increase of saturation is taking place.

Further if water or air flow occurs in the respective phases the permeability coefficients will, by similar arguments, be again unique functions of S_w. Thus for instance

$$k_w = k_w(S_w) \tag{18a}$$

and

$$k_a = k_a(S_w) \tag{18b}$$

The determination of the relationships (17) and (18) has been a subject of extensive studies (namely Liakopoulos, 1965; Neuman, 1975; Van Genuchten *et al.*, 1977; Narasimhan and Witherspoon, 1978; Safai and Pinder, 1979; Lloret and Alonso, 1980; Bear *et al.*, 1984; Alonso *et al.*, 1987). In Fig. 2 we illustrate some typical results.

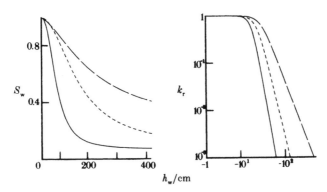

FIG. 2. Typical relations between pore pressure head, $h_w = p_w/\gamma_w$, saturation, S_w, and relative permeability, $k_r = k_w(S_w)/k_w(1)$ (Van Genuchten *et al.*, 1977). $S_w = \delta + (1-\delta)[1 + (\beta|h_w|)^\gamma]^{-1}$, $k_r = [1 + (a|h_w|)^b]^{-\alpha}$. ———, sand ($\delta = 0.0689$, $\beta = 0.0174$, $\gamma = 2.5$, $a = 0.0667$, $b = 5$, $\alpha = 1$); ---, loam ($\delta = 0.2$, $\beta = 0.00481$, $\gamma = 1.5$, $a = 0.04$, $b = 3.5$, $\alpha = 0.64$); ----, used in present San Fernando dam analysis ($\delta = 0.0842$, $\beta = 0.007$, $\gamma = 2$, $a = 0.05$, $b = 4$, $\alpha = 0.9$). β and a have units cm^{-1}.

3.1 Modification of Effective Stress

Following similar arguments to those used to establish the effective stress concept under saturated conditions it is reasonable to assume that we can define now a new effective stress as

$$\sigma'_{ij} = \sigma_{ij} + \delta_{ij}(S_w p_w + S_a p_a) \qquad (19a)$$

Certainly the term equivalent to pressure in eqn (1) is now simply given by the mean pressure of water and air exerted on the particles and indeed in the limiting case of $S_w = 1$ the original definition (1) is retrieved. Bishop (1959) appears to be the first to use and justify an effective stress thus defined.

The concept of defining the material behaviour in terms of the new effective stress is by no means universally accepted. Arguments against its use have been cogently summarised by Lloret *et al.* (1986). In particular, as pointed out by these authors, it is noted that on first 'wetting' of soil, irreversible compressive volumetric strains occur if the soil is subjected to compressive total tress – a result apparently at variance with the expected volumetric expansion which would occur due to the reduction of negative pore pressures. However, this effect decreases on subsequent wet-ting/drying cycles and although in principle it is possible to include it by a slight modification of the constitutive relation, in the present examples we shall simply assume that the relationship (19a) holds together with the appropriate, general, constitutive law valid in both saturated and non-saturated zones.

A further criticism of the relation (19a) is that it does not take into account the relative 'wettability' of the two fluid phases with the solid skeleton. Such wettability does in fact determine the effective contact areas and hence the definition of the true stresses in the solid matrix. A possible alternative form of relation (19a) is given below in which f_w and f_a are appropriate functions of the saturation.

$$\sigma'_{ij} = \sigma_{ij} + \delta_{ij}(f_w p_w + f_a p_a) \tag{19b}$$

where $f_w = f_w(S_w)$, $f_a = f_a(S_a)$ and $f_w + f_a = 1$.

Although it would be a simple matter to adjust the following formula-tions to use relation (19b), the lack of experimental data precludes its current use.

Before establishing the final equations requiring solution which in a full three phase mixture would require the consideration of air as well as water flow we introduce here a further simplifying assumption. This implies that the resistance to the flow of air is so small that at all points of the system the air pressure is zero (ambient external, atmospheric, value).

Putting thus $p_a = 0$ the effective stress can be defined in the manner analogous to that of equation (2) as

$$\sigma''_{ij} = \sigma_{ij} + \alpha \delta_{ij} S_w p_w \tag{20}$$

This definition allows us to make necessary modifications to the equations of saturated behaviour to deal with semi-saturated media. As S_w and p_w are uniquely related no additional variables are introduced in the final solution as shown in the next section.

3.2 Modification of Governing Equations
We shall, in what follows, use the simplified governing equations of section 2.3 omitting the small, underlined, terms. First we note that the overall equilibrium equation (13) remains unchanged in terms of the total

stress as

$$\sigma_{ij,j} - \rho \ddot{u}_i + \rho b_i - r_i = 0 \tag{21}$$

In the above, the new definition of effective stress (20) needs to be used before the constitutive relation via eqn (4) is employed and the density of the soil mixture ρ in eqn (7) is now rewritten, neglecting the weight of air, as

$$\rho = S_w n \rho_f + (1 - n) \rho_s \tag{22}$$

However more substantial modifications need to be introduced to the governing equation of flow. Here the starting point is provided by eqns (8)–(11), omitting the negligible terms for clarity. The fluid momentum equation (8) remains unchanged as

$$-p_{,i} - R_i - \rho_f \ddot{u}_i + \rho_f b_i = 0 \tag{23}$$

putting

$$p_w \equiv p \tag{24}$$

in the above and omitting the suffix in what follows.

Similarly no changes are introduced in the Darcy seepage law of eqn (9), i.e.

$$k_{ij} R_j = w_i \tag{25}$$

However the continuity equation (11) must now be modified. Here the most important change is the addition of the storage term due to changes of saturation S_w, i.e.

$$n \frac{\partial S_w}{\partial t} = n \frac{\partial S_w}{\partial p} \frac{\partial p}{\partial t} = C_s \frac{\partial p}{\partial t} = C_s \dot{p} \tag{26}$$

In the above C_s is the so-called specific moisture capacity which can readily be determined from the knowledge of the porosity n and the slope of the curve relating S_w and p (see Fig. 2).

With this modification the continuity equation (11) can still be written in a similar form as

$$\dot{p}/Q^* + \alpha \dot{\varepsilon}_{ii} + w_{i,i} + \dot{s}_0 = 0 \tag{27}$$

but Q^* is now redefined as

$$\frac{1}{Q^*} = C_s + n \frac{S_w}{K_f} + (\alpha - n) \frac{S_w}{K_s} \left(S_w + C_s \frac{p}{n} \right) \tag{28}$$

This value is now strongly dependent on p in semi-saturated conditions but, of course, becomes identical to that given by eqn (12) when $S_w = 1$.

The derivation of the above expression can be found in Xie (1990) though of course the origin of the first two terms is self evident. Indeed $K_s \Rightarrow \infty$ and the compressibility of the solid phase is negligible in most soils.

Elimination of w_i from eqn (27) by the use of eqns (23) and (25) results finally in a form identical to that of (14), i.e.

$$\alpha \dot{\varepsilon}_{ii} - (k_{ij} p_{,j})_{,i} + \dot{p}/Q^* + \dot{s}_0 + (k_{ij} \rho_f b_j)_{,i} - (k_{ij} (\rho_f \ddot{u}_j + 1))_{,i} = 0 \qquad (29)$$

With $\mathbf{u}$ and $p = p_w$ the only variables remaining, the determination of S_w, k_w and C_s is accomplished by supplying the physical data of the type shown in Fig. 2 in the form of approximate expressions or simply by interpolation from given curves.

The general code so extended allows both fully and partially saturated regions to be treated simultaneously.

It is of interest to remark that the highly non-linear variation of permeability with the negative pore pressures which describes a physical reality (as shown in Fig. 2) is in fact frequently used as a pure computational device to solve steady state seepage problems with a 'free' surface. Such a 'free', phreatic, surface is generally assumed to separate the 'wet', saturated, zones in which flow occurs from 'dry' regions with no flow. The drop of permeability in the semi-saturated zone is such that the flow is reduced to a negligible amount there and the contour of $S_w = 1$ is an approximation of the phreatic surface but certainly the soil zone above it is not dry.

Desai (1976), Bathe and Khoshgoftan (1979) and Desai and Li (1983) have introduced this artifice for the solution of steady state seepage problems. In a subsequent example we shall show how the present formulation achieves this naturally.

4   THE DISCRETE APPROXIMATION AND ITS SOLUTION

To obtain a numerical solution a suitable discretization process is necessary. We shall use here the finite element procedures for both spatial and time discretization. The general notation follows the text of Zienkiewicz and Taylor (1989) in which for compactness we use a vector notation in place of indices and thus replace u_i by $\mathbf{u}$, σ_{ij} by $\boldsymbol{\sigma}$ etc. (This is not necessarily the optimal form for computation in which we have in fact retained the index notation).

First we introduce a spatial approximation writing this in the form

$$\mathbf{u} \approx u(t) = \mathbf{N}^u \bar{\mathbf{u}}(t) \tag{30a}$$

and

$$p \approx p(t) = \mathbf{N}^p \bar{\mathbf{p}}(t) \tag{30b}$$

where $N^{u/p}$ are approximate 'basis' or 'shape' functions and $\bar{\mathbf{u}}$ and $\bar{\mathbf{p}}$ list a finite set of parameters. Such parameters may (but not must) correspond to nodal values of the appropriate variables.

The governing equations (21) and (29) can now be transformed into a set of algebraic equations in space with only time derivatives remaining by the use of an appropriate Galerkin (or weighted residual) statement. This permits the approximation to satisfy the equations in an integral, mean, sense.

Thus, pre-multiplying eqn (21) by $\mathbf{N}^{u^T}$ and integrating over the spatial domain results (after the use of Green's theorem and insertion of boundary conditions) in

$$\int_\Omega \mathbf{B}^T \sigma \, d\Omega + \left(\int_\Omega \mathbf{N}^{u^T} \rho \mathbf{N}^u \, d\Omega \right) \ddot{\bar{u}} = \mathbf{f}^{(1)} \tag{31}$$

where $\mathbf{B}$ is the well-known strain matrix relating, by use of expression (5), the increments of strain and displacement, that is

$$d\varepsilon = \mathbf{B} \, d\bar{u} \tag{32}$$

The 'load' vector $\mathbf{f}^{(1)}$, equal in size to vector $\mathbf{u}$, contains the body forces, boundary tractions, omitted error terms and prescribed boundary values as shown below.

$$\mathbf{f}^{(1)} = \int_\Omega \mathbf{N}^{u^T} \rho \mathbf{b} \, d\Omega - \int_\Omega \mathbf{N}^{u^T} \mathbf{r} \, d\Omega + \int_{\Gamma_t} \mathbf{N}^{u^T} \bar{\mathbf{t}} \, d\Gamma \tag{33}$$

where $\bar{\mathbf{t}}$ is the prescribed traction on part of the boundary Γ_t.

Using the definition giving the effective stresses (eqn (20)) and introducing certain abbreviations, eqn (31) can be rewritten as

$$\mathbf{M}\ddot{\bar{u}} + \int_\Omega \mathbf{B}^T \sigma'' \, d\Omega - \mathbf{Q}\bar{\mathbf{p}} - \mathbf{f}^{(1)} = 0 \tag{34}$$

where

$$\mathbf{M} \equiv \int_\Omega \mathbf{N}^{u^T} \rho \mathbf{N}^u \, d\Omega \tag{35a}$$

is the well-known mass matrix and

$$Q = \int_\Omega \mathbf{B}^T S_w \alpha \mathbf{m} N^p \, d\Omega \qquad (35b)$$

is a coupling matrix with $\mathbf{m}$ being a vector equivalent to the Kronecker δ_{ij}.

The above discrete governing equation contains implicitly the two unknown parameters $\bar{\mathbf{u}}$ and $\bar{\mathbf{p}}$ only, as the increments of stresses are given by the constitutive relation in terms of displacement increments. That relation can be rewritten in the vectorial notation as

$$d\boldsymbol{\sigma}'' = \mathbf{D}[\mathbf{B} \, d\bar{\mathbf{u}} - d\varepsilon^\circ] + \mathbf{A}^T \boldsymbol{\sigma}'' \, \mathbf{A}\mathbf{B} \, d\bar{\mathbf{u}} \qquad (36)$$

where $\mathbf{A}$ is a suitably defined matrix operator. The above allows $\boldsymbol{\sigma}''$ to be continuously integrated from the known initial values of the problem.

The second governing equation (eqn (29)) and its boundary conditions are similarly discretized using now $\mathbf{N}^{p^T}$ as the weighting function and this results in a set of ordinary differential equations of the form

$$Q^T \dot{\bar{\mathbf{u}}} + \mathbf{H}\bar{\mathbf{p}} + \mathbf{S}\dot{\bar{\mathbf{p}}} - \mathbf{f}^{(2)} = 0 \qquad (37)$$

where Q is the matrix already defined in eqn (35b), $\mathbf{f}^{(2)}$ represents a 'force' vector, with dimension of $\bar{\mathbf{p}}$, incorporating body forces, error terms, the $\dot{s}_0$ term and boundary conditions.

The remaining matrices are defined below with

$$\mathbf{H} = \int_\Omega [\nabla N^p]^T \mathbf{k} \nabla N^p \, d\Omega \qquad (38a)$$

as the permeability matrix in which $\mathbf{k}$ is the matrix of coefficients in eqns (10) and (18a).

$$\mathbf{S} = \int_\Omega \mathbf{N}^{p^T} \frac{1}{Q^*} N^p \, d\Omega \qquad (38b)$$

is the compressibility matrix (frequently taken as zero) and

$$\mathbf{f}^{(2)} = -\int_{\Gamma_w} \mathbf{N}^{p^T} \mathbf{k} \frac{\partial p}{\partial n} \, d\Gamma + \int_\Omega (\nabla N^p)^T \mathbf{k}\rho_f \, \mathbf{b} \, d\Omega - \int_\Omega (\nabla N^p)^T \mathbf{k} l \, d\Omega$$
$$- \left[\int \left(\nabla N^p \right)^T \mathbf{k}\rho_f N^u \, d\Omega \right] \ddot{\bar{u}} - \int \mathbf{N}^{p^T} s_0 \, d\Omega \qquad (39)$$

To complete the numerical solution it is necessary to integrate the ordinary differential equations (34), (36) and (37) in time by one of the many available schemes. Various forms utilize the finite element concept in the time domain but here we shall use the simplest single step schemes available (Newmark, 1959; Katona and Zienkiewicz, 1985).

In all such schemes we shall write a recurrence relation linking known values of $\bar{u}_n$, $\dot{\bar{u}}_n$ and $\bar{p}_n$ available at time t_n with the values of $\bar{u}_{n+1}$, $\dot{\bar{u}}_{n+1}$, $\bar{p}_{n+1}$, which are valid at time $t_n + \Delta t$ and are the unknowns. If we assume that eqns (34) and (37) have to be satisfied at each discrete time and $\ddot{\bar{u}}_n$ and $\dot{\bar{p}}_n$ are added to the known conditions at t_n with $\ddot{\bar{u}}_{n+1}$ as $\dot{\bar{p}}_{n+1}$ remaining as unknown, we require that

$$\mathbf{M}\ddot{\bar{u}}_{n+1} - \left[\int_\Omega \mathbf{B}^T \sigma'' \, d\Omega\right]_{n+1} - \mathbf{Q}\bar{p}_{n+1} - f^{(1)}_{n+1} = 0 \tag{40a}$$

$$\mathbf{Q}^T \dot{\bar{u}}_{n+1} + \mathbf{H}_{n+1}\bar{p}_{n+1} + \mathbf{S}_{n+1}\dot{\bar{p}}_{n+1} - f^{(2)}_{n+1} = 0 \tag{40b}$$

and that eqn (36) is satisfied.

The link between the successive values is provided by a truncated series expansion taken in the simplest case as

$$\dot{\bar{u}}_{n+1} = \dot{\bar{u}}_n + \ddot{\bar{u}}_n \, \Delta t + \Delta \ddot{\bar{u}}_n \beta_1 \, \Delta t \tag{41a}$$

$$\bar{u}_{n+1} = \bar{u}_n + \dot{\bar{u}}_n \, \Delta t + \ddot{\bar{u}}_n \, \Delta t^2/2 + \Delta \ddot{\bar{u}}_n \beta_2 \, \Delta t^2/2 \tag{41b}$$

$$\bar{p}_{n+1} = \bar{p}_n + \dot{\bar{p}}_n \, \Delta t + \Delta \dot{\bar{p}}_n \theta \, \Delta t \tag{41c}$$

where

$$\Delta \ddot{\bar{u}}_n = \ddot{\bar{u}}_{n+1} - \ddot{\bar{u}}_n \quad \text{and} \quad \Delta \dot{\bar{p}}_n = \dot{\bar{p}}_{n+1} - \dot{\bar{p}}_n$$

are as yet undetermined quantities. The parameters β_1, β_2 and θ are chosen in the range of 0–1 but for unconditional stability of the recurrence scheme we require (Zienkiewicz and Taylor, 1985; Chan, 1988)

$$\beta_2 \geqslant \beta_1 \geqslant \tfrac{1}{2} \quad \text{and} \quad \theta \geqslant \tfrac{1}{2}$$

and their optimal choice is a matter of computational convenience, the discussion of which can be found in the literature.

Insertion of the relationships (41) into eqn (40) yields a general non-linear equation set in which only $\Delta \ddot{\bar{u}}_n$ and $\Delta \dot{\bar{p}}_n$ remain as unknowns. This can be written as

$$\psi^{(1)}_{n+1} = \mathbf{M}_{n+1} \Delta \ddot{\bar{u}}_n + \mathbf{P}(\bar{u}_{n+1}) - \mathbf{Q}_{n+1}\theta \, \Delta t \, \Delta \dot{\bar{p}}_n - \mathbf{F}^{(1)}_{n+1} = 0 \tag{42a}$$

$$\psi^{(2)}_{n+1} = \mathbf{Q}^T_{n+1}\beta_1 \Delta t \Delta \ddot{\bar{u}}_n + \mathbf{H}_{n+1}\theta \Delta t \Delta \dot{\bar{p}}_n + \mathbf{S}_{n+1}\Delta \dot{\bar{p}}_n - \mathbf{F}^{(2)}_{n+1} = 0 \tag{42b}$$

where $\mathbf{F}^1_{n+1}$ and $\mathbf{F}^2_{n+1}$ can be evaluated explicitly from the information at time n and

$$\mathbf{P}(\bar{\mathbf{u}}_{n+1}) = \int_\Omega \mathbf{B}_{n+1}^\mathsf{T} \sigma_{n+1}'' \, d\Omega = \int_\Omega \mathbf{B}^\mathsf{T} \Delta\sigma_n'' \, d\Omega + \mathbf{P}(u_n) \qquad (43)$$

where $\Delta\sigma_n''$ is evaluated by integrating eqn (36) and $\bar{\mathbf{u}}_{n+1}$ is defined by eqn (41b).

The equations will generally need to be solved by a convergent, iterative process using some form of Newton procedure typically written as

$$J \begin{Bmatrix} \mathrm{d}(\Delta\ddot{\mathbf{u}}_n) \\ \mathrm{d}(\Delta\dot{\bar{\mathbf{p}}}_n) \end{Bmatrix}^{l+1} = - \begin{Bmatrix} \psi_{n+1}^{(1)} \\ \psi_{n+1}^{(2)} \end{Bmatrix}^l \qquad (44)$$

where l is the iteration number.

The Jacobian matrix can be written as

$$\mathbf{J} = \begin{bmatrix} \partial\psi^{(1)}/\partial(\Delta\ddot{\mathbf{u}}_n) & \partial\psi^{(1)}/\partial(\Delta\dot{\bar{\mathbf{p}}}_n) \\ \partial\psi^{(2)}/\partial(\Delta\ddot{\mathbf{u}}_n) & \partial\psi^{(2)}/\partial(\Delta\dot{\bar{\mathbf{p}}}_n) \end{bmatrix} = \begin{bmatrix} \mathbf{M} + \mathbf{K}_\mathsf{T}\beta_2\,\Delta t^2/2 & -\mathbf{Q}\,\theta\,\Delta t \\ \mathbf{Q}^\mathsf{T}\beta_1\,\Delta t & \mathbf{H}_{n+1}\theta\,\Delta t + S_{n+1} \end{bmatrix} \qquad (45)$$

where

$$K_\mathsf{T} = \int_\Omega \mathbf{B}^\mathsf{T}\mathbf{D}_\mathsf{T}\,\mathbf{B}\,d\Omega + \int_\Omega \mathbf{B}^\mathsf{T}\mathbf{A}^\mathsf{T}\sigma''\,\mathbf{A}\mathbf{B}\,d\Omega \qquad (46)$$

which are well-known expressions for tangent stiffness and 'initial stress' matrices evaluated in the current configuration.

Two points should be made here:

(a) that in the linear case a single 'iteration' solves the problem exactly
(b) that the matrix $\mathbf{J}$ can be easily made symmetric by a simple scalar multiplication of the second row (providing K_T is itself symmetric).

In practice the use of various approximations for the matrix $\mathbf{J}$ is often advantageous and we have found the use of 'secant' updates of the Davidon (1968) form particularly useful.

A particularly economical computation form is given by choosing $\beta_2 = 0$ and representing $\mathbf{M}$ in a diagonal form. This explicit procedure first used by Leung (1984) and Zienkiewicz et al. (1980a) is however only conditionally stable and is efficient only for short duration phenomena.

The iterative procedure allows the determination of the effect of terms neglected in the $u - p$ approximation and hence an assessment of the accuracy.

The process of the time domain solution can be amended to that of successive separate solutions of eqns (42a) and (42b) for variables $\Delta\ddot{u}$ and $\Delta\dot{p}$, respectively, using an approximation for the remaining variable. Such *staggered* procedures, if stable, can be extremely economical as shown by Park and Felippa (1983) but the particular system of equations presented here needs stabilization. This was first achieved by Park (1983) and later a more effective form was introduced by Zienkiewicz *et al.* (1988).

It should be remarked that the basic form of solution for the two unknowns u and p remains unchanged whether the solid is fully saturated or partially saturated. If the pore pressure is positive (i.e. above atmospheric) full saturation (i.e. $S_w = 1$) is assumed. If the pore pressure becomes negative during the computation, partial saturation becomes immediately operative and $S_w < 1$ is fixed by the appropriate relationship (namely Fig. 2) and simultaneous change of permeability is recorded.

Special cases of solution are incorporated in the general solution scheme presented without any modification and indeed without loss of computational efficiency.

Thus for *static or quasi-static* problems, it is merely necessary to put

$$\mathbf{M} = 0$$

and immediately the transient *consolidation equation* is available. Here time is still real and we have omitted purely the inertia effects (though with implicit schemes this *a-priori* assumption is not necessary and inertia effects will simply appear as negligible without any substantial increase of computation).

In pure statics the time variable is still retained but is then purely an artificial variable allowing load incrementation.

In static or dynamic *undrained* analysis the permeability (and compressibility) matrices are set to zero, that is

$$\mathbf{H},\ \mathbf{f}^{(2)} = 0 \quad \text{and usually} \quad \mathbf{S} = 0$$

resulting in a zero diagonal term in the Jacobian matrix of eqn (45).

The matrix to be solved in such a limiting case is identical to that used frequently in the solution of problems of incompressible elasticity or fluid mechanics and in such studies places limitations on the approximating functions $\mathbf{N}^u$ and $\mathbf{N}^p$ used in eqn (30) if the so called Babuska–Brezzi convergence conditions are to be satisfied (Babuska, 1971, 1973; Brezzi 1974). Until now we have not referred to any particular element form and indeed a wide choice is available to the user if the limiting (undrained) condition is never imposed. Due to the presence of first derivatives in space in all the equations, it is necessary to use C_0-continuous interpolation functions (Zienkiewicz and Taylor, 1989) and Fig. 3 shows some elements incorporated in the formulation. The form of the elements used satisfies the necessary convergence criteria of the undrained limit (Zienkiewicz, 1984).

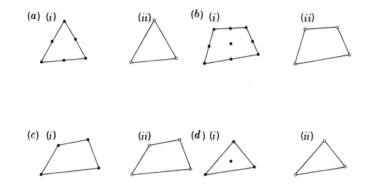

FIG. 3. Elements used for coupled analysis, displacement (u) and pressure (p) formulation. (a) (i), quadratic for u; (ii), linear for p: (b) (i), biquadratic for u; (ii), bilinear for p: (c) (i), linear for u; (ii), linear for p: (d) (i), linear (with cubic bubble) for u; (ii), linear for p. Element (c) is not fully acceptable at incompressible-undrained limits.

5 SOIL CONSTITUTIVE MODELS

Constitutive modelling of soil (or other geomaterials) is clearly a keystone of successful quantitative solution possibilities. Without a reasonable constitutive model the computations are worthless; but indeed a good constitutive model without a computation framework in which to use it is only an academic exercise.

It is not surprising therefore that much research work has been devoted to determining such models in the last quarter of the century in parallel to the development of computation (numerical analysis) procedures. This work is too extensive to report here but progress has been such that recently the behaviour of both cohesive and non-cohesive soil, rocks and concrete can be described with a reasonable amount of accuracy for most loading paths. Of course the research continues and every year new constitutive models are added to the repertoire.

Most of the soil deformation is independent of time and hence can be cast in the form of eqn (4) or (36) where the **D** matrix is defined by the current state of stress and strain, its history and, importantly, the direction of the strain or stress changes. The latter is essential if plastic (or irreversible) deformations are occurring as inevitably happens in most soils. The first important contribution to deriving constitutive models were based on the classical theory of plasticity and here the work of the Cambridge group in the early 1960s paved the way for the basis of deriving cohesive soil (clay) models. The work of Roscoe *et al.* (1958) and Schofield and Wroth (1968) is now classic in this context.

In recent years it has been realized that such modelling can be included simply in a *generalized plasticity theory* (Mroz and Zienkiewicz, 1984; Zienkiewicz and Mroz, 1984) and that this new formulation permits the definition of plastic laws in a very simple manner without the necessity of defining yield or flow surfaces. This theory has recently allowed the behaviour of sands and similar materials to be modelled effectively with a relatively small number of experimental parameters. Pastor and Zien-kiewicz (1986) and Pastor *et al.* (1988, 1990) discuss the details of this formulation and Figs 4(a)–(d) show how well the typical behaviour of sand under complex loading can be reproduced with few physical constants identified (8–10 is sufficient for most sands). We relegate the description of the detailed features of the theory to the Appendix and to the relevant papers but some important aspects must be noted

(a) The specification of the **D** is such that

$$\mathbf{D} = \mathbf{D}_{\text{load}} \quad \text{when} \quad \mathbf{n}^{\text{T}} \, d\boldsymbol{\sigma}^{\text{e}} \geqslant 0$$

and

$$\mathbf{D} = \mathbf{D}_{\text{unload}} \quad \text{when} \quad \mathbf{n}^{\text{T}} \, d\boldsymbol{\sigma}^{\text{e}} < 0$$

with $d\boldsymbol{\sigma}^{\text{e}} = \mathbf{D}^{\text{e}} \, d\boldsymbol{\varepsilon}$ and $\boldsymbol{\sigma}^{\text{e}}$ standing for the elastic part of effective stresses.

The loading and unloading matrices differ thus allowing permanent deformation to occur in a load cycle.

(b) The model includes plastic deformation both in 'loading' and 'unloading'. For sands this deformation causes a decrease of volumetric strain in both directions when a drained sample is tested or, when the sample is undrained, pore pressure rise is observed leading to soil 'liquefaction' or at least the so-called 'cyclic mobility'. This is a most important phenomenon which accounts for such failures as that illustrated in Fig. 1.

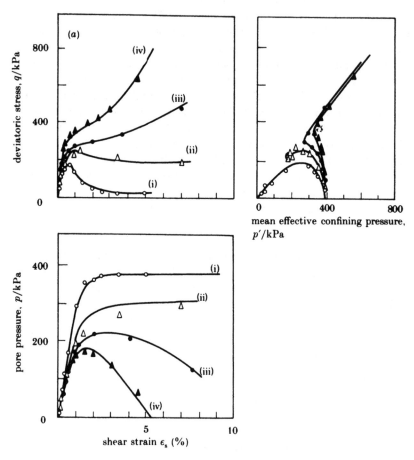

FIG. 4. (a) Undrained triaxial, monotonic load test for sands of various densities (Castro 1969). (i) ○, relative density, $D_R = 29\%$; (ii) △, $D_R = 44\%$; (iii) ●, $D_R = 47\%$; (iv) ▲, $D_R = 64\%$; ———, computational model. (b) Drained triaxial, monotonic load tests; loose and dense sands (Taylor 1984). When $D_R = 100\%$ then ●, experiment; ———, computational model. When $D_R = 20\%$ then ○, experiment; – – –, computational model. (c) Undrained behaviour of loose sand under reversal of stress (Ishihara & Okada 1978); ●, experimental; ———, computational model. (d) Undrained one-way cyclic loading of loose sand (Castro 1969). $D_R = 33\%$. (i) Experimental, (ii) computational model. (e) Undrained two-way cyclic loading of loose Niigana sand (Tatsuoka & Isihara 1974). (i) Experimental, (ii) computational model.

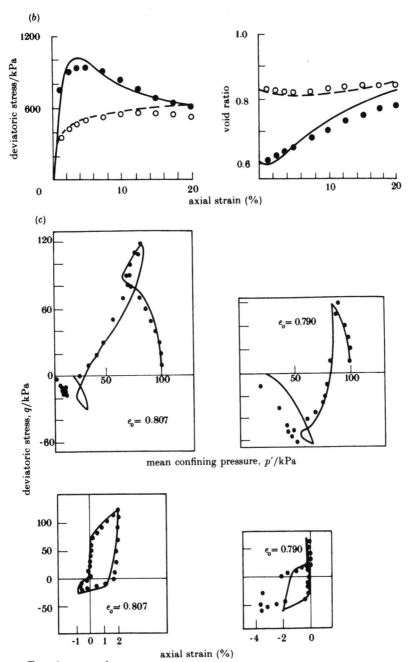

FIG. 4. — *contd.*

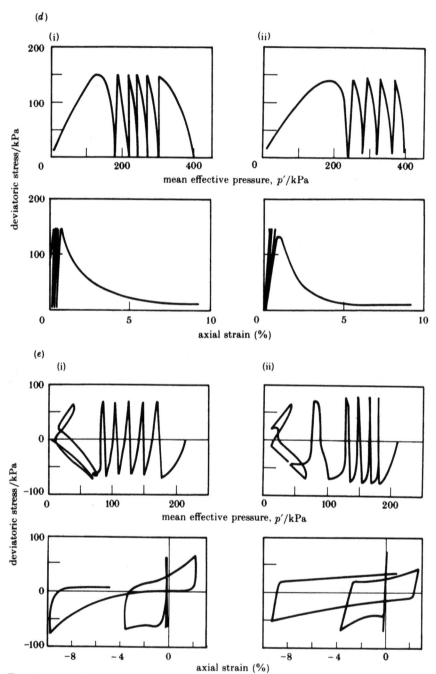

FIG. 4. — *contd.*

(c) The particular model used does not account for possible deformations caused by pure rotation of principal stress axes which is known to be able to cause liquefaction effects. The experimental evidence of this is however only quantitatively confirmed to date and in the practical applications shown, the effect can probably be disregarded. The possible extension of the model to deal with this phenomenon is discussed by Pastor *et al.* (1990) and can be put into effect when quantitative evidence is available.

(d) The models used in this paper use essentially the predictive capacity for non-cohesive soils. Modification for clays and similar soils is relatively simple and is described in the last reference.

In closing this section we must remark that no available soil model is ideal in the sense of being able to reproduce *precisely* all observed features of experimental behaviour of soils. The present one is currently optimal but new ones will doubtless be developed and, if successful, can easily be substituted in the code. However, the quest for precision should be tempered by a realisation that laboratory tests suffer from considerable error in the application of prescribed stresses or strains, and further that *in-situ* soil shows considerable variability and parameters chosen for a particular reason in practice need to show a statistical approach.

6   VERIFICATION

With the approximation involved in the numerical process and more importantly the possible deficiency of constitutive laws and variability of the quantitative parameters defining these, it is essential to be able to verify predictions on prototype or scale model computation, having assured by preliminary computation that essential phenomena are correctly modelled.

The first example represents such a qualitative test to demonstrate that the effects observed in the Niigata soil layer liquefaction of Fig. 1 can be reproduced. In Fig. 5 we model such a layer subject to a horizontal earthquake input in its base (using soil parameters and input as closely matched with observation as possible) and show the computed pore pressure development. The 'liquefaction' or at least the development of internal pore pressures to the point of balancing the soil stresses due to its own weight and thus giving zero effective mean stress is demonstrated.

Of particular interest is the pattern of 'post-earthquake' consolidation in which the excess pressures drop to zero. The fairly rapid reduction of

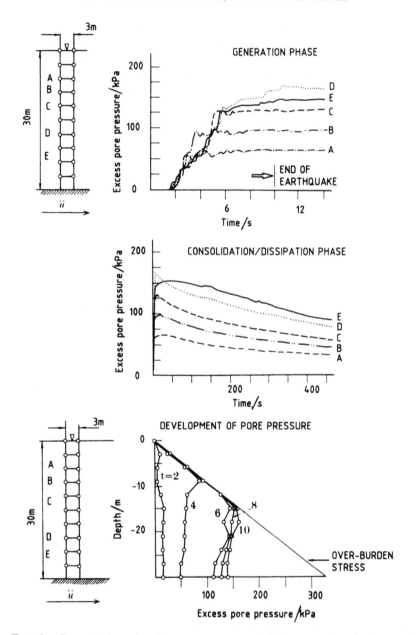

FIG. 5. Computation showing pore pressure build up to liquefaction and subsequent consolidation in a typical soil layer.

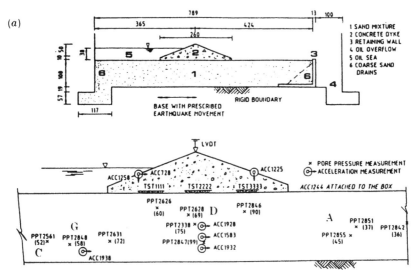

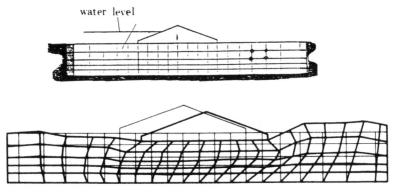

FIG. 6. (a) Centrifuge model of a dyke, computational mesh (bilinear elements) and deformation pattern after earthquake ($\times$ 10 magnification). (b) Pore pressure and displacement computation (right) against experiment (left). Excess pore pressure at (i) G, (ii) D, (iii) A, (iv) C; (v) vertical displacement of the dyke. The device type is a 6 pressure transducer. Device number (i) 2561, (ii) 2338, (iii) 2851, (iv) 2848, (v) 873. Range $0 < t < 0.16$ s. (c) As Fig. 5(b) with range $0 < t < 2.5$ s.

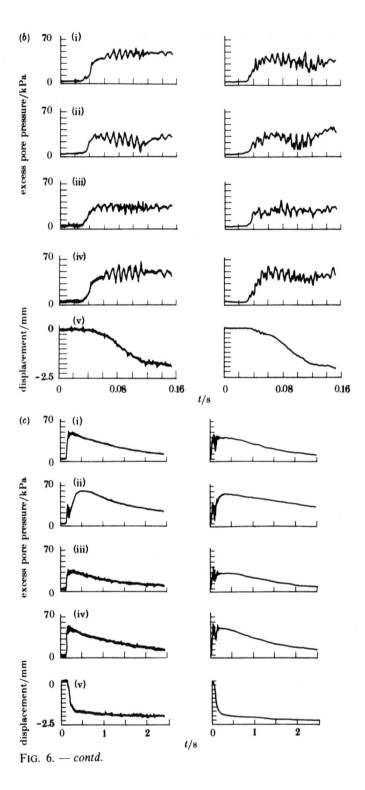

FIG. 6. — *contd.*

such pressures at points situated at considerable depth and the almost constant excess pressures persisting for a long period closer to the surface are at first glance a paradox. However, this phenomenon has been observed in many laboratory tests simulating the problem and are simply explained by observing that close to the surface, the appreciable volume expansion following the initial pore pressure rise requires more time to re-establish soil particle contact than at deeper points.

The first example is qualitative for the fairly obvious reason that any *a-posteriori* reconstruction of events taking place in nature will lack sufficient quantitative information on the event and the materials of the structure.

For this reason a series of controlled experiments on the centrifuge at Cambridge University were carried out by Professor Schofield to provide precise measurements of dynamic input, displacements and pore pressure development in some typical situations.

Such experiments are shown in Fig. 6 together with corresponding computations. The difficulties of recording displacements in the experiment limit that particular set of data but those observed together with pore pressure recorded show a very reasonable comparison with results of computation.

Tests carried out at Cambridge were recorded by Venter (1987) who also carried out tests allowing at least some of the soil parameters to be identified. Others had to be guessed from external information.

We do not show here any static computation results, although such static analyses had to be carried out to establish initial conditions of dynamics. Quasi-static consolidation validation is however well documented in Fig. 6c where the dynamic effects are insignificant.

7 COLLAPSE OF THE LOWER SAN FERNANDO DAM MODELLED

The failure of the Lower San Fernando earth dam in 1971 with nearly catastrophic consequences is typical of what can occur in a poorly consolidated soil structure due to shaking resulting from an earthquake. The problem has of course been studied extensively since but so far no quantitative analysis has succeeded in reproducing the mechanism of the failure (Zienkienwicz *et al.* (1981) and Castro *et al.* (1985)). The reconstruction of events by Seed *et al.* (1975) and Seed (1979) is however remarkable in attempting to explain why the failure occurred apparently

Cross-section through embankment after earthquake

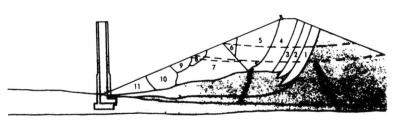

Reconstructed cross-section

FIG. 7.   Failure and reconstruction of original conditions of lower San Fernando dam after 1971 earthquake, according to Seed (1979).

some 60–90 s after the start of the earthquake which was recorded to last some 14 s.

The hypothesis made by Seed was that the important pressure build-up occurring due to cyclic loading which manifested itself first in the central portions of the dam, 'migrates' in the post earthquake period to regions closer to the 'heel' of the dam where it triggers the failure.

In the present section we use this example to illustrate the computational process although comparison with the actual occurrence is, perforce, still basically impossible due to the lack of full records. While in the last section we used a fully documented centrifuge test as a benchmark similar tests are not available here, probably due to scaling difficulties in the centrifuge test necessary to model the semi-saturated conditions in the upper region of the dam. The resistance of this region is of considerable importance in stability analysis and indeed the neglect of negative pressures there leads to unrealistic results.

The actual collapsed dam and a 'reconstructed' cross section are shown in Fig. 7 following Seed (1979). In Fig. 8 we show the material idealiz-

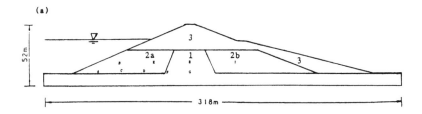

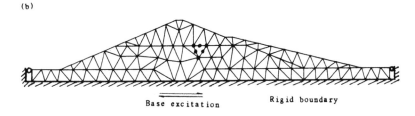

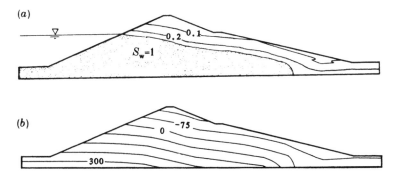

FIG. 8. Idealization of San Fernando dam for analysis. (a) Material zones (see Table 1); (b) displacement discretization and boundary conditions; (c) pore pressure discretization and boundary conditions.

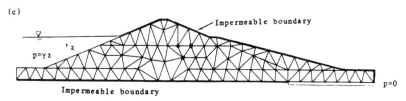

FIG. 9. Initial steady-state solution. Only saturation (a) and pressure contours (b) are shown. Contour interval in (b) is 75 kPa.

TABLE 1
MATERIAL PROPERTIES USED IN SAN FERNANDO DAM ANALYSIS

Material zone	ρ (kg/m³)	ρ_f (kg/m³)	K_s (Pa)	K_f (Pa)	v	n	$k(1)$ (m/s)	K_{evo}	K_{eso}	M_g	M_f	α_g	α_f	β_0	β_1	H_0	H_{uo} (Pa)	γ_u	γ_{DM}
1	2090·0	980·0	1·0E+12	2·0E+9	0·2857	0·375	1·0E−3	1200	180·0	1·55	1·400	0·45	0·45	4·2	0·2	700·0	6·00E+7	2·0	2·0
2a	2020·0	980·0	1·0E+12	2·0E+9	0·2857	0·375	1·0E−2	70·0	105·0	1·51	0·755	0·45	0·45	4·2	0·2	408·3	3·50E+7	2·0	2·0
2b	2020·0	980·0	1·0E+12	2·0E+9	0·2857	0·375	1·0E−2	75·0	112·6	1·51	0·906	0·45	0·45	4·2	0·2	437·5	3·75E+7	2·0	2·0
3	2020·0	980·0	1·0E+12	2·0E+9	0·2857	0·375	1·0E−3	80·0	120·0	1·51	1·133	0·45	0·45	4·2	0·2	467·0	4·00E+7	2·0	2·0

ation, finite element meshes and boundary conditions used. Figure 9 shows the initial steady state solution for the saturation and the pore pressure distribution indicating clearly the 'phreatic' line and the suction pressures developing above which give a substantial cohesion there. Preliminary computation indicates clearly that without such cohesion, an almost immediate local failure develops in the dry material on shaking.

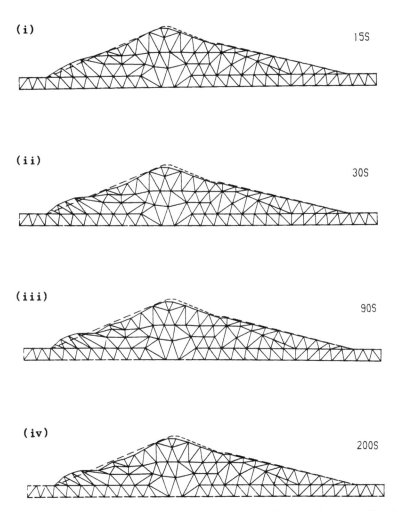

FIG. 10.   Deformed shapes of the dam at times (i) 15s (end of earthquake), (ii) 30s, (iii) 90s, (iv) 200s.

displacement/m

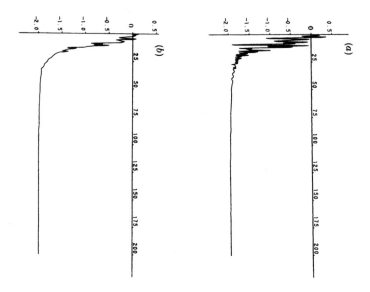

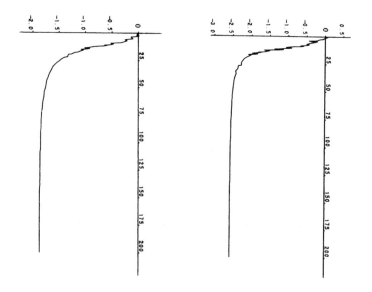

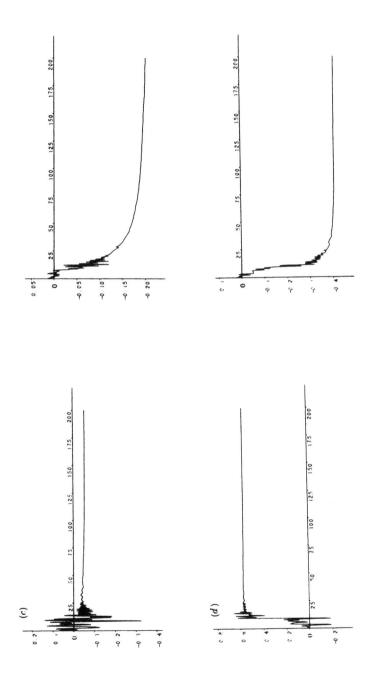

time/s

FIG. 11. Horizontal (left) and vertical (right) displacements: (a) on the crest; (b) at point E, (c) at point H; (d) at point I. See Fig. 8(a).

pressure/ Pa

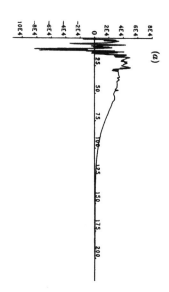

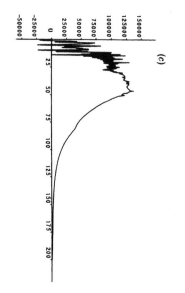

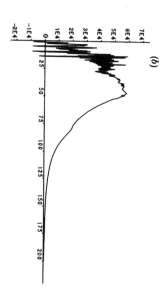

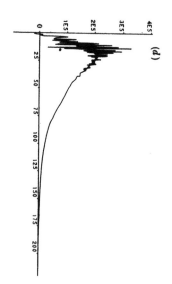

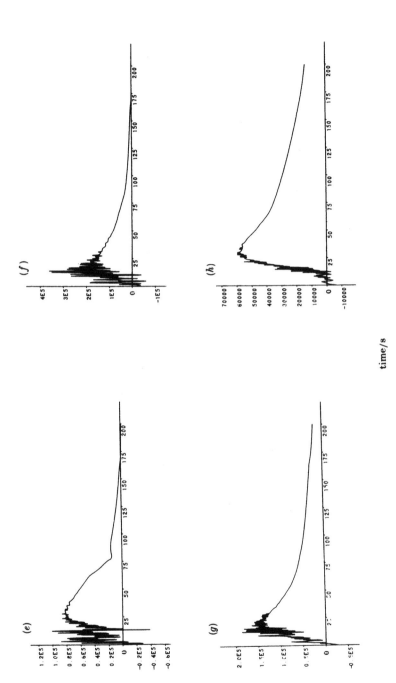

time/s

FIG. 12.   Excess pore pressure at points (a) A, (b) B, (c) C, (d) D, (e) E, (f) F, (g) G, (h) H. See Fig. 8(a).

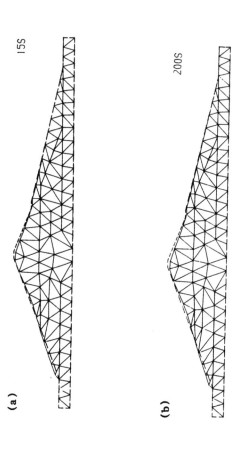

(a) 15S

(b) 20S

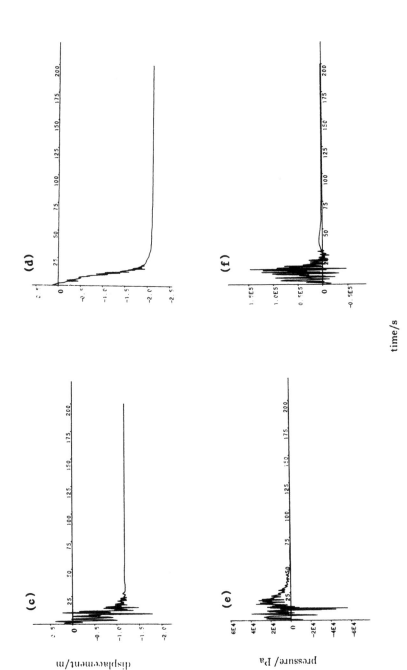

FIG. 13. Results of analysis with increased permeabilities: (a) deformed shape of the dam at 15s; (b) deformed shape of the dam at 200s; (c) horizontal displacement on the crest; (d) vertical displacement on the crest, (e) excess pore pressure at point A; (f) excess pore pressure at point D. See Fig. 8(a).

(i)

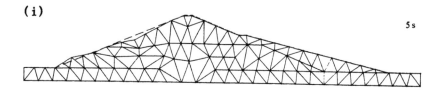

5 s

(ii)

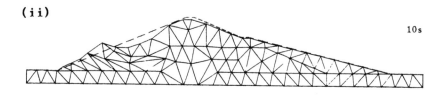

10 s

(iii)

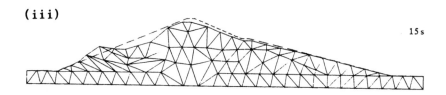

15 s

(iv)

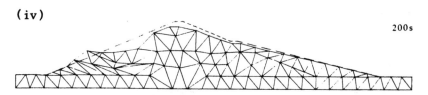

200 s

FIG. 14.   Results of analysis with softer materials. Deformed shapes of the dam
at (i) 5s, (ii) 10s, (iii) 15s, (iv) 200s.

The above static, initial computation was carried out by the full program now operating in a static mode assuming the gravity and external water pressure to be applied without dynamic effects.

Table 1 summarises the material properties assumed to describe the various zones of the dam using the constitutive model described in the Appendix.

Starting with the computed effective stress and pressure distribution the computation is carried out for the full period of the earthquake and continued for a further time of 200 s. Figures 10, 11 and 12 show the displaced form of the dam at various times, the plot of displacements at some characteristic points and the development and decay of pore pressures (here only the excess, i.e. the change from the steady-state, initial values, is shown).

The results are, we believe, quite remarkable. First, deformations are increasing for a considerable period after the end of the earthquake. This undoubtedly is aided by the redistribution of pore pressures which fall rapidly in the central portions but continue to increase or maintain nearly constant values near the upstream surface. Second, the pattern of deformation is very similar to that which occurred in the actual case showing large movements near the upstream base and indicating the motion along the failure plane.

If the permeability of the dam material is sufficiently high, it maybe impossible for an earthquake to cause any build-up of pore pressures in the embankment since the pore pressure can dissipate by drainage as rapidly as the earthquake can generate them by shaking. Figure 13 shows the results of computation with permeabilities in Table 1 being increased by a factor of 10.

In an additional calculation with the dam material assumed to be softer, which implies in the present constitutive model that the ratio of M_f/M_g is reduced, significantly larger displacements are recorded at the early stages of the earthquake shaking as shown in Fig. 14.

8 CONCLUSION

This Chapter attempts, and we believe succeeds, in showing that numerical computation and modelling can reproduce quantitatively many of the phenomena for which previously only rough rules of thumb were available. Much work has been reported by Zienkiewicz and others in the past 15 years on computational possibilities in the geomechanics area but for quite plausible reasons such computations are still infrequently used. We hope that the prediction verified here and the possibility of modelling all phases of mechanical soil behaviour in a unified manner will have an appeal to both the practitioner and researcher.

APPENDIX: GENERALISED PLASTICITY MODEL

In elasto-plasticity, permanent deformations can occur in a load cycle and all corresponding theories distinguish between loading and unloading. In the generalised plasticity formulation, at every point of the stress space σ_{ij}, a direction tensor n_{ij} is specified which serves to distinguish between loading and unloading. Thus if D^e_{ijkl} corresponds to the elastic (or direction independent) modular matrix we have

(loading) $d\sigma_{ij} = D^L_{ijkl} \, d\varepsilon_{kl}$ if $n_{ij} D^e_{ijkl} \, d\varepsilon_{kl} \leqslant 0$

(unloading $d\sigma_{ij} = D^U_{ijkl} \, d\varepsilon_{kl}$ if $n_{ij} D^e_{ijkl} \, d\varepsilon_{kl} \leqslant 0$
or reverse
loading)

To obtain uniqueness of strain changes corresponding to given stress changes, it is necessary that along the neutral loading direction

$$[D^L_{ijkl}]^{-1} = [D^U_{ijkl}]^{-1}$$

This is easily achieved by writing

$$[D^L_{ijkl}]^{-1} = [D^e_{ijkl}]^{-1} + n^{Lg}_{ij} n_{kl}/H^L$$

$$[D^U_{ijkl}]^{-1} = [D^e_{ijkl}]^{-1} + n^{Ug}_{ij} n_{kl}/H^U$$

where n^{Lg}_{ij} and n^{Ug}_{ij} are arbitrary (unit) tensors and $H^{L/U}$ are scalar plastic moduli.

The above definition results in purely elastic and hence unique deformation in neutral loading with $n_{ij} D^e_{ijkl} \, d\varepsilon_{kl}$ and $n_{ij} \, d\sigma_{ij}$ equal to zero.

Note that in the above the inverse of a tensor has the meaning

$$D_{ijkl} [D_{wxyz}]^{-1} = \delta_{kw} \delta_{lx} \delta_{iy} \delta_{jz}$$

The complete plastic deformation satisfying all consistency rules can be obtained by defining the tensor n_{ij} (loading direction), $n^{gL/U}_{ij}$ plastic flow rule tensors permitting plasticity to occur during both loading and unloading, $H^{L/U}$ the plastic moduli and D^e_{ijkl} giving the basic elastic constants.

These quantities can be defined directly without specifying yield or plasticity potential although of course the classic definitions occasionally provide a useful subset.

Details of the model used in this paper for sands can be found in Pastor and Zienkiewicz (1986) and Pastor et al. (1988, 1990).

The model is written in terms of the three stress invariants

- p mean effective stress
- q deviatoric stress
- θ the Lode angle.

Vector convection is used in specifying the directions **n** and $\mathbf{n}^q$ and transformation to the cartesian system follows the procedure given in detail by Chan et al. (1988).

The definitions and parameters which need to be determined are given below.

(1) *Loading direction vector*
$$n_p = [1 + \alpha_f][M_f - \eta]$$
$$n_q = 1$$
$$n_\theta = -\tfrac{1}{2}q \cdot M_f \cos 3\theta$$
where α_f, M_f are model parameters and η is the stress ratio defined as q/p.

(2) *Plastic flow vector*
(a) *Loading*
$$n_p^{gL} = (1 + \alpha_g)(M_g - \eta)$$
$$n_q^{gL} = 1$$
$$n_\theta^{gL} = -\tfrac{1}{2}q \cdot M_g \cos 3\theta$$
where α_g is a material parameter and M_g is the slope of the critical state line (Roscoe et al. (1958)) and it is related to the residual friction angle by

$$M_g = \frac{6 \sin \phi'}{3 - \sin \phi' \sin 3\theta}$$

(b) *Unloading*
$$n_p^{gU} = -|n_p^{gL}|$$
$$n_q^{gU} = n_q^{gL}$$
$$n_\theta^{gU} = n_\theta^{gL}$$

This is so chosen to ensure densification occurs in unloading.

(3) *Plastic moduli*
(a) *Loading*
$$H_1 = H_0 \cdot p' \cdot [1 - \eta/\eta_f]^4 \cdot [H_v + H_s] \cdot [\eta/\eta_{max}]^{-\gamma_{DM}}$$

$$\eta_f = \left[1 + \frac{1}{\alpha_f}\right] M_f$$

$$H_v = 1 - \eta/M_g, \quad H_s = \beta_0 \beta_1 \exp(-\beta_0 \xi)$$

$$\xi = \int |d\varepsilon_q|$$

where H_0, γ_{DM}, β_0, β_1 are model parameters, η_{max} is the maximum stress ratio reached and ξ is the cumulative deviatoric plastic strain.

(b) *Unloading*

where $H_u = H_{uo}[\eta_u/M_g]^{-\gamma_u}$

H_{uo} and γ_u are material parameters and η_u is the stress ratio from which unloading commenced.

(4) *Elastic constants*

   (a) *Bulk modulus*: $K = H_{evo} p'$

   (b) *Shear modulus*: $G = H_{eso} p'/3$ where

both H_{evo} and H_{eso} are model constants.

The vectors are transformed using appropriate transformation (Chan *et al.* 1988) to the cartesian stress space for numerical calculations.

REFERENCES

ALONSO, E.E., GENS, A. and HIGHT, D.W. (1987). Special problems of soil: general report, *Proc. 9th Europ. Conf. Int. Soc. Soil Mech. Found. Eng.*, Dublin.

BABUSKA, I. (1971). Error bounds for finite element methods, *Numer. Math.*, **16**, 322–33.

BABUSKA, I. (1973). The finite element method with Lagrange multipliers, *Numer. Math.*, **22**, 179–92.

BATHE, K.J. and KHOSHGOFTAN, M.R. (1979). Finite element for surface seepage analysis without mesh iteration, *Int. J. Numer. Anal. Methods Geomech.*, **3**, 13–22.

BEAR, J., CORAPCIOGLU, M.Y. and BALAKRISHNA, J. (1984). Modelling of centrifugal filtration in unsaturated deformable porous media, *Adv. Water Resources*, **7**, 150–67.

BIOT, M.A. (1941). General theory of three-dimensional consolidation, *J. Appl. Phys.*, **12**, 155–64.

BIOT, M.A. (1955). Theory of elasticity and consolidation for a porous anisotropic solid, *J. Appl. Phys.*, **26**, 182–5.

BIOT, M.A. (1956). Theory of propagation of elastic waves in a fluid-saturated porous solid, part I: low-frequency range, Part II: higher frequency range, *J. Acoust. Soc. Am.*, **28**, 168–91.

BIOT, M.A. and WILLIS, P.G. (1957). The elastic coefficients of the theory of consolidation, *J. Appl. Mech.*, **24**, 594–601.

BISHOP, A.W. (1959). The principle of effective stress, *Teknisk Ukeblad*, **39**, 859–863.

BOWEN, R.M. (1976). In *Theory of Mixtures in Continuum Physics*, A.C. Eringen (Ed.), Vol. III, Academic Press, New York.

BREZZI, F. (1974). On the existence, uniqueness and approximation of saddle point problems arising from Lagrange multipliers, *R.A.I.R.O.*, **8**, 129–51.

CASTRO, G. (1969). Liquefaction of sands, Ph.D. thesis, Harvard University, U.S.A.

CASTRO, G., POULOS, S.J. and LEATHERS, F.D. (1985). Re-evaluation of slide of Lower San Fernando dam, *J. Geotech. Eng. Div. ASCE*, **111**, (GT9), 1093–1107.

CHAN, A.H.C. (1988). A unified finite element solution to static and dynamic problems in geomechanics, Ph.D. thesis, University College of Swansea, U.K.

CHAN, A.H.C., ZIENKIEWICZ, O.C. and PASTOR, M. (1988). Transformation of incremental plasticity relation from defining space to general Cartesian stress space, *Commun. Appl. Numer. Methods*, **4**, 577–81.

DARCY, H. (1856). *Les Fonteines*, Publiques de le villa de Dijon, Delmont, Paris.

DAVIDON, W.C. (1968). Variance algorithms for minimization, *Computer J.*, **10**, 406–10.

DESAI, C.S. (1976). Finite element, residual schemes for unconfined flow, *Int. J. Numer. Methods Eng.*, **10**, 1415–18.

DESAI, C.S. and LI, G.C. (1983). A residual flow procedure and application for free surface flow in porous media, *Adv. Water Resources*, **6**, 27–35.

GREEN, A.E. and ADKIN, J.E. (1960). *Large Elastic Deformations and Nonlinear Continuum Mechanics*, Oxford University Press, London.

GREEN, A.E. (1969). On basic equations for mixtures, *Quart. J. Mech. Appl. Math.*, **22**, 428–38.

ISHIHARA, K. and OKADA, S. (1978). Yielding of overconsolidated sand and liquefaction model under cyclic stresses, *Soils and Foundations*, **18**, 57–72.

KATONA, M.G. and ZIENKIEWICZ, O.C. (1985). A unified set of single step algorithms, Part 3: The Beta-m method a generalization of the Newmark scheme, *Int. J. Numer. Methods Eng.*, **21**, 1345–59.

LEUNG, K.H. (1984). Earthquake response of saturated soils and liquefaction, Ph.D. thesis, University College of Swansea, U.K.

LI, X., ZIENKIEWICZ, O.C. and XIE, Y.M. (1990). A numerical model for immiscible two-phase fluid flow in a porous medium and its time domain solution, *Int. J. Numer. Methods Eng.*, **30**, 1195–212.

LIAKOPOULOS, A.C. (1965). Transient flow through unsaturated porous media, D. Eng. dissertation, University of California, Berkeley.

LLORET, A. and ALONSO, E.E. (1980). Consolidation of unsaturated soils including swelling and collapse behaviour, *Géotechnique*, **30**, 449–77.

LLORET, A., ALONSO, E.E. and GENS, A. (1986). Undrained loading and consolidation analysis for unsaturated soils, *Proc. 2nd Euro. Conf. Numer. Methods Geomech.*, Stuttgart.

MROZ, Z. and ZIENKIEWICZ, O.C. (1984). Uniform formulation of constitutive equations for clays and sands. In *Mechanics of Engineering Materials*, C.S. Desai and R.H. Gallagher (Eds.), Wiley, New York, Ch. 22.

NARASIMHAN, T.N. and WITHERSPOON, P.A. (1978). Numerical model for saturated-unsaturated flow in deformable porous media 3. Applications, *Water Resources Res.*, **14**, 1017–34.

NEUMAN, S.P. (1975). Galerkin approach to saturated-unsaturated flow in porous media. In *Finite Elements in Fluids*, R.H. Gallagher *et al.* (Eds.), Wiley, London, Vol. 1, pp. 201–17.

NEWMARK, N.M. (1959). A method of computation for structural dynamics, *Proc. Am. Soc. Civ. Eng.*, **8**, 67–94.

PARK, K.C. and FELIPPA, C.A. (1983). Partitioned analysis of coupled systems. In *Computational Methods for Transient Analysis*, T. Belytschko and T.J.R. Hughes (Eds.) Elsevier, Amsterdam, Ch. 3.

PARK, K.C. (1983). Stabilization of partitioned solution procedure for a pore fluid-soil interaction analysis, *Int. J. Numer. Methods Eng.*, **19**, 1669–73.

PASTOR, M. and ZIENKIEWICZ, O.C. (1986). A generalised plasticity, hierarchical model for sand under monotonic and cyclic loading. In *Numerical Methods in Geomechanics*, G.N. Pande and W.F. van Impe (Eds.), Jackson and Son, London, pp. 131–50.

PASTOR, M., ZIENKIEWICZ, O.C. and CHAN, A.H.C. (1988). Simple models for soil behaviour and applications to problems of soil liquefaction. In *Numerical Methods in Geomechanics*, G. Swoboda (Ed.), A.A. Balkhema, pp. 169–80.

PASTOR, M., ZIENKIEWICZ, O.C. and CHAN, A.H.C. (1990). Generalized plasticity and the modelling of soil behaviour, *Int. J. Numer. Anal. Methods Geomech.*, **14**, 151–90.

ROSCOE, K.H., SCHOFIELD, A.N. and WROTH, C.P. (1958). On the yielding of soils, *Géotechnique*, **8**, 22–53.

SAFAI, N.M. and PINDER, G.F. (1979). Vertical and horizontal land deformation in a desaturating porous medium, *Adv. Water Resources*, **2**, 19–25.

SCHOFIELD, A.N. and WROTH, C.P. (1968). *Critical State Soil Mechanics*, McGraw-Hill, New York.

SEED, H.B., LEE, K.L., IDRISS, I.M., and MAKDISI, F.I. (1975). Analysis of slides of the San Fernando dams during the earthquake of February 9, 1971, *J. Geotech. Eng. Div. ASCE*, **101**(GT7), 651–688.

SEED, H.B. (1979). Consideration in the earthquake resistant design of earth and rockfill dams, *Géotechnique*, **29**, 215–63.

SKEMPTON, A.W. (1961). *Effective Stress in Soils, Concrete and Rocks, Pore Pressure and Suction in Soils*, The British National Society of the International Society of Soil Mechanics and Foundation Engineering, pp. 4–16.

TATSUOKA, T. and ISHIHARA, K. (1974). Yielding of sand in triaxial compression, *Soil and Foundations*, **14**, 63–76.

TAYLOR, D. (1984). *Fundamentals of Soil Mechanics*, Wiley, New York.

TERZAGHI, K. (1943). *Theoretical Soil Mechanics*, Wiley, New York.

VAN GENUCHTEN, M. Th., PINDER, G.F. and SAUKIN, W.P. (1977). Modeling of leachate and soil interactions in an aquifer, *Proc. 3rd Annual Municipal Solid Waste Res. Symp.*, EPA-600/9-77-026, pp. 95–103.

VENTER, K.V. (1987). Modelling the response of sand to cyclic loads, Ph.D. thesis, Department of Engineering, Cambridge University, U.K.

XIE, Y.M. (1990). Finite element solution and adaptive analysis for static and dynamic problems of saturated-unsaturated porous media, Ph.D. thesis, University College of Swansea, UK.

ZIENKIEWICZ, O.C. HINTON, E. LEUNG, K.H. and TAYLOR, R.L. (1980a). Staggered time marching schemes in dynamic soil analysis and selective explicit extrapolation algorithms, *Proc. 2nd Symp. on Innovative Numerical Analysis for the Engineering Science*, R. Shaw *et al.* (Eds.), University of Virginia Press.

ZIENKIEWICZ, O.C., CHANG, C.T. and BETTESS, P. (1980b). Drained, undrained consolidating dynamic behaviour assumptions in soils, *Géotechnique*, **30**, 385–95.

ZIENKIEWICZ, O.C., LEUNG, K.H., HINTON, E. and CHANG, C.T. (1981). Earth dam analysis for earthquakes: numerical solution and constitutive relations for non-linear (damage) analysis, *Proc. Int. Conf. Dams and Earthquake*, London, pp. 179–94.

ZIENKIEWICZ, O.C. (1982). Basic formulation of static and dynamic behaviour of soil and other porous material. In *Numerical Methods in Geomechanics*, J.B. Martins (Ed.), D. Riedel, London.

ZIENKIEWICZ, O.C. and BETTESS, P. (1982). Soils and other saturated media under transient dynamic conditions: general formulation and the validity of various simplifying assumptions. In *Soil Mechanics-Transient and Cyclic Loads*, G.N. Pande and O.C. Zienkiewicz (Eds.), Wiley, New York, Ch. 1.

ZIENKIEWICZ, O.C., LEUNG, K.H. and HINTON, E. (1982). Earthquake response behaviour of soils with drainage, III, *Proc. 4th Int. Conf. on Numerical Methods in Geomechanics*, Edmonton, Canada, pp. 983–1002.

ZIENKIEWICZ, O.C. (1984). Coupled problems and their numerical solution. In *Numerical Methods in Coupled Systems*, R.W. Lewis, P. Bettess and E. Hinton (Eds.), Ch. 1, pp. 35–8.

ZIENKIEWICZ, O.C. and SHIOMI, T. (1984). Dynamic behaviour of saturated porous media: the generalised Biot formulation and its numerical solution, *Int. J. Numer. Anal. Methods Geomech.*, **8**, 71–96.

ZIENKIEWICZ, O.C. and MROZ, Z. (1984). Generalised plasticity formulation and applications to geomechanics. In *Mechanics of Engineering Materials*, C.S. Desai and R.H. Gallagher (Eds.), Wiley, New York, Ch. 33.

ZIENKIEWICZ, O.C. and TAYLOR, R.L. (1985). Coupled problems – a simple time stepping procedure, *Commun. Appl. Numer. Methods*, **1**, 233–9.

ZIENKIEWICZ, O.C., PAUL, D.K. and CHAN, A.H.C. (1988). Unconditionally stable staggered solution procedure for soil-pore fluid interaction problems, *Int. J. Numer. Methods Eng.*, **26**, 1039–55.

ZIENKIEWICZ, O.C. and TAYLOR, R.L. (1989). *The Finite Element Methods*, 4th edn., McGraw-Hill, New York.

ZIENKIEWICZ, O.C., CHAN, A.H.C., PASTOR, M., PAUL, D.K. and SHIOMI, T. (1990a). Static and dynamic behaviour of soils: a rational approach to quantitative solutions, Part I: Fully saturated problems, *Proc. R. Soc. London*, Ser. A, **429**, 285–309.

ZIENKIEWICZ, O.C., XIE, Y.M., SCHREFLER, B.A., LEDESMA, A. and BICANIC, N. (1990b). Static and dynamic behaviour of soils: a rational approach to quantitative solutions, Part II: Semi-saturated problems, *Proc. R. Soc. London, Ser. A*, **429**, 311–21.

ZIENKIEWICZ, O.C., SCHREFLER, B.A., SIMONI, L., XIE, Y.M. and ZHAN, X.Y. (1991). Two and three-phase behaviour in semi-saturated soil dynamics, *Nonlinear Computational Mechanics, A State of the Art*, P. Wriggers and W. Wagner (Eds.), Springer-Verlag, Berlin, to appear.

Chapter 2

FINITE ELEMENT ANALYSIS OF COMPACTION PROBLEMS

R.B. SEED[a], J.M. DUNCAN[b] and C.Y. OU[c]

[a]*Department of Civil Engineering, University of California at Berkeley, USA*
[b]*Department of Civil Engineering, Virginia Polytechnic Institute and State University, USA*
[c]*Department of Construction Engineering and Technology, National Taiwan Institute of Technology, Taiwan*

ABSTRACT

This chapter describes analytical models and a finite element analysis methodology for evaluation of compaction-induced soil stresses and resulting soil-structure interaction effects. These analytical methods model the incremental placement and compaction of soil, and are based on a hysteretic model for residual soil stresses induced by multiple cycles of loading and unloading. Compaction loading is realistically considered as a transient, moving, surficial load of finite lateral extent, which passes one or more times over some specified portion of the fill surface at each stage of backfill compaction. These analytical tools are well-supported by field case history studies.

1 INTRODUCTION

Geotechnical projects routinely involve the placement and compaction of soil fill in layers or 'lifts'. Compaction of the soil is accomplished by means of one or more 'passes' by compaction equipment operating on the surface of the most recently placed layer of fill. These repeated passes

FIG. 1.   Reinforced concrete cantilever retaining wall sections that failed during
placement and compaction of backfill.

represent a process of repeated application and removal of a travelling,
transient surface load. This process introduces stresses within the backfill,
both during and after completion of compaction, which are not amenable
to reliable analysis by conventional methods.

The stresses induced by compaction have potentially important implications in virtually all areas of geotechnical engineering because the strength and stress-strain behavior of a soil depend largely on the levels of stresses within the soil mass. An ability to analyze such compaction effects is thus necessary in order to model properly the response of compacted soils to both static and dynamic loads. In addition, compaction-induced earth pressures, compaction-induced loads on structures and compaction-induced deformations are potentially important in the design and analysis of many types of soil-structure systems such as retaining walls, bridge abutments, basement walls, buried structures and pipes, reinforced-soil walls, etc. Compaction-induced stresses can also influence resistance to both liquefaction and hydraulic fracturing within a soil mass, the strength and stiffness of pavement subgrades and railroad ballast, and internal stress distributions and deformations of compacted earth and rockfill dams and embankments.

Figure 1 shows an example of the type of adverse effect compaction-induced stresses can exert on soil-structure systems; retaining wall sections have been toppled by lateral earth pressures during placement and compaction of backfill behind the wall sections. The wall failure mechanism in these cases was a flexural failure of the reinforced concrete at the base of the wall.

In recent years new behavioral models and finite element analysis techniques have been developed for modelling and analysis of compaction-induced stresses and deformations. This chapter presents a brief description of these models and analysis techniques, as well as a review of full-scale field case studies used to refine and verify the accuracy and usefulness of these analytical methods. Special emphasis is given to the modelling parameters and to the field compaction procedures which most significantly affect the influence of compaction effects on overall soil behavior. In addition, the principal lessons learned during the development and implementation of these analytical methods are summarized.

2 CONCEPTUAL MODEL

The behavioral models and analytical procedures developed for analysis of compaction-induced soil stresses and resulting deformations are based on a relatively simple conceptual model originally proposed by Broms (1971). Broms considered the process of fill placement and compaction to be an incremental process represented by the placement of successive layers of fill, with compaction accomplished by the application and subsequent removal of a moving load acting at the surface of each successive fill layer, as illustrated schematically in Fig. 2(a).

Broms proposed an empirical analytical procedure based on the concept of hysteretic loading and unloading behavior, in which compaction is represented as a process resulting in cyclic overconsolidation of soils as a result of application and removal of surface (compaction) loads. This empirical procedure, which was limited to consideration of placement and compaction of horizontal layers of soil adjacent to a non-deflecting vertical wall, was the first to provide reasonable qualitative agreement with field data for conditions to which it could be applied.

The behavioral model upon which Broms' method is based is illustrated in Fig. 2(b). An element of soil at some depth (e.g. Point A, in Fig. 2(a)) is considered to exist at some initial stress state (Point 1). An increase in vertical effective stress (loading) results in no lateral stress increase unless and until the K_0-line is reached (Point 2), after which further loading results in an increase in horizontal stress as $\sigma'_h = K_0 \sigma'_v$. A

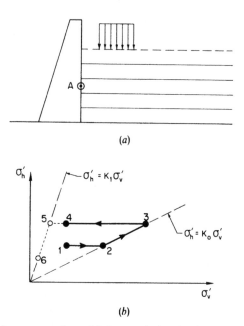

(a)

(b)

FIG. 2. Broms' conceptual model for analysis of compaction-induced lateral stresses. (a) Idealized incremental fill placement and compaction. (b) Simple hysteretic stress-path model for loading and unloading.

subsequent decrease in vertical effective stress (unloading) results in no decrease in lateral stress unless and until a limiting condition is reached (Point 5) after which further unloading results in a decrease in horizontal stress as $\sigma_h' = K_1 \, \sigma_v'$. The K_1-limiting condition is essentially that of passive failure, as lateral stresses become very large relative to the vertical stresses upon significant unloading.

Calculation of compaction-induced lateral stresses by Broms' method involves incremental analysis of the stresses at any given point in the ground which result from placement and compaction of each successive layer of fill. Compaction at any stage is modelled as application of the peak, transient increase in vertical effective stress ($\Delta\sigma_v'$) caused by the closest approach of the compaction vehicle, as determined by simple Boussinesq (1885) elastic analyses, followed by subsequent removal of this transient vertical load. The peak and residual horizontal effective stresses due to this transient compaction loading, as well as those due to fill placement, are then determined by the simple hysteretic stress-path model shown in Fig. 2(b).

This conceptual model and analytical technique represented a significant breakthrough in the development of methods for analysis of compaction-induced stresses. Indeed, this simple conceptual model is the underlying basis for the more complex behavioral models and finite element analysis techniques described in this chapter. These more advanced models and analysis techniques were developed in response to three principal shortcomings in Broms' original methodology:

(1) Broms' model for stresses generated by hysteretic loading and unloading was unrealistic inasmuch as it (a) provided for no lateral stress increase with 'reloading' unless and until the K_0-line was reached, and (b) provided for no relaxation (decrease) in lateral stress with unloading unless and until the passive-failure line was reached. Improved behavioral stress-path models have since been developed which correctly allow for some lateral stress increase with 'reloading', and some degree of lateral stress relaxation with 'unloading'.

(2) More recent investigations have shown that cyclic loading and unloading due to compaction at the fill surface cannot be reliably modelled on the basis of the peak, transient vertical stress increase at a point ($\Delta\sigma_v'$). As a result, an alternate technique based on consideration of the directly calculated peak *lateral* stress increase ($\Delta\sigma_{h,vc,p}'$) has been developed.

(3) Broms' early model is applicable only to 'K_0-conditions'; condi-

tions represented by placement of horizontal fill layers adjacent to a vertical, frictionless, non-displacing wall face. The more advanced analytical models and procedures described herein are based on finite element analysis techniques, and are applicable to a much broader range of problems.

The sections which follow briefly describe (a) the improved behavioral models developed for modelling stresses generated by hysteretic loading and unloading, (b) an improved basis for modelling of the peak stresses induced during compaction, and (c) the incorporation of these models and techniques into a finite element analysis methodology for analysis of compaction-induced stresses and deformations. These analytical methods are applicable to problems of arbitrary soil geometry and boundary conditions, including non-level fill layers and ground surface (e.g. embankments), irregularly shaped and/or deflecting (yielding) soil-structure interfaces, etc. Compaction loading is realistically considered as a transient moving surficial load of finite lateral extent which may pass over either the full surface of a given fill layer, or over specific areas only. Several case studies are presented in which analytical results are compared with field measurements in order to verify the accuracy and usefulness of the models and analytical methods proposed. These case studies also serve to illustrate some of the important lessons learned regarding factors affecting compaction-induced stresses and deflections, and lead to recommendations for minimizing the potential adverse effects of compaction-induced earth pressures by means of appropriate control of field compaction procedures.

3   HYSTERETIC STRESS PATH MODELLING OF MULTI-CYCLE LOADING AND UNLOADING UNDER K_0-CONDITIONS

Seed and Duncan (1983) proposed a hysteretic stress path model for the stresses generated by multiple cycles of one-dimensional loading and unloading of soil under K_0-conditions. Figure 3 illustrates some of the principal features of this hysteretic K_0-stress-path model. The model is somewhat complex, and is described in detail by Duncan and Seed (1986). The principal features of this hysteretic stress path model are described in the following paragraphs.

Primary or 'virgin' loading (defined as loading to higher stress magnitudes than had been achieved previously) follows a K_0 stress path as

$$\sigma'_h = K_0 \cdot \sigma'_v \qquad (1)$$

where K_0 is the coefficient of at-rest lateral earth pressure. Unloading results in relaxation of lateral stresses, but some fraction of the peak lateral stress induced during loading is retained. The first unloading cycle follows a non-linear stress path (AB in Fig. 3) as

$$\sigma'_h = K_0(\text{OCR})^\alpha \cdot \sigma'_v \tag{2}$$

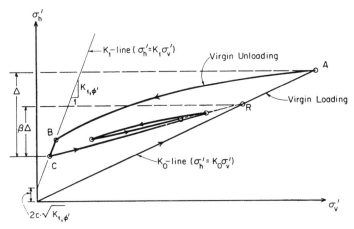

FIG. 3.   Typical stress paths of the non-linear hysteretic stress-path model for K_0-loading and unloading.

where OCR = overconsolidation ratio, and α = the K_0-unloading coefficient (a model parameter). At some stage, if unloading progresses sufficiently far, an apparent limiting condition is reached (e.g. Point B in Fig. 3) as σ'_h becomes significantly greater than σ'_v. This limiting condition is assumed to be similar to passive failure, and is governed by the equation

$$\sigma'_{h,\,\text{lim}} = K_1 \cdot \sigma'_v \tag{3}$$

where K_1 is based on the familiar Mohr-Coulomb failure criteria. The first 'reloading' cycle follows a linear stress path (CR in Fig. 3) until the virgin K_0-line is regained, after which further loading is considered as primary loading. 'Reloading' always results in a lesser increase in lateral stress than does primary loading for the same amount of vertical stress increase.

The model is more complex with regard to the second and subsequent unloading/reloading cycles, as all stress paths are a function of the complete prior stress history. For all subsequent cycles, however, the following remain true: (a) unloading follows a curvilinear unloading stress path similar in form to the first unloading cycle and subject to the same K_1-limiting condition, and (b) reloading results in a lesser increase in lateral stress than does primary loading.

This model has been shown to provide excellent agreement between predicted stresses and the stresses measured in laboratory tests for one-dimensional loading and unloading under K_0-conditions. Two examples are shown in Fig. 4 which presents comparisons between model predictions and K_0-test data for types of stress paths of particular interest in analysis of compaction-induced stresses.

This hysteretic stress path model is applicable only to the modelling of vertical and lateral stresses resulting from one-dimensional loading and

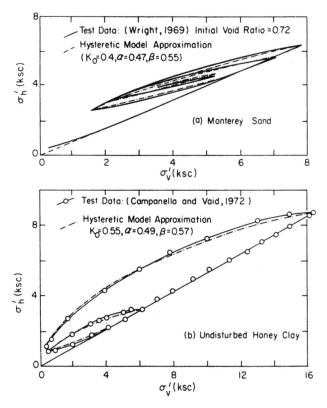

FIG. 4. Comparisons between the non-linear stress-path model and actual K_0-test data.

unloading under K_0-conditions. It is thus specifically limited to consideration of conditions wherein (a) no lateral deformations occur, (b) the major and minor principal stresses remain vertical and horizontal, and (c) the surface loading is uniform and of infinite lateral extent.

4   GENERALIZED HYSTERETIC STRESS PATH MODELLING OF MULTI-CYCLE LOADING AND UNLOADING

4.1   The Special Case of K_0-Conditions

The general hysteretic loading (and unloading) model used to control the introduction of compaction-induced stresses, as well as the subsequent interaction between geostatic and compaction-induced stresses, under non-K_0-conditions represents both an extension and a simplification of the relatively complex non-linear hysteretic model for K_0-loading conditions illustrated in Figs 3 and 4. The two models are closely interrelated, and the simplified general model can best be understood by briefly examining its evolution from a simplified model for K_0-conditions.

Figure 5 illustrates the principal features of the simplified general (non-K_0) hysteretic stress-path model in its simplest form: a stress path model for K_0-loading and unloading. In this form, the similarity between the general model and the more complex non-linear K_0-model illustrated previously in Fig. 3 is readily apparent. Primary loading again follows a

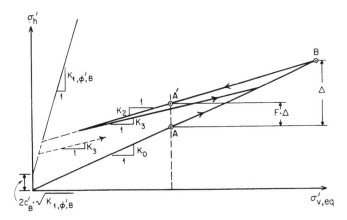

FIG. 5.   Basic components of the simplified general hysteretic loading/unloading stress-path model.

linear stress path according to eqn (1). Unloading follows a linear stress path of constant slope, regardless of previous stress history, according to the equation

$$\Delta\sigma'_h = K_2 \cdot \Delta\sigma'_v \tag{4}$$

where K_2 = the incremental coefficient of lateral earth pressure decrease for K_0-unloading. Unloading results in relaxation of lateral stresses, but some fraction (F) of the peak lateral stresses induced during loading are retained. As illustrated in Fig. 5, the fraction (F) of peak lateral stress increase retained as residual lateral stress increase after loading and then unloading to the same initial vertical stress (e.g. path ABA') may be expressed as

$$F = 1 - \frac{K_2}{K_0} \tag{5}$$

where F is always between zero and one. This type of stress path, representing loading followed by unloading to the initial vertical stress, is of particular importance because it is closely analogous to the type of loading/unloading cycles represented by surface compaction operations. Unloading is subject to a K_1-type of limiting condition as with the hysteretic K_0-model described in the previous section. This K_1-type of limiting condition is again considered to be controlled by passive failure (as σ'_h becomes very large relative to σ'_v after significant 'unloading'), and is governed by eqns (6) and (7) as

$$\sigma'_{h,lim} = K_{1,B} \cdot \sigma'_v \tag{6}$$

$$K_{1,B} = K_{1,\phi',B} + \frac{2c'_B}{\sigma'_3} K_{1,\phi',B} \tag{7}$$

where σ'_3 = the minor principal effective stress, and $K_{1,\phi',B}$, c'_B = model parameters to be described later. Reloading follows a linear stress path of constant slope, regardless of previous stress history, as

$$\Delta\sigma'_h = K_3 \cdot \Delta\sigma'_v \tag{8}$$

where K_3 = the incremental coefficient of lateral earth pressure for K_0-reloading. Reloading follows this stress path unless and until the K_0-line is reached, after which further loading is considered as virgin loading. Reloading thus results in a lesser increase in lateral stress than does virgin loading for the same amount of vertical stress increase.

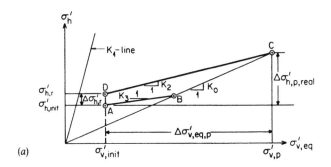

(a)

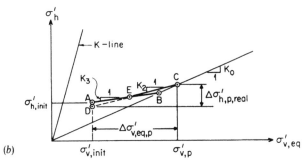

(b)

FIG. 6. Hysteretic stress-path modelling of loading followed by unloading to the same initial equivalent vertical stress. (a) Compaction loading/unloading cycle resulting in a positive residual horizontal stress increase. (b) Compaction loading/unloading cycle resulting in no residual change in horizontal stress.

Figure 6 illustrates the single exception to the rule that unloading follows a stress path either as defined by eqn (4) or controlled by a K_1-type of limiting condition. In Fig. 6(a), soil with an initial stress condition represented by Point A is loaded to B (reloading) and then C (primary loading), and is then unloaded to its initial σ'_v, resulting in a net increase in residual lateral stress of $\Delta\sigma'_{h,r}$. In Fig. 5(b), the soil is loaded from the same initial stress condition, but to a lesser peak load at C. Subsequent unloading to the initial σ'_v according to eqn (4) would result in a net decrease in residual lateral stress (Point D). An exception is made to eqn (4) such that no loading/unloading cycle to the same σ'_v results in a decrease in σ'_h. Unloading for the case shown in Fig. 2(b) thus returns the soil to the initial stress condition at A.

4.2   A General Model for Non-K_0-Conditions

In order to extend this simple hysteretic model to consideration of the more general case of compaction-induced stresses under non-K_0-conditions, it is useful to begin by defining two distinct types of horizontal stresses as follows:

(1) *Geostatic* lateral effective stresses ($\sigma'_{h,o}$) are the horizontal effective stresses not directly due to compaction, and these result from either (a) increased overburden, or (b) deflections which cause an increase in the overall horizontal effective stress (σ'_h) at any point.

(2) *Compaction-induced* lateral effective stresses ($\sigma'_{h,c}$) are the additional lateral stresses present at any point in excess of the geostatic lateral stresses, and are the direct result of surficial compaction loading. This derivation of the compaction-induced stress fraction by subtraction of the geostatic lateral stress fraction from the overall horizontal stress is illustrated (for K_0-conditions) in Fig. 7.

The overall lateral effective stress (σ'_h) at any point is then the sum of the geostatic and compaction-induced lateral stresses as

$$\sigma'_h = \sigma'_{h,o} + \sigma'_{h,c} \tag{9}$$

It should be noted that there is no evidence that these two 'types' of stresses are in fact truly different in nature; this division of horizontal stresses into two types is made simply for analytical convenience. With respect to the finite element analysis procedures described herein, geostatic lateral stresses arise due to any calculated increase in lateral stress at any stage during an incremental analysis, and thus include all components of lateral effective stress except for lateral stresses directly input as residual 'compaction-induced' stresses during the initial phase of a compaction increment, as described later.

Having thus defined geostatic and compaction-induced stresses, the simple generalized hysteretic stress path model for K_0-conditions illustrated in Figs 5 and 6 can be reformulated in terms of $\sigma'_{h,c}$ and $\sigma'_{h,o}$ by assuming $\sigma'_{h,o} = K_0 \sigma'_v$, and that $\sigma'_{h,c} =$ all horizontal stresses (σ'_h) in excess of $\sigma'_{h,o}$, as illustrated in Fig. 7. This greatly simplifies extension of this hysteretic model to the more general case of non-K_0-conditions.

The final step in transforming this hysteretic stress path model into a general model suitable for non-K_0-conditions is to substitute an 'equivalent' vertical effective stress ($\sigma'_{v,eq}$) axis for the true vertical stress axis (σ'_v) as shown in Figs 5 and 6. The model then 'operates' in terms of the true horizontal stress (σ'_h), and $\sigma'_{v,eq}$ is inferred based on σ'_h. It should be noted that $\sigma'_{v,eq}$ is exactly equal to σ'_v for the specific case of K_0-conditions, but

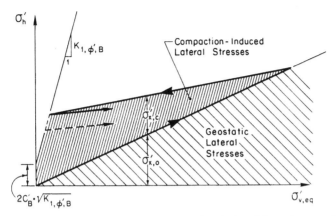

FIG. 7. Definition of geostatic and compaction-induced lateral stress fractions under K_0-conditions.

not for non-K_0-conditions. For all conditions (K_0 or otherwise), this general hysteretic stress path model provides a basis for evaluating the actual lateral effective stress (σ'_h) at any stage of loading or unloading. Although it does not reproduce hysteretic stress paths for loading and unloading under K_0-conditions with the level of accuracy provided by the more complex non-linear hysteretic stress-path model described in the previous section, this simpler model does retain the main characteristics of the more complex model.

The fully non-linear hysteretic stress-path model described in Section 3 provides a higher level of accuracy for modelling loading and unloading under K_0-conditions, and is recommended for such applications. Compaction loading, however, results from discrete, concentrated vehicle loads, and thus does not represent K_0-loading. The complex, fully non-linear stress-path model described in Section 3 has been successfully adapted to analysis of compaction-induced stresses under non-K_0-conditions, but the algorithms required are complex and the improved level of accuracy over that achieved by implementing the simpler general hysteretic stress described in this section has been shown to be small (Ou and Seed, 1987). As there is no significant improvement in overall accuracy of the analysis, and as this simpler model can be relatively simply incorporated into an incremental finite element analysis algorithm, this simplified model is recommended for analysis of compaction-induced stresses and deformations. The sections which follow describe the use of this model in performing this type of analysis.

5 GENERAL FINITE ELEMENT ANALYSIS APPROACH

The incremental finite element analysis methodology developed for analysis of stresses and deformations resulting from placement and compaction of layers of fill simulates the actual sequence of field operations in a number of sequential steps or increments. Four types of increments may be modelled: (a) placement of new soil elements, (b) compaction operations of the current fill surface, (c) placement of new structural elements, and (d) application of various types of loads to the completed fill and/or structure. These analytical procedures permit modelling of multiple 'passes' (cycles) of compaction loading at any given fill stage with a single solution increment, greatly enhancing computational efficiency. In these analyses, compaction loading is realistically considered as a transient moving concentrated surficial load which may pass one or more times over some specified portion of the current fill surface at each stage of backfill placement and compaction.

Two soil behavior models are employed in these analyses. Non-linear stress-strain and volumetric strain behavior of soil is modelled using the hyperbolic formulation proposed by Duncan et al. (1980), as modified by Seed and Duncan (1983). It should be noted that the methods for analysis of compaction effects described herein do not depend on the use of this hyperbolic model, and any other well-formulated non-linear soil behavior model might serve as well in this role.

The second soil behavior model employed is the simplified general stress path model for stresses generated by hysteretic loading and unloading of soil described in the previous section. This hysteretic model performs two roles during analyses: (a) it provides a basis for the controlled introduction of compaction-induced soil stresses at the beginning of each compaction increment, and (b) it acts as an overriding 'filter', controlling and modifying the compaction-induced fraction of soil stresses during all stages of analysis. The following sections describe the use of this model in both of these roles. In all of the sections which follow, the terms σ'_x and σ'_h will be used interchangeably. The use of both terms is for clarity and to conform, when possible, to common conventions.

5.1 Introduction of Compaction-Induced Stresses
Compaction-induced lateral stresses may be introduced into an analysis during 'compaction' increments. Both the peak and residual compaction-induced lateral stresses at any point are modelled based on the peak, virgin compaction-induced horizontal stress increase ($\Delta\sigma'_{h,vc,p}$) which is

defined as the maximum (temporary) increase in horizontal stress which would occur at any given point as a result of the most critical positioning of the surficial compaction plant loading if the soil mass was previously uncompacted (virgin soil) with no 'locked in' residual stresses due to prior compaction. The hysteretic model is then 'driven' during a given compaction loading/unloading cycle by an 'equivalent' peak vertical load increment calculated as

$$\Delta\sigma'_{v,e,p} = \frac{\Delta\sigma'_{h,vc,p}}{K_0} \tag{10}$$

During a given compaction increment, $\Delta\sigma'_{v,e,p}$ is first applied and then removed, as compaction is considered to result in no net increase in residual vertical stresses.

This modelling of compaction on the basis of $\Delta\sigma'_{h,vc,p}$ is convenient because it is a value which has been shown to be readily calculated by simple, linear-elastic analyses (Seed and Duncan, 1983). Moreover, case studies based on actual full scale field data (e.g. Seed and Duncan, 1986a,b; Seed et al., 1986; Ou and Seed, 1987) indicate that transforming the value to $\Delta\sigma'_{v,e,p}$ by eqn (10), and then using $\Delta\sigma'_{v,e,p}$ to 'drive' the hysteretic stress-path model, results in calculation of appropriate true peak and residual lateral stress increases ($\Delta\sigma'_{h,p}$ and $\Delta\sigma'_{h,r}$, respectively).

It is important to note that the modelling of peak compaction loading must be based on directly calculated *lateral* stress increases rather than directly calculated vertical stress increases multiplied by some constant (e.g. K_0 or K_A), because for surficial loading of finite lateral extent the relationship between $\Delta\sigma'_h$ and $\Delta\sigma'_v$ is far from constant. This is illustrated in Fig. 8, which schematically shows $\Delta\sigma'_h$ and $\Delta\sigma'_v$ at several depths due to a surficial point load. It may be noted that $\Delta\sigma'_h$ is negative where $\Delta\sigma'_v$ is greatest, and that it exceeds $\Delta\sigma'_v$ at other locations.

$\Delta\sigma'_{hvc,p}$, which is independent of previous hysteretic stress history effects, can be readily calculated by simple linear-elastic analyses (as described in Section 7) and is directly input for each soil element during each compaction increment. The general hysteretic model then accounts for the effects of previous hysteretic loading cycles (e.g. previous compaction increments) and calculates both the actual peak and residual lateral stress increases based on $\Delta\sigma'_{h,vc,p}$. The resulting compaction-induced residual stress increases are added to each soil element at the beginning of each compaction increment, and associated compaction-induced nodal point forces are used to 'drive' a two-iteration finite element solution process during the compaction increment.

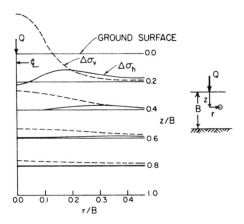

FIG. 8.   Schematic illustration of $\Delta\sigma_h$ and $\Delta\sigma_v$ due to a vertical surface point load.

The initial residual horizontal effective stress at any point (in any soil element) during the initial phase of a compaction increment (prior to the occurrence of nodal point deflections) is modelled as

$$\sigma'_{h,r} = \max \left\{ \begin{array}{l} \sigma'_{h,o} + F(\sigma'_{h,p} - \sigma'_{h,o}) \\ \text{or} \\ \sigma'_{h,o} + \sigma'_{h,c,old} \end{array} \right\} \qquad (11)$$

where

$\sigma'_{h,o}$	= the geostatic fraction of the lateral effective stress
$\sigma'_{h,c,old}$	= the compaction-induced fraction of the lateral effective stress prior to the compaction increment
F	= the fraction of peak stress increase retained as residual stress (see eqn. 5) and
$\sigma'_{h,p}$	= the peak (transient) horizontal effective stress during compaction which may be calculated as

$$\sigma'_{h,p} = \max \left\{ \begin{array}{l} \sigma'_{h,o} + \Delta\sigma'_{h,vc,p} \\ \text{or} \\ \sigma'_{h,o} + \sigma'_{h,c,old} + (K_3/K_0) \cdot \Delta\sigma'_{h,vc,p} \end{array} \right\} \qquad (12)$$

The value of $\sigma'_{x,r}$ as determined by eqn (11) is subject to two overriding limiting conditions. The directly input compaction-induced residual increase in lateral stress in any soil element is limited such that the final principal stress ratio (σ'_1/σ'_3); (1) is not caused to exceed $K_{1,B}$ (see eqns (4) and (7)) and (2) does not represent mobilization of greater than 85% of the shear strength of the soil based on Mohr–Coulomb failure criteria. When these two limiting conditions conflict, the more strictly limiting are the two controls. These limiting conditions will not, however, cause the residual lateral stress to decrease prior to the occurrence of nodal displacements.

The resulting modelled increase in horizontal stress, in turn, affects the normal stresses of all orientations except the vertical stress. In order to maintain vertical and horizontal equilibrium at all stages of an analysis, it is assumed that: (a) the initial (pre-nodal displacement) portion of a compaction increment (after addition of the new compaction-induced residual lateral stress) results in no net change in σ'_y or τ_{xy} at any point, in order to maintain vertical equilibrium, and (b) the residual increases in σ'_x which result from compaction are accompanied by horizontal nodal forces exerted at the boundaries of the soil mass, in order to maintain horizontal equilibrium.

5.2   Control of Compaction-Induced Lateral Stresses During All Increments

In addition to establishing the magnitude of residual compaction-induced lateral stresses introduced during the initial stages of compaction increments (prior to nodal displacements), the general hysteretic model also acts as a 'filter', controlling and modifying the compaction-induced fraction of lateral stresses in soil elements at all stages of analysis. This involves monitoring two types of lateral stress changes as follows:

(1) *Increase in geostatic horizontal stress*: All calculated increases in σ'_x at any stage during an analysis (except for the directly input compaction-induced increases) are considered to represent an increase in geostatic lateral stress.

Although the contribution of $\sigma'_{x,c}$ to σ'_x is continuously monitored, the global finite element solution procedures used to calculate incremental lateral stress increases are not affected by the natures of the lateral stresses involved (geostatic vs compaction induced) and thus do not account for the fact such increases constitute hysteretic 'reloading' if a compaction-induced stress component is present. Subsequent to the solution of the global stiffness and displacement equations for any increment, therefore, the resulting calculated increases in σ'_x (assumed to represent an increase in $\sigma'_{x,o}$) are in turn used to calculate, using the general hysteretic stress-path model, an associated decrease in the compaction-induced fraction of lateral stress ($\sigma'_{x,c}$) to a new value defined by the equation

$$\sigma'_{x,c,new} = \max \begin{cases} \sigma'_{x,c,old} - \Delta\sigma'_{x,o} \dfrac{(K_0 - K_3)}{K_0} \\ \text{or} \\ 0 \end{cases} \tag{13}$$

Because σ'_x at any point is the sum of $\sigma'_{x,o}$ and $\sigma'_{x,c}$, this progressive erasure or 'overwriting' of compaction-induced lateral stresses by increased geostatic lateral stresses results in an overall increase of σ'_x less than the calculated increase in $\sigma'_{x,o}$ for soil with some previously 'locked-in' compaction-induced lateral stress component, and corresponds to 'reloading'. In order to maintain equilibrium at all boundaries where nodal point forces were previously applied to reflect directly input compaction-induced stress increases, appropriate nodal forces of opposite sign must be applied to reflect this reduction in lateral stress ($\sigma'_{x,c}$).

(2) *Decrease in lateral stress*: When solution of the global stiffness and displacement equations results in a calculated decrease in σ'_x, it is assumed that this decrease is borne by both the geostatic and compaction-induced fractions of the pre-existing lateral stress in direct proportion to their contributions to the overall lateral effective stress. The magnitude of the net change (decrease) in the overall effective lateral stress (σ'_x) is thus unaffected by the relative proportions of geostatic and compaction-induced lateral stresses. Because the hysteretic stress-path model results in no overall lateral stress modifications, no additional nodal forces are applied.

6 MODELLING MULTIPLE CYCLES OF COMPACTION LOADING WITH A SINGLE SOLUTION INCREMENT

Compaction-induced lateral stress increases in a soil mass can exert increased pressure against adjacent structures, resulting in structural deflections which may in turn partially alleviate the increased lateral stresses. Multiple passes of a surficial compaction plant, however, continually re-introduce the lateral stresses and result in progressive displacements at shallow depths. This results in a situation in which compaction-induced lateral stresses resulting from multiple passes of a compaction plant behave essentially as 'following' loads (being undiminished as a result of deflections) in the region near the ground surface where compaction-induced lateral stresses are most significant in magnitude relative to geostatic stresses. At greater depths, compaction-induced structural deflections can cause reduction in both the geostatic and compaction-induced lateral stresses to such an extent that the overall lateral effective stress is reduced as a result of compaction.

In order to approximate this complex cyclic process with a single solution increment, both compaction-induced lateral stresses and the corresponding nodal point forces for a given compaction increment are assumed to represent 'following' loads from the current ground surface down to a second surface, the depth of which is specified for each compaction increment. All soil elements above this second surface are assigned very small values of modulus during the two-iteration solution of the compaction increment. This modelling procedure results in calculation of displacements at *all* nodal points as a result of compaction-induced lateral forces. However, (a) no changes in soil stresses result from displacements in soil elements above the specified depth of 'following' compaction loading, and (b) no changes in compaction-induced nodal forces result in this upper region.

This procedure is illustrated schematically in Fig. 9. Figure 9(a) shows

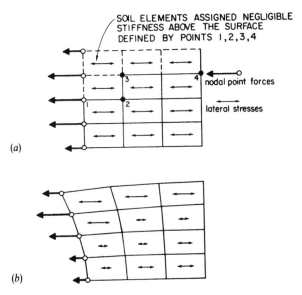

FIG. 9. Schematic illustration of the use of a single solution increment to model multiple compaction loading cycles. (a) Conditions after assignment of compaction induced lateral stresses and nodal forces, prior to the occurrence of nodal deflections. (b) Conditions after the occurrence of nodal deflections.

the modelled conditions following the initial portion of a compaction increment (prior to the occurrence of nodal deflections). The figure corresponds to conditions within a fill adjacent to a flexible wall (located at the left boundary of the mesh). Using the hysteretic model, residual compaction-induced lateral stress increases (represented by double-ended arrows) have been assigned to the soil elements, and associated nodal point forces (represented by large arrows) have been assigned to the nodes at the soil/structure interfaces. All soil elements above the surface defined by points 1, 2, 3, and 4 have been assigned negligible stiffness in order to model 'following' compaction loading in this near-surface region. In Fig. 9(b), the compaction-induced nodal forces have caused displacement of the structure to the left, resulting in a decrease in lateral stresses in soil elements below the depth of following loading, and an associated decrease in the lateral nodal forces exerted at the soil/structure interface in this zone. In the (upper) zone of 'following' loading, however, neither the lateral soil stresses nor the interface nodal forces have been diminished by the occurrence of deflections.

The modelled depth of following loading need not represent a horizontal plane. Instead, this depth may vary across a finite element mesh. Based on analyses of a number of well documented field studies, it appears that good results can be obtained by modelling 'following' compaction loading from the surface to a depth (h) defined by the expression

$$h = \Delta\sigma'_{h, vc, p}/(K_0 \cdot \gamma_m) \tag{14}$$

where h = depth below the current fill surface and γ_m = the average unit weight of the overlying soil.

7   PERFORMING AN INCREMENTAL ANALYSIS INCLUDING MODELLING OF COMPACTION-INDUCED STRESSES

The steps involved in actually using these models and techniques for analysis of a problem involving soil compaction are: (1) definition of overall problem geometry; (2) development of a suitable finite element mesh, with appropriate consideration of the incremental sequence of fill placement and compaction; (3) evaluation of non-linear stress-strain and bulk modulus soil model parameters for the various soils involved; (4) evaluation of modelling parameters for structural elements (and soil/structure interfaces), if present; (5) evaluation of model parameters for the general hysteretic stress path model described in Section 4; (6) evaluation of $\Delta\sigma'_{h,vc,p}$ in *each* soil element for *each* compaction increment; and (7) performance of the incremental finite element analysis. Steps 1, 2, 3, 4 and 7 are common to all incremental non-linear geotechnical finite element analyses, and require no discussion. Steps 5 and 6, development of hysteretic model parameters and evaluation of $\Delta\sigma'_{h,vc,p}$, are unique to the compaction problem and are discussed in this section. These analysis techniques are also illustrated by analysis of a series of full-scale field case studies, described in Section 8.

7.1 Evaluation of Hysteretic Stress-Path Model Parameters

Six material property parameters are required for the hysteretic model: K_0, $K_{1,\phi',B}$, c'_B, K_2 (or F), and K_3. These parameters are described in Table 1, and were illustrated previously in Fig. 5. The hysteretic stress-path model was developed in a manner which lends itself to relatively simple derivation of these model parameters by means of empirical correlations with conventional soil parameters. These model parameters can be determined by either of two procedures: (a) based on data for a single-cycle K_0-loading and unloading test, or (b) based on empirical correlations with ϕ', c' (and compaction water content (w_c) or degree of saturation (S) for clayey soils), as described in Table 1. The empirical correlations suggested in Table 1 are based on laboratory K_0-test data as well as analysis of full-scale field case studies. Seed and Duncan (1983) provide a detailed description of parameter evaluation techniques for cohesionless soils, and a similar description for cohesive soils is provided by Ou and Seed (1987).

It is suggested that best results with regard to analysis of compaction-induced stresses can generally be achieved by setting $K_3 \approx (0.8)(K_2)$. In addition, K_2 and F are directly related by eqn (5), so that choosing one automatically 'fixes' the value of the other. It is thus necessary to determine only the following four parameters: K_0, $K_{1,\phi',B}$, c'_B, and F. Model parameters used should be representative of the post-compaction properties of the soils in question, as each lift may be subjected to multiple cycles (passes) of compaction loading, and the as-compacted soil properties at the end of compaction will dominate overall behavior.

Analysis of cohesionless soils is performed using effective stress modelling parameters, and the parameter evaluation techniques presented in Table 1 are well supported for cohesionless soils. It should be noted that a high degree of compaction results in an increased friction angle (ϕ'), and this in turn increases F, the fraction of peak compaction-

TABLE 1
GENERALIZED HYSTERETIC LOADING/UNLOADING STRESS PATH MODEL PARAMETERS

Parameter	Name or description	Recommended range of values	Recommended method of estimation based on correlation with ϕ' and c'
K_0	Coefficient of at-rest lateral earth pressure for virgin loading	$0.3 \leqslant K_0 \leqslant 0.7$	$K_0 \simeq 1 - \sin \phi'$
$K_{1,\phi',B}$	Frictional component of the limiting coefficient of at-rest lateral earth pressure	$K_0 \leqslant K_1 \leqslant K_p$	$K_{1,\phi',B} \simeq (2/3)\tan^2(45 + \phi'/2)$
c'_B	Cohesive component of the limiting coefficient of at-rest lateral earth pressure	$c'_B \geqslant 0$	$c'_B \simeq 0.8c'$
F	Fraction of peak lateral compaction stress retained as residual stress for virgin soil	$0 \leqslant F \leqslant 0.8$	*For cohesionless soils:* $F = 1 - \dfrac{(OCR - OCR^\alpha)}{(OCR - 1)}$ with a recommended value of overconsolidation ratio (OCR) equal to 5, and α selected as a function of ϕ' based on Fig. 11
or			*For cohesive soils:* $F \approx 0.3$–0.6 for $w_c < w_{opt}$ $F \approx 0$–0.3 for $w_c > w_{opt}$
K_2	Incremental coefficient of at-rest lateral earth pressure for unloading $[K_2 = K_0(1-F)]$	$K_0 \geqslant K_2 \geqslant 0$	
K_3	Incremental coefficient of at-rest lateral earth pressure for reloading	$0 \leqslant K_3 \leqslant K_0$	$K_3 \simeq (0.8)K_2$

Note: (1) $K_p = \tan^2(45 + \phi'/2)$; (2) w_c = compaction water content; (3) w_{opt} = the Modified AASHTO (ASTM D-1557) optimum water content.

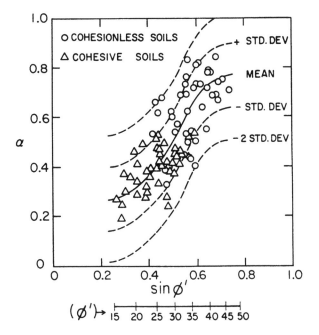

FIG. 10.   Suggested relationship between sin ϕ' and α.

induced lateral stress which is retained in the soil as a residual 'locked in' stress after removal of the compaction equipment loads. This increase is relatively minor, however, so that residual compaction-induced stresses are controlled mainly by the stresses induced by the compaction equipment and are only moderately influenced by degree of compaction. It will be demonstrated later, in discussing the Case Studies, that the dominant factors influencing the magnitude of compaction induced stresses are the equipment and procedures used to place and compact the fill, and *not* the degree of compaction achieved.

Analyses of cohesive soils are performed using total stress analysis and modelling parameters. There are less data available for evaluation of parameters for 'rapid' (undrained) loading and unloading of these soils, so the recommendations which follow are not yet well supported and may change as more data become available. The total stress at-rest lateral earth pressure coefficient can be estimated as

$$K_{0,\text{total stress}} \approx \frac{K_0(\sigma_v - u) + u}{\sigma_v} \tag{15}$$

where

$\sigma_v =$ total vertical stress

$u =$ pore pressure and

$K_0 =$ the coefficient of earth pressure at rest for 'virgin' loading conditions (based on effective stresses), which can be approximated as

$$K_0 \approx 1 - \sin \phi' \qquad (16)$$

Fully saturated cohesionless soils have been shown to retain little or none of the peak lateral compaction-induced lateral stress as residual stress, and should thus be modelled with $F \approx 0$. Similarly, cohesive soils compacted wet of the Modified AASHTO optimum water content have low retention values ($F \approx 0$–0.3). Cohesive soils compacted dry of the Modified AASHTO optimum water content can retain a significant fraction of peak compaction-induced stresses, and recommended values of F are on the order of 0.3 to 0.6 for these conditions. As an example, Fig. 11 illustrates the variation of F as a function of as-compacted density

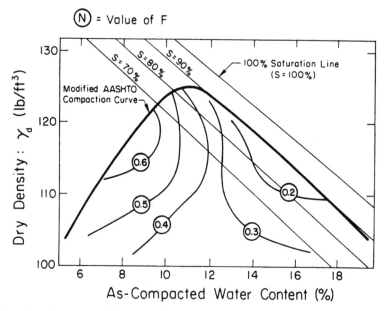

FIG. 11. Variation of the fraction (F) of peak lateral stress increase retained as residual lateral stress in a cohesive soil as a function of compaction conditions.

and water content for a silty clay (Ou and Seed, 1987). Samples were compacted to a given density and water content, and were then rapidly loaded and unloaded under one-dimensional K_0-conditions in a rigid-walled oedometer designed to measure total horizontal stresses without permitting horizontal displacement. The F-values shown in Fig. 11 represent the fraction of total peak lateral stress increase retained as 'locked in' residual lateral stress increase after removal of the vertical loads applied.

7.2 Evaluation of $\Delta\sigma'_{h,vc,p}$

The most important factors controlling the magnitude of compaction-induced stresses at any point are the footprint geometry, weight and closest proximity achieved by any compaction or construction vehicle. These factors control $\Delta\sigma'_{h,vc,p}$, as illustrated schematically in Fig. 12 for the case of $\Delta\sigma'_{h,vc,p}$ against a vertical wall.

The calculation of $\Delta\sigma'_{h,vc,p}$ for each soil element at each stage of fill placement and compaction is thus of significant importance to the overall analysis of compaction effects. Seed and Duncan (1983) present a study of techniques suitable for analysis of $\Delta\sigma'_{h,vc,p}$, which is defined as the peak, transient horizontal soil stress increase due to application of loads at the surface of a soil mass with no 'locked-in' residual compaction-induced stresses. The following is a brief summary of their findings:

(1) $\Delta\sigma'_{h,vc,p}$ cannot be correctly calculated as some constant (e.g. K_0 or K_A) multiplied by the peak *vertical* stress increase (as illustrated in Fig. 8). Instead the peak *horizontal* stress increase must be *directly* calculated.

(2) Calculation of $\Delta\sigma'_{h,vc,p}$ is a three-dimensional problem.

(3) Values of $\Delta\sigma'_{h,vc,p}$ can be calculated with good accuracy using simple linear-elastic analyses. Suitable procedures include three-dimensional finite element analyses, as well as closed form solutions such as the Boussinesq (1885) solution. For saturated clays, Poisson's ratio should be taken as 0·5. For cohesionless soils, as well as non-saturated cohesive soils, Poisson's ratio (v) should be taken as

$$v = 0·5(v_{K_0}) + 0·25 \qquad (17)$$

where

$$v_{K_0} = \frac{K_0}{1 + K_0} \qquad (18)$$

and K_0 is the virgin loading at-rest earth pressure coefficient.

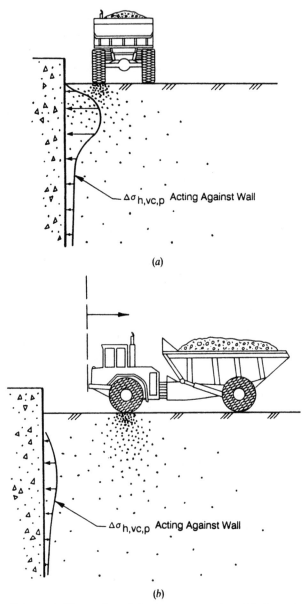

(a)

(b)

FIG. 12.   Schematic illustration of the importance of vehicle footprint proximity in determining $\Delta\sigma_{h,vc,p}$. (a) Vehicle in close proximity to wall. (b) Vehicle not closely approaching wall.

(4) $\Delta\sigma'_{h,vc,p}$ acting against a vertical wall can be calculated with excellent accuracy using 'free field' solutions applicable to analysis of an infinite elastic half-space, by simply doubling the 'free field' value of $\Delta\sigma'_{h,vc,p}$ at the wall face. This applies *whether or not* the wall face deflects somewhat under the resulting earth pressure increase.

(5) The peak, transient vertical surface loading induced by a vibratory roller operating efficiently is typically between two and four times the static weight of the roller (e.g. Toombs, 1972). In the absence of actual vertical thrust measurements during vibratory operation, a multiplier of three is recommended for vibratory rollers.

Figure 13 illustrates the accuracy with which elastic analyses can be used to calculate $\Delta\sigma'_{h,vc,p}$. It is not currently possible to make reliable in-situ measurements of earth pressures in the free field. Accordingly, the only reliable measurements of $\Delta\sigma'_{h,vc,p}$ due to surface loads are measurements of $\Delta\sigma'_{h,vc,p}$ against walls or other structures. Figure 13 shows measured values of $\Delta\sigma'_{h,vc,p}$ against a vertical wall due to application of a 1000 lb (454 kg) surface point load at various distances from the wall face. The data shown were developed by Gerber (1929) and Spangler (1938), and are of significant interest because the clever techniques used to measure earth pressures against the walls avoided the types of measurement problems which will be discussed later. As a result, these early data are of unusual value and accuracy.

The solid lines in Fig. 13 show $\Delta\sigma'_{h,vc,p}$ calculated by doubling the Boussinesq (1885) calculated values of $\Delta\sigma_h$. It can be seen that the level of agreement is very good.

Despite this, however, there remains considerable confusion and disagreement among geotechnical engineers as to how best to calculate $\Delta\sigma'_{h,vc,p}$. Indeed, a number of 'empirical' procedures have been proposed for calculating $\Delta\sigma'_{h,vc,p}$ against walls. Some of these can lead to erroneous results.

This confusion within the geotechnical profession is the result of a large body of misleading data regarding field measurements of $\Delta\sigma'_{h,vc,p}$ due to concentrated surface loads. This, in turn, is the result of two sources of error: (a) errors in earth pressure measurements, and (b) 'locked-in' compaction-induced stresses, which cause measured values of $\Delta\sigma_h$ due to surface loading to represent reloading and thus to underestimate $\Delta\sigma'_{h,vc,p}$.

The two most common types of systematic earth pressure measurement system error are illustrated in Fig. 14. An earth pressure measurement cell

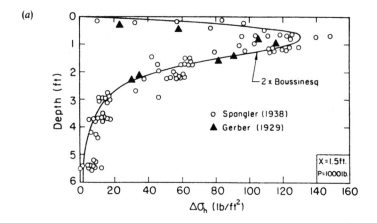

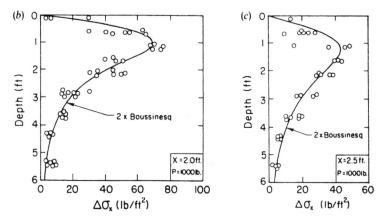

FIG. 13. Comparison between calculated and measured lateral earth pressure increases against a wall due to application of a 454 kg (1000 lb) surface point load. (a) Point load 0·5 m (1·5 ft) from the wall. (b) Point load 0·6 m (2·0 ft) from the wall. (c) Point load 0·75 m (2·5 ft) from the wall.

which protrudes from a wall face, as in Fig. 14(a), is typically much less compressible than the soil which it replaces. The protruding cell thus represents a rigid inclusion, and attracts more than its share of the earth pressures, resulting in overregistration. When the earth pressure cell is inset into the wall with its face flush with the wall face, as shown in Fig. 14(b), a second problem can arise. Most earth pressure cells require some

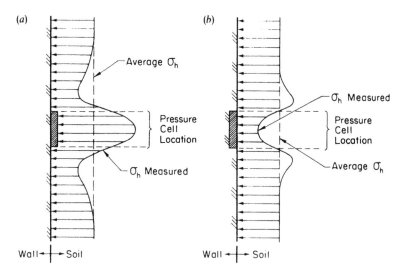

FIG. 14. Schematic illustration of potential errors in measurement of earth pressures acting against a wall. (a) Earth pressure cell protruding from wall face. (b) Earth pressure cell inset flush with wall face.

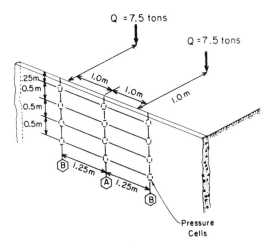

FIG. 15. Peak point loading configuration and pressure cell layout: Stockholm test wall.

small deflection of their faces in order to register some measurement. Unfortunately, very small pressure cell face deflections are sufficient to cause soil arching, which results in under-registration of earth pressures. The best technique for avoiding these problems is to employ a very stiff (essentially non-displacement) type of earth pressure cell inset into the wall with its face flush with the wall face.

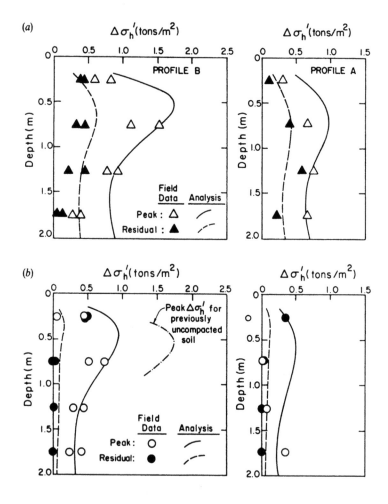

FIG. 16.   Measured peak and residual lateral earth pressure changes, Stockholm test wall: gravelly sand backfill. (a) Loosely dumped fill. (b) Fill placed and compacted in layers.

The problem of 'locked-in' compaction-induced stresses is well illustrated by the studies performed by Rehnman and Broms (1972) who applied and removed a pair of point loads at the surface of a fill against a wall, and measured the resulting peak increase in earth pressures against the wall during load application, and the residual earth pressure increases after load removal. Figure 15 shows the wall and backfill

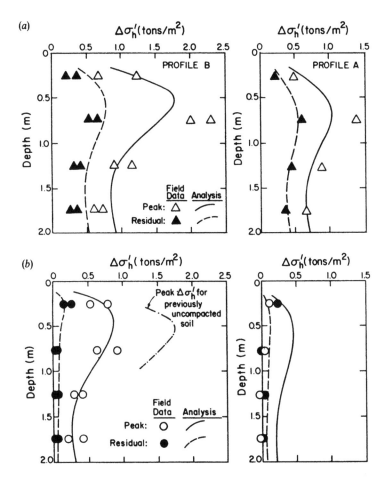

FIG. 17. Measured peak and residual lateral earth pressure changes, Stockholm test wall: silty fine sand backfill. (a) Loosely dumped fill. (b) Fill placed and compacted in layers.

configuration, the load application points, and the locations of the pressure cells used to measure peak and residual pressure increases.

The tests were performed using two different types of backfill: a gravelly sand and a fine silty sand. Initially, both backfills were loosely dumped. The resulting peak lateral pressure increases in the loosely dumped fill are shown by the open triangles in Figs 16(a) and 17(a). The solid lines in these figures are values of $\Delta\sigma'_{h,vc,p}$ calculated by simple elastic analyses. Allowing for scatter in the measurements, the agreement is fairly good. The solid triangles show the residual earth pressure increases after load removal, and the dashed lines the residual compaction-induced lateral stresses calculated using the analytical techniques described in this chapter (Duncan and Seed, 1986). Agreement between these analyses and the field measurements can be seen to be good.

Rehnman and Broms then repeated these tests, this time placing and compacting the backfill in layers so as to deliberately induce compaction-induced stresses. When the loads were subsequently applied to the surface of the compacted fill, the resulting measured peak earth pressure increases were much smaller than for the uncompacted fills, as shown by the open circles in Figs 16(b) and 17(b). This clearly illustrates the fact that application of loads to a soil which already has significant compaction-induced lateral stresses constitutes 'reloading', and results in a smaller increase in lateral stress than would virgin loading. Failure to account for this reloading effect can result in misinterpretation of test results.

8 FULL-SCALE FIELD CASE STUDIES

It is important to calibrate and verify new analysis techniques and behavioral models, such as those described in this chapter, by comparison with full-scale field case studies. A number of such case study comparisons have been made in which predictions developed using these analytical tools were compared with stresses and deformations measured in the field. These studies have involved measurements and analyses of compaction effects on both rigid and deformable walls, bridge abutments, long-span flexible metal culvert structures, and reinforced-soil walls (e.g. Seed and Duncan, 1986a,b; Seed et al., 1986; Seed and Ou, 1987; Ou and Seed, 1987). The good levels of agreement achieved between analyses and field measurements in these cases provide strong support for the accuracy and usefulness of these new analytical techniques.

This section presents three such case studies as an illustration of the uses of these new methods for analysis of compaction-induced stresses and deformations. These case studies also serve to illustrate the principal factors affecting compaction-induced stresses, and lead to recommendations for minimizing the potential adverse effects of compaction-induced soil stresses by means of appropriate control of field compaction procedures.

8.1 TRRL Test Walls

Carder *et al.* (1977) provided an excellent set of field measurements of compaction-induced earth pressures and resulting wall deflections in a test bin at the TRRL Experimental Retaining Wall Facility in Crowthorne, England. Figure 18 shows the TRRL test facility. A reinforced concrete trough 22 m long and 3 m deep was backfilled with a compacted clean medium sand, and the resulting earth pressures against the two side walls were measured. Backfill was placed in 0.15 m lifts, and each lift was compacted with a 1.3 mg twin-roll vibratory roller operating parallel to the walls and approaching to within 0.15 m of the walls.

As shown in Fig. 18(a), the wall at one test section was a 1 m thick reinforced concrete wall 2 m high, and at the other test section the wall was a 2 m high steel wall braced with hydraulic jacks. Earth pressures on the faces of both walls were measured using arrays of flush-mounted pressure cells of three types: hydraulic, pneumatic, and stiff strain-gauged diaphragm. The resulting pressure distributions measured by each cell type on the face of the metal wall were compared with the total lateral force exerted on the wall as measured by means of load cells on the bracing jacks. Based on this comparison, all hydraulic and diaphragm cell measurements were scaled by factors of 1·01 and 0·85 respectively, and the pneumatic measurements were discarded as unreliable. The remaining scaled earth pressure measurements represent a set of compaction-induced lateral earth pressure measurements of unusually high reliability. The massive reinforced concrete wall at the first test section was assumed to be essentially non-deflecting (non-yielding) under the types of earth pressures exerted by the backfill, and deformations of the braced steel wall were carefully measured during placement and compaction of the backfill.

The finite element mesh shown in Fig. 18(b) was used to model placement and compaction of fill at the metal test wall section. Advantage was taken of the approximate symmetry of the test trough, and nodal points at the right-hand boundary of the mesh were free to displace

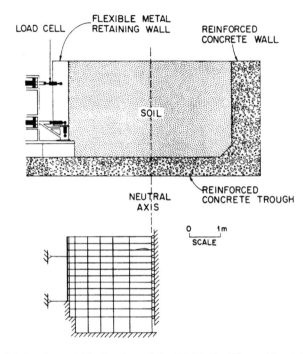

FIG. 18.   Finite element idealization of the TRRL flexible metal retaining wall.

vertically, but were fixed against lateral deflection. Beam elements were used to model the metal test wall, and bar elements were used to represent the supports. A similar mesh, but with a laterally non-yielding soil/wall boundary, was used to model placement and compaction to fill at the reinforced concrete test wall section. The following hyperbolic stress-strain model parameters were used to model the compacted sand backfill: $\gamma_m = 19 \cdot 6 \, \text{kN/m}^2$ (125 pcf), $K = 500$, $K_{ur} = 700$, $n = 0 \cdot 4$, $R_f = 0 \cdot 7$, $K_B = 140$, $m = 0 \cdot 2$, $\phi' = 38 \cdot 7°$, $\Delta\phi' = 0°$, and $c' = 0$. Based on the reported post-compaction friction angle of $38 \cdot 7°$, and the empirical parameter determination procedures suggested in Table 1, the following hysteretic stress-path model parameters were used to model the compacted sand backfill: $K_0 = 0 \cdot 37$, $K_{1,\phi',B} = 2 \cdot 88$, $c'_B = 0 \cdot 0$, and $K_2 = K_3 = 0 \cdot 10$.

Having thus established finite element meshes and appropriate model parameters, the next step in analyzing the placement of backfill against either wall was to determine suitable values of $\Delta\sigma'_{h,vc,p}$ for modelling peak compaction loading at all points during each compaction increment.

Several profiles of $\Delta\sigma'_{h,vc,p}$ vs depth below the current fill surface were calculated because the compaction plant only approached to within 0·15 m (0·5 ft) of each wall, and the walls themselves influenced the three-dimensional peak compaction-induced stress fields. Figure 19(a)

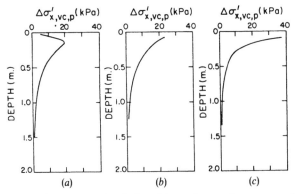

FIG. 19.   Peak virgin, compaction-induced lateral stress increase profiles used in analyses of the TRRL wall. (a) Against wall. (b) $x=0·15$ m. (c) Free field.

shows $\Delta\sigma'_{h,vc,p}$ acting at the soil/wall interfaces, and Fig. 19(b) shows $\Delta\sigma'_{h,vc,p}$ at a distance of 0·15 m (0·5 ft) from the walls. Farther from the walls, 'free field' conditions prevailed, and the compaction plant passed fully over the underlying soil. Figure 20(c) illustrates the profile of $\Delta\sigma'_{h,vc,p}$ used in this 'free field' region. All three profiles of $\Delta\sigma'_{h,vc,p}$ were based on three-dimensional linear elastic analyses, using $v\approx0·38$.

  Having thus established model parameters and a basis for modelling peak compaction loading ($\Delta\sigma'_{h,vc,p}$), the computer program SSCOMP (Seed and Duncan, 1984) was used to model incremental placement and compaction of backfill adjacent to the non-deflecting concrete wall. Based on the empirical guidelines proposed in eqn (14), compaction loading was modelled as 'following' loading to a depth of two full element layers below the current fill surface during all compaction increments. The final results of this analysis are presented in Fig. 20, which shows a comparison between the calculated and measured final lateral earth pressures. The degree of agreement between the calculated and measured values is very good. Also shown by the dashed line in Fig. 20 is the conventional at-rest earth pressure distribution which would have been calculated if compaction-induced stresses had not been included in the analysis. Inclusion of compaction-induced stresses nearly doubled the overall final

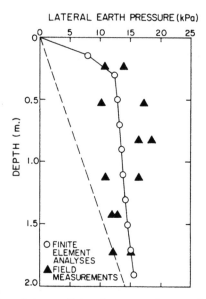

FIG. 20.   Comparison between finite element analyses and measured pressures:
TRRL concrete retaining wall.

lateral force acting against the wall, and tripled the soil-induced overturning moment about the base of the wall.

The same procedures were used to model placement and compaction
of backfill against the 'flexible' metal wall section, and the results are
presented in Fig. 21. The solid lines represent the final calculated wall
deflections and lateral earth pressures acting against the wall. Once again,
agreement with the actual field measurements is good. The dashed lines
in Fig. 21 represent earth pressures and wall deflections calculated by
means of conventional incremental finite element analyses without including compaction-induced stresses. It can be seen that modelling
compaction-induced stresses again resulted in greatly increased wall
deflections and bending moments, caused an 80% increase in the overall
final lateral force acting against the wall, and caused a three-fold increase
in pressures acting on the upper part of the wall.

In addition to providing an illustration of the use of the analytical
techniques described in this chapter, the TRRL case study also serves to
illustrate a number of fundamental aspects of compaction-induced soil
stresses and their influences on soil/structure systems. Figures 20 and 21

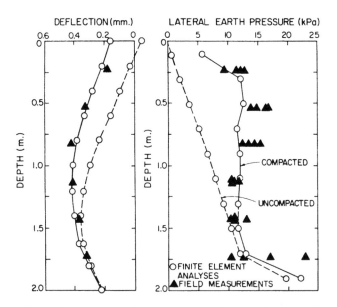

FIG. 21.   Comparison between finite element analyses and measured pressures
and deflections: TRRL metal retaining wall.

show that compaction causes the largest increase in lateral soil stresses
at shallow depths, and that the magnitude of the compaction-induced
earth pressure decreases with depth. This is because (a) the peak, transient
stress increase is largest at relatively shallow depth, as shown in Fig. 19,
and (b) the fraction of this transient stress increase which remains
'locked-in' as a residual, compaction-induced lateral stress is progressively 'erased' or 'overwritten' by increasing geostatic stresses as the depth of
burial increases. This process represents 'reloading', as described previously in Sections 3 and 4. At some depth of burial, all of the compaction-
induced stresses are overwritten, and earth pressures are the same as
would occur without compaction.

In this TRRL case study, the compaction-induced stresses induced by
a light, twin-drum hand-operated vibratory compactor were significant
to a depth of approximately 1·5 m (about 5 ft). For typical trucks, tractors
and self-propelled rollers (not hand compactors), the depth of influence is
typically on the order of 3–4·7 m (10–15 ft), though very heavy compaction equipment has been shown to produce residual compaction-induced
soil stresses to depths as great as 20 m (Seed and Duncan, 1983).

8.2 The Promontory Culvert Overpass Structure

This and the next case study represent field measurements and finite element analyses of compaction-induced deformations and bending moments for a pair of nearly identical long-span metal culvert overpass structures. For the first of these structures, field compaction procedures were carefully controlled, allowing only light hand-operated compaction equipment to operate in close proximity to the structure in order to minimize compaction-induced structural deformations and bending stresses. For the second structure, such field procedural controls were not enforced, and the operation of large compaction and construction vehicles in close proximity to the structure resulted in large compaction-induced structural deformations and the onset of plastic yield in the culvert structure. As a result, these two cases provide an excellent illustration of the importance of field procedures (type, size, and proximity of compaction equipment) in determining the influence of compaction-induced stresses on soil/structure interaction and performance.

The first case involves the Promontory culvert structure, located in Mesa, California. Figure 22(a) shows a cross section through the structure. The arched culvert has a span of 11·7 m (38 ft 5 in), a rise of 4·8 m (15 ft 9 in) and a length of 24·4 m (80 ft). The culvert is founded on 0·9 m (3 ft) high reinforced concrete walls with a reinforced concrete base slab. The culvert itself consists of corrugated aluminium structural plate, and the crown section is reinforced with externally attached (bolted) aluminium stiffener ribs spaced 23 cm (9 in) apart.

The foundation soil at the site was a stiff, silty sandy clay of low plasticity (CL-SC). Chemical tests indicated that this sandy clay was potentially corrosive with respect to the culvert structure. As a result, a crushed basalt material (select fill) was imported for use as a protective backfill envelope within 0·9–1·2 m (3–4 ft) of the culvert. This crushed basalt was an angular silty sand (SM), and was placed to a minimum width of 1·2 m (4 ft) at both sides of the culvert, and was continued to the final fill surface. The existing sandy clay was used as backfill outside of this select fill zone. Both materials were compacted to a minimum of 95% of the maximum dry density determined by a Standard Proctor Compaction Test (ASTM 698-D).

Backfill operations were well controlled and measured deformations at two sections (Sections A-A and B-B) were nearly identical at all backfill stages. Figure 23 shows the final deformed culvert shapes at both sections. In this figure, deformations are exaggerated by a scaling factor of 5 for clarity. From this point on, all 'measured' deformations discussed

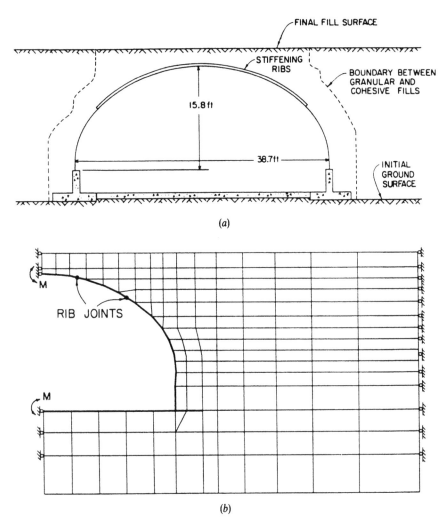

FIG. 21.   The promontory overpass culvert structure. (a) Cross section showing structural and backfill configurations. (b) Finite element mesh used for analyses.

will represent averaged deformations for the two measured sections.

The general pattern of culvert deformations measured during backfill placement and compaction can be well characterized by monitoring the vertical deflection of the crown point and the inward radial deformation of the quarter point, as shown in Figs 23 and 24. The general pattern of

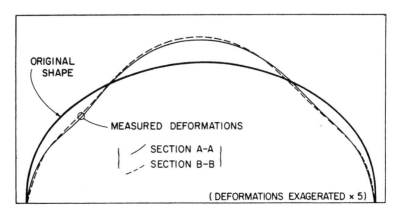

FIG. 23. Final deformed shapes of sections A-A and B-B of the promontory overpass culvert structure.

culvert deformations consisted of inward flexure of the upper quarter points and upward movement of the crown ('peaking'). In Fig. 24, in which the solid triangles represent the measured crown and quarter point deflections as a function of backfill level, it can be seen that as backfill was placed above the crown of the structure, peaking reversed and the crown began to descend slightly under the weight of the cover fill.

The most important factors affecting the magnitude of compaction-induced soil stresses at any given point in the ground are the contact pressure, footprint geometry and closest proximity to the point of interest achieved by the compaction equipment (or other construction vehicles). This is because these factors control the magnitude of $\Delta\sigma'_{h, vc, p}$. In order to model compaction-induced earth pressures acting against the culvert, it was thus necessary to continuously monitor the closest proximity to the culvert achieved by each construction vehicle at each stage of backfill placement and compaction. A team of field observers maintained detailed records of these vehicle movements.

Six types of construction equipment were used during backfill operations: a small pan or scraper, a tracked bulldozer, a front loader with four rubber wheels, a 2000-gallon water truck, a two-drum vibratory hand roller, and a medium-sized single-drum vibratory roller. During backfill placement and compaction, large equipment and vehicles were not allowed to operate in close proximity to the culvert structure. Only the hand compactor was permitted to operate within 1·2 m (4 ft) of the structure. As a result, compaction-induced earth pressures acting against

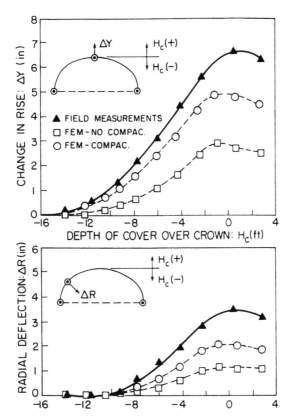

FIG. 24. Deformations versus fill height (crown cover depth) for the promontory overpass culvert structure.

the culvert were very sensitive to the closest proximity to the culvert achieved by each piece of compaction equipment at each backfill stage.

Two types of finite element analyses were performed in order to evaluate the significance of compaction effects on culvert deformations and stresses: (a) conventional analyses without any provision for compaction-induced stresses, and (b) analyses incorporating consideration of compaction-induced soil stresses and associated deformations. Both sets of analyses used the same soil and structural modelling parameters. Soil properties were defined by laboratory testing, and structural stiffnesses had been defined by previous studies, so that all parameters needed for stress–strain and volumetric strain modelling of the soil/structure system were well established. It should be noted that calculation of $\Delta\sigma'_{h,vc,p}$ for *each* element at *each* stage of backfill placement and compaction was a time-consuming task for a problem of this complexity with the locations of each piece of compaction equipment varying at each stage of construction.

The open squares in Fig. 24 show the deflections calculated by 'conventional' incremental finite element analyses, without consideration of compaction-induced soil stresses. As shown in this figure, these analyses underestimate the measured crown deflection by a factor of 2 to 2·5, and underestimate the inward radial deflections of the upper quarter point by a factor of 3 or more. The open circles in Fig. 24 represent the deflections calculated using the analysis techniques described in this chapter. It can be seen that inclusion of consideration of compaction effects resulted in much larger calculated crown and quarter point deflections, and significantly improved agreement with observed field behavior.

Figure 25 shows the structural bending moments calculated with and without modelling of compaction effects. As shown in this figure, inclusion of compaction effects resulted in significantly increased calculated bending moments. This is important, because the possibility of flexural failure is a major design consideration for this type of structure. Con-

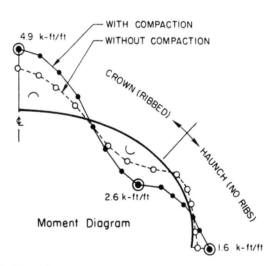

FIG. 25. Final bending moments in the promontory overpass culvert structure calculated with and without modelling of compaction effects.

sideration of compaction effects resulted in a decrease in the calculated factor of safety with respect to the formation of a plastic hinge at the top of the haunch region (which had no stiffening ribs) from $FS = 7.4$ to only $FS = 1.4$ as a result of overall moment increases and a shift in moment distribution.

8.3 The Vista Culvert Overpass Structure

The Vista culvert overpass structure is virtually identical in configuration to the Promontory culvert structure described in the previous section. Installed at Vista, California (less than 48 km (30 miles) from the Promontory structure), the Vista culvert has the same span and rise as the Promontory culvert, and rests atop an identical reinforced concrete base slab and wall system. One difference between the two structures was the use of thicker corrugated aluminium plate and more reinforcing ribs on the Vista structure, so that this second structure had approximately twice the flexural stiffness and almost twice the flexural capacity of the Promontory structure.

A second difference between the two structures was that the backfill for the Vista culvert was a non-plastic silty sand (SM), which was compacted to a relative compaction of between 95% and 98%, based on the Standard Proctor compaction test, at a range of water contents near optimum. This backfill represents a slightly higher overall quality than the two backfills used for the Promontory culvert, and might have been expected to result in smaller structural deflections and bending moments.

A third point of difference between the Promontory and Vista culverts was the equipment and procedures used to achieve backfill compaction. Whereas large vehicles were proscribed from operating in close proximity to the Promontory culvert, and backfill near the culvert was compacted with hand-operated vibratory rollers, such constraints were not enforced for the Vista culvert structure. As a result, the contractor elected to allow heavy equipment (including loaded scrapers, dump trucks, and a water truck) to operate on the fill in close proximity to the culvert structure. This, in turn, resulted in extremely large compaction-induced earth pressures at the soil/culvert interface, and led to very large structural deformations and bending stresses. Indeed, the structure suffered plastic yielding in flexure and had to be excavated and reinstalled using more careful compaction procedures, with large vehicles prevented from operating near the structure.

Figure 26 shows a comparison between the measured deformations and those calculated with and without modelling of compaction-induced

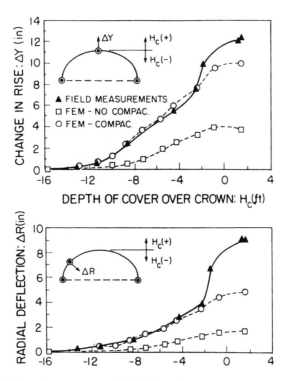

FIG. 26. Deformations versus fill height (crown cover depth) for the Vista overpass culvert structure.

stresses and deformations. It may be seen that analyses performed without consideration of compaction effects led to significant under-estimation of deformations. On the other hand, the analyses performed with modelling of compaction effects provided excellent agreement with field measurements until the backfill was 0·3 m (1 ft) below the crown of the structure, at which point plastic yielding of the structure began. This excellent level of agreement between analyses and the field measurements for this complex soil-structure interaction problem provides good sup-port for the accuracy and usefulness of the analytical methods described in this chapter. Moreover, the large differences in calculated defor-mations, as well as bending moments (as shown in Fig. 27), calculated by analyses with and without modelling of compaction effects shows very clearly the importance of including these effects in soil/structure interac-tion analyses of this type.

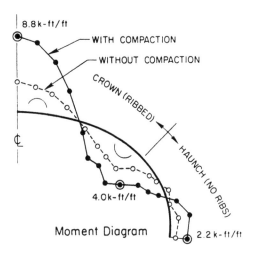

FIG. 27. Final bending moments in the Vista overpass culvert structure calculated with and without modelling of compaction effects.

Comparison of the measured deformations of the Vista culvert (Fig. 26)
with those of the Promontory culvert (Fig. 24), shows that the Vista
culvert underwent approximately twice as much deformation during
backfill placement and compaction as the Promontory structure, despite
the fact that the Vista culvert structure was almost twice as stiff as the
Promontory structure. This difference in performance is attributable to
the difference in field compaction procedures, as large vehicles operated
in close proximity to the Vista structure but not the Promontory
structure. The significant impact of this difference in compaction procedures provides a very clear illustration of the importance of compaction
procedures on overall soil-structure interaction and system performance.

9 SUMMARY AND CONCLUSIONS

The analytical models and procedures described in this chapter provide
a basis for analysis of compaction-induced stresses and deformations. The
three-dimensional nature of the peak, transient compaction-induced
stress increases induced by construction vehicles is taken into account by
the use of $\Delta\sigma'_{h,vc,p}$ as a basis for introducing compaction effects. This, in
turn, permits modelling of compaction loading as a transient, moving
surficial load of finite lateral extent which may pass one or more times

over either the full surface of a given fill layer, or over specific areas only.

The accuracy and usefulness of these analytical procedures are supported by analyses of a number of full-scale field case studies. In the course of performing these analyses, it has become apparent that for most problems the dominant factors controlling the magnitude of compaction-induced stresses are the footprint geometry, the contact force, and the closest proximity of the compaction or construction equipment. This is true because these are the factors that control the magnitude of the peak, transient virgin stress increase $\Delta\sigma'_{h,vc,p}$. For this reason, in performing analyses of compaction-induced stresses, particular attention should be devoted to the calculation of $\Delta\sigma'_{h,vc,p}$ at each stage of backfill placement and compaction. As a corollary, the most effective way to minimize potential adverse effects of compaction-induced earth pressures against a structure is to minimize $\Delta\sigma'_{h,vc,p}$. This is done by keeping large, heavy vehicles from operating in close proximity to structures and *not* by decreasing the final degree of backfill compaction or density required.

REFERENCES

BOUSSINESQ, J. (1885). *Application Des Potentials a L'Etude de L'Equilibre et du Mouvement des Solides Elastiques*, Gauthier-Villars, Paris.

BROMS, B. (1971). Lateral earth pressures due to compaction of cohesionless soils, *Proc. 4th Budapest Conf. on Soil Mech. and Found. Engrg.*, pp. 373–84.

CARDER, D.R., POCOCK, R.G. and MURRAY, R.T. (1977). Experimental retaining wall facility-lateral stress measurements with sand backfill, Transport and Road Research Laboratory Report No. LR 766.

DUNCAN, J.M., BRYNE, P., WONG, K.S. and MABRY, P. (1980). Strength, stress-strain and bulk modulus parameters for finite element analyses of stresses and movements in soil masses, Geotechnical Engineering Research Report No. UCB/GT/80-01, University of California, Berkeley.

DUNCAN, J.M. and SEED, R.B. (1986). Compaction-induced earth pressures under K_0-conditions, *J. Geotech. Engng. Am. Soc. Civil Engrs*, 112(1) 1–22.

GERBER, E. (1929). Untersuchunger Uber die Druckverteilung Im Ortlich Belasteten Sand, Diss. A.-G. Gerb. Leemann, Zurich.

OU, C.Y. and SEED, R.B. (1985). The Promontory long-span culvert overpass structure: field measurements and finite element analyses, Geotechnical Research Report No. SU/GT/85-01, Stanford Universtiy.

OU, C.Y. and SEED, R.B. (1987). Finite element analysis of compaction-induced stresses and deformations, Geotechnical Research Report No. SU/GT/87−03, Standford University.

REHNMAN, S.E. and BROMS, B.B. (1972). Lateral pressures on basement wall. Results from full-scale tests, *Proc. 5th European Conf. on Soil Mech. and Found. Engrg.*, Vol. 1, pp. 189–97.

SEED, R.B. and DUNCAN, J.M. (1983). Soil-structure interaction effects of compaction-induced stresses and deflections, Geotechnical Engineering Research Report No. UCB/GT/83-06, University of California, Berkeley.

SEED, R.B. and DUNCAN, J.M. (1984). SSCOMP: A finite element analysis program for evaluation of Soil-Structure interaction and COMPaction Effects, Geotechnical Engineering Research Report No. UCB/GT/84-02, University of California, Berkeley.

SEED, R.B. and DUNCAN, J.M. (1986a). FE analyses: compaction-induced stresses and deformations, *J. Geotech. Engng. Am. Soc. Civil Engrs*, **112**(1) 23–43.

SEED, R.B. and DUNCAN, J.M. (1986b). Analysis of compaction-induced stresses and deformations, *Proc. 2nd International Conf. on Numerical Methods in Geomechanics*, Belgium, pp. 439–50.

SEED, R.B., COLLIN, J.G. and MITCHELL, J.K. (1986). FEM analyses of compacted reinforced-soil walls, *Proc. 2nd International Conf. on Numerical Methods in Geomechanics*, Belgium, pp. 553–62.

SEED, R.B. and OU, C.Y. (1987). Measurement and analysis of compaction effects on a long-span culvert, Transportation Research Record, No. 1087, Transportation Research Board, National Research Council, pp. 37–45.

SPANGLER, G. (1938). Lateral pressures on retaining walls caused by superimposed loads, *Proc. Highway Research Board*, Part II, England.

TOOMBS, A.F. (1972). The performance of Bomag BW 75S and BW 200 double vibrating rollers in the compaction of soil, Transport and Road Research Laboratory Report No. LR 480.

Chapter 3

FINITE ELEMENT ANALYSIS OF THE STABILITY OF A VERTICAL CUT USING AN ANISOTROPIC SOIL MODEL

P.K. BANERJEE[a], A.S KUMBHOJKAR[b] and N.B. YOUSIF[c]

[a]Department of Civil Engineering, State University of New York at Buffalo, Buffalo, New York, USA
[b]Department of Civil and Architectural Engineering, University of Miami, Coral Gables, Florida, USA
[c]Department of Civil Engineering, University of Basra, Basra, Iraq

ABSTRACT

This chapter describes further developments of an anisotropic soil model and a finite element (FE) analysis of the field test excavation in Welland Clay performed using the developed anisotropic soil model. It presents the details of the model, the FE formulation, and the transient effective stress stability analysis, and compares the FE results with the field measurements. The analysis reflects the post-excavation decrease in the factor of safety with time and predicts the failure of the slope along the observed failure surface. The parametric study shows that the time to failure is a function of the pore pressure boundary conditions at the excavation surface which affects the transient factor of safety.

1 INTRODUCTION

Stability of unsupported excavations in saturated clays is routinely analyzed using the total stress analysis which provides an estimate of the factor of safety on the assumption that the soil is in an undrained condition. The excavation changes traction and pore pressure boundary conditions of the previous site configuration and the unloading (removal of overburden and lateral restraint) generates excess negative pore pressure in the soil around the excavated trench. Dissipation of the excess negative pore pressure starts immediately, and usually continues well beyond the completion of the excavation. Thus, if it exists at all, the idealized undrained condition is short lived. Many case histories available in the literature (for example, Lambe and Turner, 1970; Kwan, 1971; DiBagio and Roti, 1972; Dysli and Fontana, 1982) have demonstrated this fact. The change in effective stresses due to the dissipation of excess negative pore pressure alters the strength of the soil, and as a consequence, the factor of safety obtained on the basis of undrained shear strength becomes unrepresentative of the stability of the excavation during the post-excavation period. Critical analysis of known slope failures in overconsolidated clays (Duncan and Dunlop, 1969) has shown that the total stress approach, in many cases, has given a factor of safety greater than one, when the slopes in fact have failed. Effective stress analysis provides a rational approach to solve this problem, as it makes it possible to take account of phenomena such as progressive failure, swelling, and change in the strength (Simpson et al., 1979), provided a realistic soil behavior model is available. With the help of the finite element method or other numerical methods, it is then possible to deal with other modeling complexities such as excavation simulation, transient pore pressure development, changes in the boundary condition, variation in in-situ stresses and nonhomogeneities.

The finite element method has been employed in the past to perform excavation analysis. However, these efforts are limited to the simulation of only some specific aspects of the soil response to excavation construction. For example, Osaimi and Clough (1979) simulated the pore pressure dissipation during a hypothetical excavation, Dysli and Fontana (1982) attempted to predict the displacements around an excavation, and Popescu (1982) performed stability analysis using the stresses obtained from the finite element simulation of an excavation. All these analyses use rather simple models for soil behavior (elastic, nonlinear elastic, and Von Mises' law with an isotropic strain hardening, respectively by Popescu, Osaimi and Clough, and Dysli and Fontana) which can neither account for stress anisotropy nor consider sequential changes in the effective stress distribution. Dysli and Fontana (1982) in fact realized the limitation of their model (one of the better models employed for excavation analysis) and concluded that more realistic and generalized soil behavior models were necessary for the analysis of excavations in common soils. The anisotropic model presented in this paper attempts to fulfill this requirement. The work presented here is a synthesis of efforts previously reported by Yousif (1984), Kumbhojkar (1987) and Banerjee et al. (1988).

In the following two sections the model and the finite element formulation are described and examples of model prediction and accuracy of the finite element formulation are given. The behavior of a vertical cut is then examined using the case study provided by Kwan (1971), and the results of the analysis are compared with actual field measurements.

2 THE ANISOTROPIC MODEL

Field and laboratory investigations over the last 35 years have now established that inherent and induced stress anisotropy significantly influence the subsequent stress–strain behavior of the soils. Some discussions of the stress history – anisotropy – soil fabric relationship is available in Banerjee et al. (1984) and Wroth and Houlsby (1985). The plasticity based anisotropic model described below provides a description of these anisotropy effects on clay behavior by upgrading the memory of stress history. The development and change in anisotropic character depend on the magnitude and symmetry of the current stress tensor (σ_{ij}) in relation to the maximum stress tensor (σ_{ij}^0). The tensor σ_{ij}^0 is a dynamic quantity and is not only a static memory of the initial consolidation stress state. During initial consolidation, both the tensors increase identically until unloading occurs. The memory of σ_{ij}^0 is retained during the unloading in the form of the yield surface, and a new yield surface is developed whenever the current stress state (σ_{ij}) reaches beyond the existing yield surface.

Details of the laboratory investigation on anisotropically consolidated clays, which provide the basis for the formulation of the model, are available in Banerjee and Stipho (1978, 1979), Stipho (1978), Banerjee et al. (1984) and the theoretical plasticity concepts related to the present elastoplastic constitutive relationship can be found in Banerjee et al. (1984, 1988), Banerjee and Yousif (1986), and Kumbhojkar (1987). Only salient features of the formulation are briefly described here.

2.1 State of Stress
The tensor σ_{ij} represents the effective state of stress at a point, which can be conveniently described in the form of stress invariants p, J_2 and J_3,

where

$$p = \tfrac{1}{3}\sigma_{ii} \tag{1}$$

$$J_2^2 = \tfrac{1}{2}(s_{ij}s_{ij}) \tag{2}$$

$$J_3^3 = \tfrac{1}{3}(s_{ij}s_{jk}s_{ki}) \tag{3}$$

and s_{ij} is the deviatoric stress tensor.

$$s_{ij} = \sigma_{ij} - \tfrac{1}{3}\delta_{ij}\sigma_{kk} = \sigma_{ij} - \delta_{ij}p \tag{4}$$

The third stress invariant can also be expressed as the load angle, θ,

$$\theta = \tfrac{1}{3}\sin^{-1}\left[-\frac{3\sqrt{3}}{2}\frac{J_3^3}{J_2^3}\right], \quad \text{for} \quad \frac{\pi}{6} \geqslant \theta \geqslant -\frac{\pi}{6} \tag{5}$$

Similarly, the deviatoric stress tensor at maximum stress state (s_{ij}^0) is given by

$$s_{ij}^0 = \sigma_{ij}^0 - \tfrac{1}{3}\delta_{ij}\sigma_{kk}^0 = \sigma_{ij}^0 - \delta_{ij}p_0 \tag{6}$$

where δ_{ij} is the Kronecker delta, and p_0 is the maximum mean effective stress.

2.2 Yield Function

The anisotropic yield function $f(\sigma_{ij}, s_{ij}^0, p_0)$, represents a family of yield surfaces given by (Banerjee et $al.$ 1984)

$$f(\sigma_{ij}, s_{ij}^0, p_0) = \frac{3}{2}\frac{1}{g^2(\bar{\theta})}\left[\bar{s}_{ij}\bar{s}_{ij} - \frac{1}{9}\frac{p}{p_0}s_{ij}^0 s_{ij}^0\right] - pp_0 + p^2 = 0 \tag{7}$$

where

$$g(\bar{\theta}) = \frac{2kg(\pi/6)}{(1+k) - (1-k)\sin(3\bar{\theta})}$$

$$k = \frac{3 - \sin\phi'}{3 + \sin\phi'}$$

$g(\pi/6) = M$, the slope of the critical state line as in Roscoe and Burland (1968)

$$\bar{\theta} = \sin^{-1}\left(-\frac{3\sqrt{3}}{2}\frac{\bar{J}_3^3}{\bar{J}_2^3}\right)$$

$$\bar{J}_2^2 = \tfrac{1}{2}\bar{s}_{ij}\,\bar{s}_{ij}$$

$$\bar{J}_3^3 = \tfrac{1}{3}\bar{s}_{ij}\bar{s}_{jk}\bar{s}_{ki}$$

$$\bar{s}_{ij} = s_{ij} - \frac{2}{3}\frac{p}{p_0}s_{ij}^0$$

The reduced state of the deviatoric stress tensor, $\bar{s}_{ij}$, is the algebraic difference between s_{ij}, and the weighted value of s_{ij}^0 and the $\bar{\theta}$, $\bar{J}_2$ and $\bar{J}_3$ are reduced stress invariants defined in the same way as the stress invariants θ, J_2 and J_3 in terms of $\bar{s}_{ij}$.

These yield surfaces are distorted ellipsoids asymmetrically oriented along the initial consolidation line (K_0-line for natural soils) in the principal stress space (Fig. 1) which appear as distorted ellipsi in a $p-q$

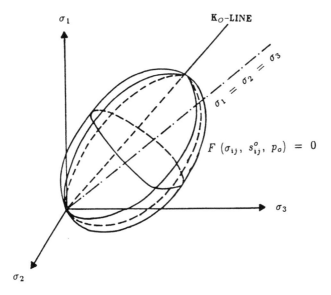

FIG 1.   A schematic representation of the yield surface in principal stress space.

space (Fig. 2). Natural clay deposits are known to display this type of yield surface (for example, Tavenas and Lerouiel, 1977; Graham *et al.*, 1983; Wroth and Houlsby, 1985).

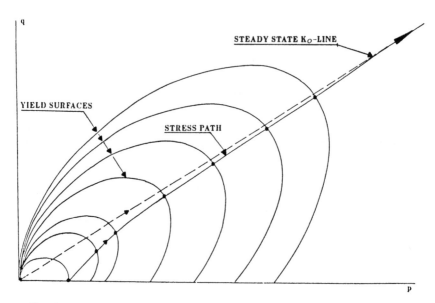

FIG 2.   Development of the orientation of the yield surface in $p-q$ space.

2.3   Hardening Rule

The anisotropic hardening rule used in this model is a combination of isotropic and kinematic hardening rules, and is given by

$$dp_0 = \frac{1+e_0}{\lambda-\kappa} p_0 d\varepsilon_{ii}^p \qquad (8)$$

$$ds_{ij}^0 = \frac{2}{3} \frac{1+e_0}{\lambda-\kappa} p_0 \left[g(\bar{\theta}) - \frac{\sqrt{3 J_2}}{p} \right] dE_{ij}^p \qquad (9)$$

where

$dE_{ij}^p \quad = d\varepsilon_{ij}^p - \dfrac{d\varepsilon_{ii}^p}{3},$ the deviatoric plastic strain rate

$\varepsilon_{ij}^p \quad$ = plastic strain tensor

$\varepsilon_{ii}^p \quad$ = volumetric plastic strain

$e_0 \quad$ = initial void ratio

$\lambda, \kappa \quad$ = slopes of consolidation and swelling lines in e-ln(p) space, and

$p_0, s_{ij}^0 \quad$ = hardening parameters.

Since the yield criterion (eqn (7)) is a function of p_0 and s_{ij}^0, the increments in the current state of stress ensure the translation and expansion of the yield surface in the general stress space when the current stress state reaches the yield surface, since the loading criterion demands that the state of stress must always remain on the yield surface. The rotation of the yield surface then becomes the function of the relative magnitudes of tensors s_{ij}^0 and p_0. The deviatoric hardening rule satisfies the conditions at failure, that is when $\sqrt{3J_2}/p$ equals $g(\theta)$, the strain hardening ceases. (This of course means that the contribution of constitutive equations in phenomenon such as localized yielding or shear band formation are excluded.) Similarly, during steady state K_0-consolidation, it ensures continuous expansion of the yield surface only along the K_0-line without any rotation as ds_{ij}^0/dp_0 becomes a constant tensor.

For axisymmetric problems such as conventional triaxial and one-dimensional (K_0) consolidation tests $(\sigma_2 = \sigma_3)$, the yield function $f(\sigma_{ij}, s_{ij}^0, p_0)$ can be simplified to

$$f(p, q, p_0, q_0) = \frac{1}{g^2(\theta)}\left[\left(q - \frac{2}{3}\frac{p}{p_0}q_0\right)^2 - \frac{1}{9}\frac{p}{p_0}q_0^2\right] - pp_0 + p^2 = 0 \quad (10)$$

where

$$q_0 = \sigma_1^0 - \sigma_3^0$$

and

$$g(\theta) = g(\bar{\theta}) = g\left(\frac{\pi}{6}\right) = \frac{6\sin\phi'}{3 - \sin\phi'} = M$$

As mentioned earlier these yield surfaces are distorted ellipsoids asymmetrically oriented along the initial consolidation line in the principal stress space, which happens to be the K_0-consolidated line for natural deposits. Figure 3 shows a typical state boundary surface in $e-p-q$ space. The q_0 specifies orientation of the yield surface with respect to the p-axis (or the space diagonal in a three-dimensional stress space), and the slope q_0/p_0 is the measure of instantaneous anisotropy at any state of stress during initial consolidation. The state of stress lies sequentially on successive yield surfaces defined by the increasing values of p_0 and q_0. For isotropically consolidated soils, q_0 becomes zero, and eqn (10) reduces once again to the modified cam-clay (Roscoe and Burland, 1968) yield surface in $p-q$ space.

Equations (8) and (9) can be simplified for the biaxial condition (as in

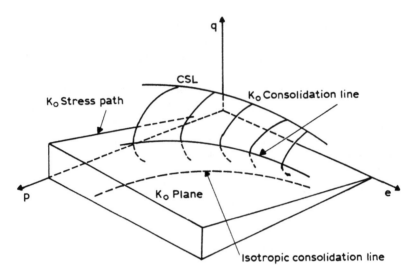

FIG 3.   The anisotropic stress boundary surface in $e-p-q$ space.

a standard triaxial test) as

$$dp_0 = \frac{1+e_0}{\lambda-\kappa} p_0 \, d\varepsilon_v^p \qquad (11)$$

$$dq_0 = \frac{1+e_0}{\lambda-\kappa} p_0 \left(g(\theta) - \frac{q}{p} \right) d\varepsilon_d^p \qquad (12)$$

Since (10) is a function of p_0 and q_0, dp_0 and dq_0, the increments in p_0 and q_0, ensure the translation and expansion of the yield surface in the $p-q$ stress space (Fig. 2). Similarly, during steady state consolidation ($q/p =$ constant), it ensures continuous expansion of the yield surface only along the K_0-line without any rotation. This definition of hardening allows us to develop the incremental elastoplastic constitutive relationship for the steady state consolidation which will help establish the rationality of the choice of hardening parameters defined by (11) and (12).

Hooke's law gives the elastic strains as

$$d\varepsilon_v^e = dp_0/K \qquad (13)$$

$$d\varepsilon_d^e = dp_0/3G \qquad (14)$$

where

$$K = \frac{1+e_0}{\kappa} p = \text{the bulk modulus}$$

$$G = \frac{3(1-2v)}{1+v} K = \text{the shear modulus}$$

Noting that in-situ stresses and their increments in consolidating soils are respectively equal to the maximum consolidation stresses and their increments $(p=p_0, \; q=q_0, \; dp=dp_0, \; \text{and} \; dq=dq_0)$; elastic and plastic strain components from eqns (11), (12) and (13), (14) can be added together to give

$$\frac{d\varepsilon_d}{d\varepsilon_v} = \frac{\dfrac{2(1+v)p_0}{9(1-2v)(1+e_0)} + \dfrac{(\lambda-\kappa)p_0}{\left(M-\dfrac{q_0}{p_0}\right)(1+e_0)}}{\dfrac{\lambda p_0}{1+e_0}} \frac{dq_0}{dp_0} \tag{15}$$

During a steady state K_0 consolidation, lateral strains (ε_2 and ε_3) are zero, and the stress ratio (σ_3/σ_2) is constant. Therefore, substituting

$$\frac{q_0}{p_0} = \frac{dq_0}{dp_0} = n_0, \quad \text{a constant}$$

and

$$\frac{d\varepsilon_d}{d\varepsilon_v} = 2/3$$

in eqn (15) and rearranging the terms we get

$$\frac{2\kappa}{\lambda}(1+v)n_0^2 - \left\{6(1-2v) + \frac{2\kappa}{\lambda}(1+v)M + 9(1-2v)\left(1-\frac{\kappa}{\lambda}\right)\right\}n_0$$

$$+ 6(1-2v)M = 0 \tag{16}$$

Provided κ/λ is non-zero, eqn (7) is a quadratic in n_0, and one of its roots is

$$n_0 = \frac{-b-(b^2-4ac)^{1/2}}{2a}$$

where

$$a = \frac{2\kappa}{\lambda}(1 + v)$$

$$b = -\left\{6(1 - 2v) + \frac{2\kappa}{\lambda}(1 + v)M + 9(1 - 2v)\left(1 - \frac{\kappa}{\lambda}\right)\right\}$$

$$c = 6(1 - 2v)M$$

We can express K_0^{nc} (K_0 for a normally consolidated soil) in terms of n_0 as

$$K_0^{nc} = \frac{3 - n_0}{2n_0 + 3} \tag{17}$$

Substituting the value of n_0 in equation (17) we get

$$K_0^{nc} = \frac{6a + b + (b^2 - 4ac)^{1/2}}{6a - 2b - 2(b^2 - 4ac)^{1/2}} \tag{18}$$

Equation (18) gives K_0^{nc} as an explicit function of the parameters λ, κ, ϕ' and v. Since the values of κ/λ, ϕ' and v lie in a finite range, it is possible to make predictions for K_0^{nc}, for virtually any soil. If the soil is assumed to have a perfectly elastic skeleton, κ/λ becomes equal to 1 and the K_0^{nc} expression reduces to the well-known elastic relationship

$$K_0^{nc} = \frac{v}{1 - v} \tag{19}$$

If a soil is assumed to be perfectly plastic, κ/λ becomes zero, and the K_0^{nc} expression obtained from eqn (18) simplifies to

$$K_0^{nc} = \frac{5 - 3\sin\phi'}{5 + \sin\phi'} \tag{20}$$

It is obvious that for a perfectly plastic material K_0^{nc} is independent of Poisson's ratio, an elastic property, and should become a function of ϕ' alone. The limiting cases, namely soil either as an elastic or plastic material, have some practical significance. Since unloading during over-consolidation can often be regarded as elastic, eqn (19) provides a good estimate of the ratio of the incremental values of the reduction of horizontal principal stress to that of the vertical principal stress during the unloading. Thus at any instant of the unloading process during overconsolidation the K_0 can be expressed as

$$K_0 = R \cdot K_0^{nc} + \frac{v}{1 - v}(1 - R) \tag{21}$$

where R is the overconsolidation ratio.

It may also be noted that Jaky's (1944) expression

$$K_0^{nc} = \left(1 + \frac{2}{3}\sin\phi'\right)\frac{1-\sin\phi'}{1+\sin\phi'} \simeq 1 - \sin\phi' \qquad (22)$$

which gives K_0^{nc} as a function of ϕ' alone, is based on the assumption that soil is a rigid plastic material.

Any similar expression which gives K_0^{nc} as a function of ϕ' alone can be said to consider only plastic deformations. Equation (18) therefore provides a common basis for comparison of K_0^{nc} predictions with different theoretical and empirical $K_0^{nc} - \phi'$ relations. Figure 4, a plot comparing

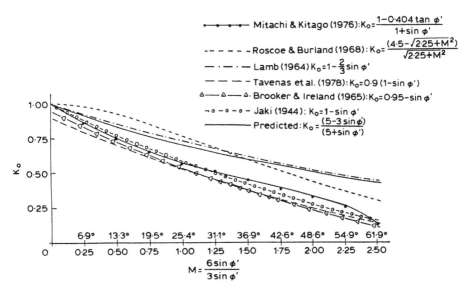

FIG 4.   Comparison of available $K_0 - \phi'$ relationships for normally consolidated soils.

different available expressions, shows two distinct trends. The most widely known approximation of Jaky's expression (eqn (22)), expressions of Brooker and Ireland (1965), Mitachi and Kitago (1976) and Tavenas et al. (1978) predict lower K_0^{nc} than those of Lambe (1964) and the special cases ($\kappa = 0$) of the proposed anisotropic, and modified cam-clay models. The difference in the two trends becomes significant at high values of ϕ'. The K_0^{nc} values using eqn (20) are close to those by Lambe (1964) over the entire range of ϕ', although the validity of the latter is said to be limited to a range of 20–54°. The modified cam-clay model, however, gives much higher K_0^{nc} values for ϕ' lower than 25°, and considerably lower K_0^{nc} for ϕ' greater than 45°.

The proposed elastoplastic K_0^{nc} expression can be used to explain observed variations in K_0^{nc} measurements for different soils as a consequence of differences in values of κ/λ, and v, apart from ϕ'. Forty-nine K_0^{nc} values reported (Mayne and Kulhaway, 1982) in the literature are plotted in Fig. 5 as a function of ϕ'. All data points lie within the bounds

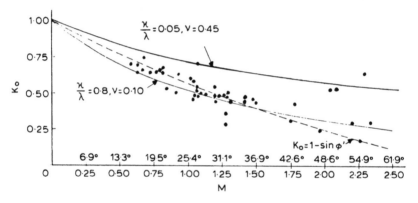

FIG 5. Comparison of $K_0 - \phi'$ experimental data for normally consolidated soils with theoretical results.

provided by the proposed expression (eqn (20)). For comparison, Jaky's expression is also plotted. Although $(1 - \sin \phi')$ represents the mean trend well, it is unable to explain the rather large scatter. Reported K_0^{nc} measurements are, of course, likely to be affected by differences in equipment, personnel, sampling disturbances and other factors including experimental errors. However, it is likely that one needs to take account of the effects of κ/λ and v, which are given by the bounds calculated using eqn (20).

Yousif (1984) proposed a different anisotropic hardening rule to develop the elastoplastic constitutive relationship, in which the increment in the hardening parameter s_{ij}^0 was given by

$$ds_{ij}^0 = \frac{\lambda - \kappa}{1 + e_0} s_{ij}^0 \, d\varepsilon_{kk}^p \qquad (23)$$

Banerjee and Yousif (1986) employed this relationship for the prediction of K_0^{nc}. Their formulation, which takes account of the shear distortion through s_{ij}^0, also provides elastoplastic K_0^{nc} estimates which are generally lower than those by the proposed and the cam-clay models.

2.4 Incremental Stress–Strain Relationship

The model uses the associated flow rule, hence the yield surface (eqn (1)) also serves as a plastic potential surface. The incremental stress–strain relationship is obtained by decomposing total strains into elastic and plastic components, using Hooke's law and the consistency condition in the usual manner (Banerjee et al., 1984; Yousif, 1984; Kumbhojkar, 1987). Accordingly, the constitutive equation in a general stress state is given by

$$\{d\sigma_{ij}\} = \left[[D^e] - \frac{[D^e]\left\{\dfrac{\partial f}{\partial \sigma_{ij}}\right\}\left\{\dfrac{\partial f}{\partial \sigma_{ij}}\right\}^{\mathrm{T}}[D^e]}{H_p + H_e} \right]\{d\varepsilon_{ij}\} \qquad (24)$$

where

$\{d\sigma_{ij}\}$ = stress increment tensor

$(d\varepsilon_{ij}\}$ = total strain increment tensor

$[D^e]$ = elastic constitutive matrix

$\left\{\dfrac{\partial f}{\partial \sigma_{ij}}\right\}$ = vector of partial derivatives of the yield function with respect to the stress components

H_e, H_p = the elastic and plastic components of the hardening modulus, where

$H_e \quad = \left\{\dfrac{\partial f}{\partial \sigma_{ij}}\right\}^{\mathrm{T}}[D^e]\left\{\dfrac{\partial f}{\partial \sigma_{ij}}\right\}$

$H_p \quad = \dfrac{1+e_0}{\lambda-\kappa}p_0\left[\operatorname{tr}\left\{\dfrac{\partial f}{\partial \sigma_{ij}}\right\}\dfrac{\partial f}{\partial p_0} + \left(g(\theta) - \dfrac{\sqrt{3J_2}}{p}\right)\left\{\dfrac{\partial f}{\partial s_{ij}^0}\right\}^{\mathrm{T}}\right.$

$\left. \left\{\dfrac{\partial f}{\partial \sigma_{ij}} - \tfrac{1}{3}\operatorname{tr}\dfrac{\partial f}{\partial \sigma_{ij}}\right\}\right]$

This formulation places no restriction on the elastic behavior which can either be anisotropic or isotropic. While analyzing the behavior of cross anisotropic materials such as K_0-consolidated soils, three additional elastic parameters, namely the ratio of moduli in horizontal and vertical planes ($N = E_h/E_v$), one additional Poisson's ratio with respect to the horizontal plane (ν_{hh}), and shear modulus (G_{vh}) are required. Since the determination of these parameters is usually difficult, an assumption of isotropic elasticity can be made. With this assumption eqn (24) requires only five characteristic soil parameters, namely λ, κ, ϕ', ν and e_0.

2.5 Model Predictions

Overall the model predictions are not substantially different from those shown in Banerjee et al. (1984) and Banerjee and Yousif (1986). For the present version Banerjee et al. (1988) and Kumbhojkar (1987) provide extensive comparisons of the model predictions with the experimental results and Figs 6–10 provide a few examples pertaining to loading/unloading responses and the simulation of anisotropy. The aforementioned references provide further validations of the developed model.

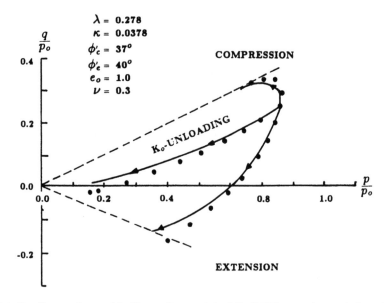

FIG 6. Comparison with Koutsoftas and Ladd's (1985) experimental data for K_0 consolidated clay.

2.6 K_0-Loading and Unloading

Koutsoftas and Ladd (1985) report the results of undrained compression, extension and K_0-unloading tests on Plastic Holocene (Atlantic Gener-

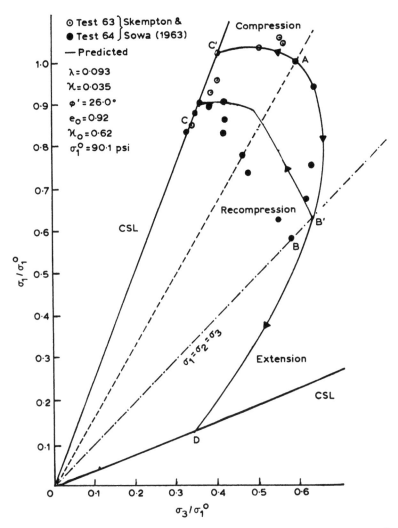

FIG 7.   Comparison with Skempton and Sowa (1963) for perfect sampling of K_0 consolidated clay.

ation Station) clay. Undisturbed samples of marine clay were one dimensionally consolidated in the laboratory to more than twice the initial preconsolidation pressure. Exact details of sample extraction, transport and the steps of reconsolidation are not available, therefore the tests are simulated assuming that the samples have only one-phase

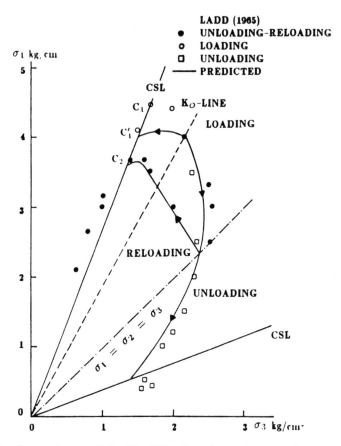

FIG 8.   Comparison with Ladd's (1965) data for perfect sampling of K_0 consolidated clay.

K_0−consolidation history. The nondimensionalized plot of the results shown in Fig. 6 uses a different convention for p and q (as in the test results), i.e., $p = 0.5(\sigma_1 + \sigma_3)$ and $q = 0.5(\sigma_1 - \sigma_3)$. These tests were simulated using the maximum vertical effective stress of 620 kPa (kg/cm^2), $\phi' = 37°$ and $40°$ respectively in compression and extension, $\lambda = 0.278$ and $\kappa = 0.0378$. The predictions follow the experimental trend very well, although there is a slight difference between the observed and predicted unloading stress paths. This difference is expected since the deformation history of the soil up to its shearing is simplified due to the reconsolidation.

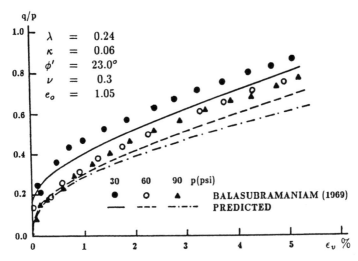

FIG 9.   Comparison with Balasubramnian's (1969) data for isotropic consolidation of K_0 consolidated samples (q/p vs ε_v).

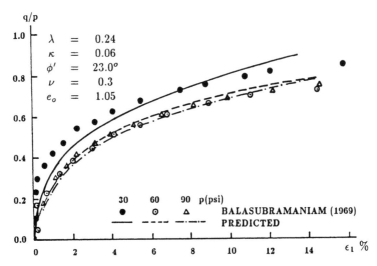

FIG 10.   Comparison with Balasubramnian's (1969) data for isotropic consolidation of K_0 consolidated samples (q/p vs ε_1).

2.6 Effect of Sampling

Skempton and Sowa (1963) studied the effect of change of state of stress due to sampling on the subsequent behavior of K_0-consolidated soils. They concluded that the undrained shear strength (C_u) for triaxial compression of 'perfect samples' was equal to that of in-situ natural deposits. The observed and predicted stress paths are shown in Fig. 7. AB, the experimental stress path represents the effect of taking a naturally K_0-consolidated sample out of the ground by a perfect sampling operation. When the undrained compression test was carried out on the sample from this state (B), the stress path rose initially elastically until it approached the yield surface, and then the soil continued to yield until it reached failure at point C. When an undrained compression test was carried out from the in-situ state (A), the sample again failed to close to C. The stress paths corresponding to AB, BC and AC obtained using the model are given by AB', $B'C$ and AC'. The model predictions follow the trend well, but there is a small difference between the predicted and observed compressive undrained strengths of the sample from its undisturbed and disturbed state. Ladd (1965) also performed experiments similar to those by Skempton and Sowa (1963), but observed some difference in C_u of natural samples (Fig. 8). Skempton and Sowa did not perform the unloading tests, but Fig. 7 also shows the simulation results of an extension test on the same sample. Ladd (1965) actually performed such an extension test and the prediction matches very well with the experimental results (Fig. 8). It should be noted that the strengths of these anisotropically consolidated samples in extension tests (after a perfect sampling operation) are significantly lower (about a half) than those in compression tests. Cairncross and James (1977) also observed this influence of anisotropy on the soil response when they studied the behavior of Kaolin, Fulford, Haney and London clay samples.

2.7 Fabric Development

The above model predictions and experimental data clearly demonstrate the influence of anisotropy on the soil behavior. However, the fabric development not only depends upon the initial and final states but also on another aspect of stress history, i.e., the stress path followed. Balasubramanian (1969) provides an interesting set of results to demonstrate this aspect. Reconsolidation of a K_0-consolidated sample to an isotropic state is one of the simplest ways to generate a complete stress history in the laboratory. Hence he also adopted the same approach, and generated different stress histories by employing different stress paths to reach the final isotropic stress states. Balasubramanian consolidated three identical Kaolin samples one dimensionally. The final vertical (σ_1) and lateral (σ_3) stresses for all the three samples were, respectively, 22·0 psi and 14·08 psi. These specimens were then brought to an isotropic stress state at mean effective pressures of 30, 60 and 90 psi, and were then sheared under a p-constant condition.

These tests were simulated by the present model in two stages. The first stage of the analysis used the end of K_0-consolidated state as the initial state and loaded the samples to reach the isotropic stress states. The positions of the yield surfaces and the void ratios corresponding to each state were determined. These results were used as an input for the next part of the analysis in which the samples were loaded under a p-constant condition until failure. Figures 9 and 10 which represent the plots of the normalized deviatoric stress (q/p) respectively against volumetric (ε_v) and axial strain (ε_1) show excellent agreement between the experimental and predicted stress–strain behavior. If the specimens were consolidated only isotropically, they would have shown an identical normalized response. Banerjee et al. (1984) and Banerjee and Yousif (1986) also simulated these tests but the present results are in better agreement with the test data. The soil parameters used for the models were $\lambda = 0.24$, $\kappa = 0.06$, $e_0 = 1.05$, $\phi = 23°$ and $v = 0.3$.

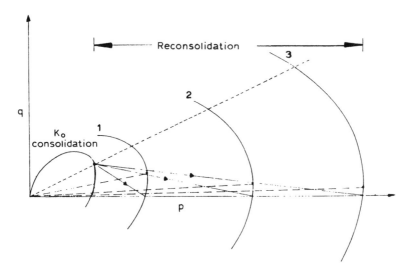

FIG. 11. Modifications of the yield surface of a K_0 consolidated clay due to isotropic consolidation.

The ability of the model to simulate the effect of the development of fabric can be very well explained using the results of these tests. The initial K_0−consolidation results in a well defined fabric characterized by the magnitude of q_0/p_0. During reconsolidation, the magnitude of q_0 reduces and p_0 increases in all three samples but by different proportions. As a consequence, all the yield surfaces rotate towards the isotropic consolidation line (p-axis) but the amount of rotation is different in each case. The longer the path length to reach the final state, the more the reduction in q_0 occurs, and the closer the samples go to the isotropic state, as shown schematically in Fig. 11. With different stress paths, the samples develop different 'memories', and without these memories the normalized results of all three samples would have been identical.

3   FINITE ELEMENT FORMULATION

The anisotropic model described above has been incorporated in a finite element program capable of modeling two-dimensional (plain strain, axisymmetric) geotechnical problems such as consolidation, multistage excavation, embankment construction and stability analyses. In addition to the elastoplastic model, the program allows one to assign linear elastic behavior to any part of the problem geometry, thus making it possible to analyze a wide ranging soil structure interaction problem or perform a complete elastic analysis. Its other important features include the provision to assign linearly varying in-situ stresses and material properties to the domain, add or subtract a region in a multiregion problem with a different set of properties assigned to each region; and perform drained, undrained or transient analyses. Description of all the program features is beyond the scope of this chapter, and a brief summary of the features relevant to the present study is given below.

4   NONLINEAR TIME DEPENDENT RESPONSE

The formulation for the consolidation type transient problems is based on the usual assumptions of saturated soil, principle of effective stress, incompressibility of soil solids and pore water, validity of Darcy's law and small deformation. Following the procedure developed by Sandhu and Wilson (1969) and Zienkiewicz et al. (1975), the unknown displacements and excess pore pressures generated by incremental loading are calculated at each time step using the following relationship:

$$\begin{bmatrix} \mathbf{K} & \mathbf{L} \\ \mathbf{L}^T & -0{\cdot}5\Delta t\,\mathbf{H} \end{bmatrix} \begin{Bmatrix} d\mathbf{u}^n \\ d\mathbf{U}^n \end{Bmatrix} = \begin{Bmatrix} df(\Delta t) \\ \Delta t\,\mathbf{H}\,\mathbf{U}^n(t) \end{Bmatrix} \tag{25}$$

where

$\quad\quad$ $\mathbf{K}$ = the elastoplastic stiffness matrix
$\quad\quad$ $\mathbf{L}$ = coupling matrix
$\quad\quad$ $\mathbf{H}$ = flow matrix
$\quad$ $d\mathbf{u}^n, d\mathbf{U}^n$ = displacement and pore pressure increment vectors in the nth step
$\quad$ $df(\Delta t)$ = algebraic sum of increments in the load vectors due to initial effective stresses, body forces and boundary tractions and
$\quad\quad$ Δt = time interval for the nth step.

The matrices $\mathbf{K}$ and $\mathbf{H}$ change with respect to time, and reflect the changes in the material properties allowing modeling of nonlinear behavior. It should be noted that for heavily overconsolidated clays it is sometimes possible to lose the positive definiteness of the tangent stiffness matrix $\mathbf{K}$. In such a case the $\mathbf{K}$ is replaced by the elastic stiffness matrix and the effects of nonlinearities are transferred to the right-hand side via the normal initial stress algorithm. Solution of eqn (25) gives the incremental changes in the pore pressure (and displacements), which are used to calculate the changes in the effective stresses. The excellent agreement between Baligh and Levadoux's (1978) results of one-dimensional consolidation under cyclic step loading based upon Terzaghi's theory and its Finite Element Simulation (Fig. 12) demonstrates the accuracy of the implementation for the linear transient problems.

4.1 Excavation

The multistage excavation is simulated by using a technique similar to that proposed by Zienkiewicz (1971). For each step of excavation, the excavation surface is made traction-free by applying nodal forces which are equal but opposite in direction to the existing forces. The equivalent nodal forces are calculated using the following expression

$$df = \sum_{i=1}^{m} \int_V \mathbf{B}^T \sigma^t \, dV \tag{26}$$

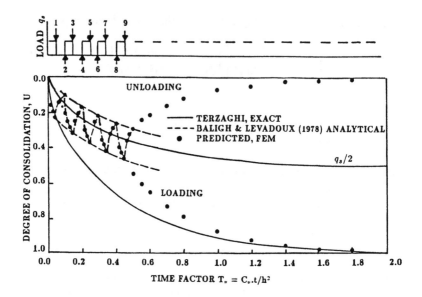

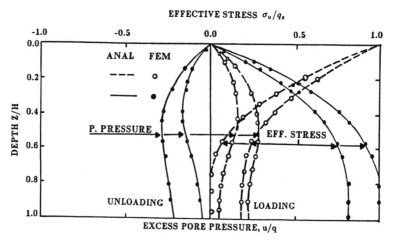

FIG 12. Validation of the finite element results for one-dimensional consolidation under transient cyclic loading.

where

df = the vector of equivalent and nodal forces
B = strain displacement relation matrix
σ^t = the total stress vector and

m = the number of elements which have a common boundary with unexcavated elements

The integral in eqn (26) is evaluated numerically using values of total stresses σ^t and $\mathbf{B}$ at the Gaussian integration points.

4.2 Stability Analysis

In the context of excavation problems, provision is made for performing the conventional stability analysis along predefined slip surfaces using the current effective stresses. For convenience, it is assumed that the centroidal values of the stresses are constant throughout the element. This assumption is similar to that used in the conventional method of slices where the stress variation within individual slices is neglected. In a fine mesh, such an assumption produces little error. The advantage of using the effective stresses obtained from the FE analysis is that the state of stress within the soil mass is known and no assumptions for the stress distribution along the slip surface are required. To perform the stability, the forces relevant to each of the trial surfaces are obtained using the elemental stresses and the factor of safety is calculated using the general limit equilibrium method. In a multistage excavation, additional trial slip surfaces are provided at each succeeding excavation step and analyses are performed for all the slip surfaces.

5 ANALYSIS OF AN EXCAVATION

The case study of an excavation of a vertical cut in Welland clay (Kwan, 1971) is used for the present analysis. The comprehensively instrumented test excavation in the Haldimand clay plain was undertaken as a full scale pilot study to investigate the stability of the side slopes of the (then) proposed realignment of the Welland canal. It was planned to continue the excavation until failure, and the behavior of the cut was to be observed. Kwan (1971) provides a detailed description of the site conditions, excavation of the test and the field instrumentation, and only a brief summary is given here.

5.1 Site and Subsurface Conditions

The Haldimand clay plain is composed of several lacustrine and glacial till beds, average 90 ft (27·4 m) thick soil deposit at the test site is underlain by lime dolomite bedrock. The uppermost 5–10 ft (1·5–3 m) thick portion of the bedrock is water-bearing, and the pore pressure in

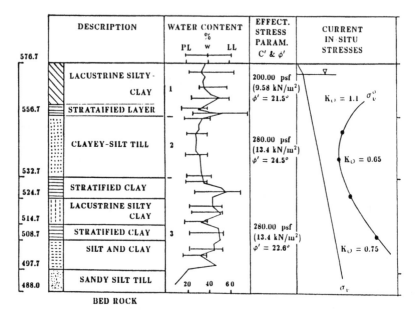

FIG 13.   Soil profile for the Welland Canal excavation site.

the rock at the test site was hydrostatic. Figure 13 shows the soil profile, preconsolidation pressures, Atterberg limits, effective strength parameters, in-situ vertical effective stresses and the values. The effective strength parameters, c' and ϕ' (effective cohesion and friction angle) are the peak values obtained from consolidated-undrained (CU) tests.

5.2   Excavation and Failure of the Slope

Excavation was carried out in stages to allow time for the instrumentation. Figure 14 shows the excavation profile and locations of piezometer stations, and Fig. 15 provides the time rate of vertical cut excavation along the centerline of the 50 ft (15 m) wide slope. The 17 ft (5·2 m) thick overburden was removed in the first 9 days and the following 12 days were utilized for the installation and stabilization of the piezometers. In the final stage of the excavation, a 32 ft (9·75 m) deep trench was cut over a period of 4 days. One side of the cut was vertical and the other side was excavated with a 1:1 slope. Excavation was stopped at this stage to avoid the possibility of bottom uplift. During the entire excavation period precipitation was low, and there was no precipitation during the excava-

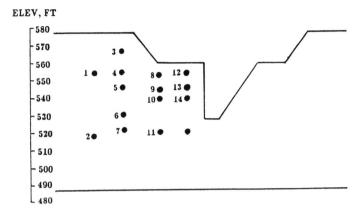

FIG 14.   The excavation geometry and the piezometer locations.

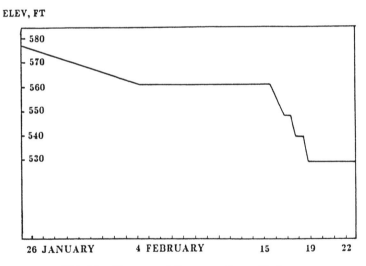

FIG 15.   Progress of the excavation with time.

tion of the vertical cut. A potential failure surface was identified at a distance of about 21 ft (6·4 m) from the vertical face (SS2 in Fig. 16) after about 3 days when external water was pumped in, on the top of the vertical block to induce failure. However, the slope failed the next day along an entirely new surface (SS1). The collapse of the slope was rapid after the appearance of the SS1.

POTENTIAL FAILURE SURFACE   ACTUAL FAILURE SURFACE

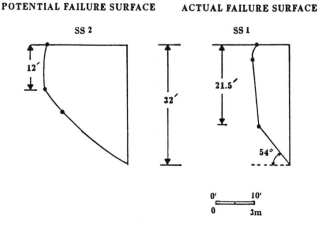

• FIELD MEASUREMENT

FIG 16.   The geometry of the slip surfaces.

5.3   Idealization and FE Simulation

The sequential behavior of the vertical cut was analyzed as a plane strain problem using the 350 six-noded linear strain triangular element-mesh shown in Fig. 17. Although the analysis required only the eastern half of the geometry pertaining to the vertical cut, the entire excavation was modeled so that the boundary fixity should not affect the stresses and displacements around the excavation. Figure 17 also shows the three stages of the excavation and boundary conditions.

A free drainage boundary (FDB) implies free flow across the boundary (excess pore pressure is equal to zero). Since excavation induces excess negative pore pressures, FDB implies the presence of an external source of water in case of excavation problems. The presence of a water table near ground level, and water-bearing bedrock therefore ensured free drainage across the restrained boundaries. Pore pressure boundary conditions for the ground level and excavation boundary were, however, uncertain in the absence of any definite information. These boundaries were certain to exert influence on the pore pressure dissipation, hence, three conceivable combinations, namely fully permeable boundary throughout the excavation and post-excavation period (PBC-1), un-drained excavation followed by free flow across the boundaries after

PLANE STRAIN

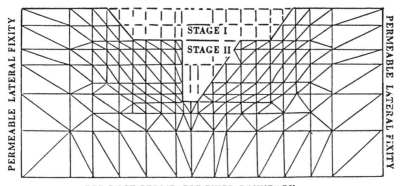

NO. OF NODES = 750
NO. OF ELEMS = 350

6 – NODED LST TRIANGLES

FIG 17.   The finite element mesh for excavation simulation.

excavation (PBC-2) and continuously sealed boundary until pumping in of water at the top of the vertical block (PBC-3), were attempted. The condition of PBC-3 appeared realistic because the record of the climatic conditions indicated the possibility that the ground and excavation surface might not have served as a free drainage until the pumping in of the water. Response of the cut was obtained for all three conditions.

The soil properties used in the analysis are summarized in Table 1. The soil deposits, although showing many distinct layers (Fig. 13), primarily consisted of two major types, namely the medium to high plasticity lacustrine clay and the medium plastic till. The clay layers 1 and 3 gave different effective friction angles and permeabilities, but their Atterberg limits, which were used to estimate the λ and κ values, are quite close. Hence, their λ and κ values were assumed to be the same. The properties of layer 2 are similar to those of Weald clay (Schofield and Wroth, 1968), hence, λ and κ of Weald clay were adopted for layer 2. The field data indicate that the soil deposit was overconsolidated and the maximum consolidation stresses would affect the soil behavior during excavation. However, in the absence of any data regarding elastic anisotropy, elastic isotropy was assumed.

TABLE 1

SOIL PARAMETERS USED IN THE FINITE ELEMENT ANALYSIS

No.	Types of soil	Elevation (ft (m))	Elastoplastic deformation parameters		Effective strength parameters		Initial void ratio,[a]	Per-mea-bility
			λ	κ	ϕ'	c' (psf (kN/m²))	e_0	K (cm/s)
1	Medium to high plasticity silty clay	556·7–576·7 (169·7–175·8)	0·11	0·04	21·5	200 (9·58)	0·96	5×10^{-6}
2	Medium plasticity clayey-silt	532·7–556·7 (162·4–169·7)	0·093	0·036	24·5	280 (13·4)	0·77	5×10^{-6}
3	Medium to high plasticity silty clay	497·7–522·7 (157·7–162·4)	0·11	0·04	22·6	280 (13·4)	1·10	5×10^{-7}

[a]The initial void ratio is calculated using the specific gravity of soil solids = 2·75 and the natural water content.

Excavation simulation closely followed the field construction sequence. The first stage of overburden removal was followed by the 12 days of the nonconstruction period. In the second stage, the 32 ft (9·75 m) deep excavation was performed in five time steps totaling about 4 days; and additional time steps were employed to study the post-excavation behavior.

6 COMPARISON OF THE RESULTS AND DISCUSSION

The finite element results pertaining to pore pressure response, stability of the cut and ground movements were compared with the available field measurements.

6.1 Pore Pressure Response

Increase in the depth of excavation increased the magnitude of excess negative pore pressure around the excavation. Accordingly, the pore pressure around the excavation at any time during and after excavation was observed to be the function of the duration (i.e., type of pore pressure boundary condition) for which the exposed surface provided free drainage. For the period up to the pumping in of water, PBC-1 and PBC-3 provided the limiting pore pressure responses. The results obtained using the condition PBC-2 were close to those due to PBC-1 after the first stage of excavation, while they were close to those due to PBC-3 on 22 February (i.e., the measurements prior to failure). These predicted pore pressure distributions around the excavation for the boundary conditions PBC-1 and PBC-3 were also compared with piezometric measurements (Kwan, 1971) recorded on 15 and 22 February (Figs 18 and 19). Prediction of pore pressure distribution on both occasions is similar to the field measurements, however, their magnitudes are different. As seen in Figs 18 and 19, the reduction in piezometric heads due to excavation is maximum around the excavation, and the effect rapidly decreases with the distance from the excavation boundary. For example, the 17 and 32 ft (5·2 and 9·75 m) deep excavations changed the piezometric head only

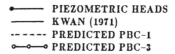

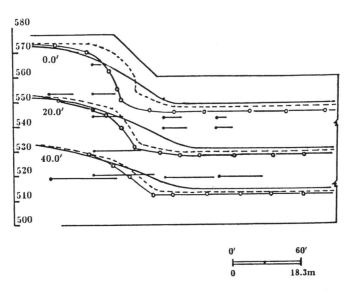

FIG 18.   Pore water pressures before the second stage excavation.

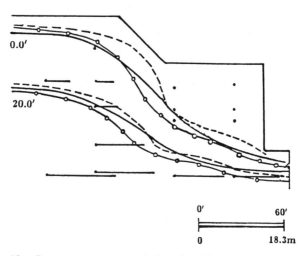

PIEZOMETRIC HEADS
KWAN (1971)
PREDICTED PBC-1
PREDICTED PBC-3

FIG 19.   Pore water pressures before the failure of the excavation.

by about 3·0 and 1·0 ft (0·9 and 0·3 m) at a distance of 50 ft (15·20 m) away
from the cut (piezometer no. 2); in contrast, the fall in the head of
piezometer 10 was, respectively, about 36 and 12 ft (11·0 and 3·65 m) after
the overburden removal and excavation of vertical cut.

6.2   Stability Analysis

Kwan (1971) performed a number of total and effective stress analyses to
obtain a reasonable estimate of undrained shear strength and predict the
failure. Different approaches and assumptions regarding the location of
failure surface, the homogeneity of the soil and methods of analysis were
used for this purpose. The estimates of undrained shear strength varied
between 800 and 1860 psf (38·3 and 89 kN/m^2). Stability analysis,
performed using Bishops's simplified method and peak strength parameters on a circular slip surface, gave a factor of safety equal to 0·9. The
choice of peak effective friction angle was appropriate for the effective
stress analysis, since the failure occurred at very small strain levels. Even

in fissured clays, use of residual strength parameters is justifiable only if failure occurs at large strains (Simpson *et al.*, 1979). Clearly, this total stress analysis was unable to take account of the changes in the state of stress, and hence should not be used in such situations. On the other hand, the effective stress analysis discussed below was able to reflect these changes and gave a more realistic result.

The effective stress analysis using the FE simulated stress-field leads to more precise prediction of the failure of the slope. The effect of the pore pressure boundary conditions was particularly evident while analyzing the transient stability of the vertical cut. Analyses were performed for both the failure surfaces (SS1 and SS2), for both boundary conditions (PBC-1 and PBC-3) after each time step following the completion of the second stage of the excavation. In all the cases, the excess negative pore pressures and the factor of safety continued to decrease with time, and the potential failure surface (SS2), gave a much higher factor of safety than SS1, the real failure surface. Although the previous excavation boundary appeared inconsistent with the observed climatic conditions, the computed pore pressures were closer to the field measurements and the corresponding effective stresses gave a factor of safety of 0·97 along the observed failure surface at the time step which fortunately coincided with the collapse in the field. At the same time, the factor of safety for SS2 was 1·89 and that for SS1 using the sealed pore pressure boundary (PBC-3) was 1·26. Continuing the time steps for PBC-3 showed that the factor of safety was reduced to less than unity along SS1 after 45 days, and the factor of safety for surface SS2 was reduced to 1·41. The finite element analysis, of course, could not take account of the effect of opening up of fissures due to excavation, the possible stress relief due to the development of SS2 and the consequent increase in the permeability. The increase in permeability probably accelerated the negative pore pressure dissipation and subsequent collapse of the slope in the field. It is likely that this effect was compensated by the continuous suction along the boundary in the PBC-1 case, even though it appeared inconsistent with the climatic data.

It has been frequently observed that the excavations fail much before the soil reaches a critical state, and this phenomenon is attributed to reduction in the effective stresses (Simpson *et al.*, 1979). The present analysis using the anisotropic model confirmed this observation, showed that the predicted changes in the state of stress were realistic and predicted the transient behavior and failure of the vertical cut.

6.3   Ground Movements Around Excavation

During the test excavation, apparently only the lateral movement of the top edge of the vertical cut was recorded. The net lateral displacement prior to the appearance of the tension crack was only 0·67 in (17 mm), or 0·18% of the depth. Clearly, the falure occurred at small strain levels. The predicted lateral displacement of the top edge at failure was 0·56 in (14 mm) which surprisingly is in excellent agreement with the field observation. Kwan did not provide any other observations of the displacements in and around the excavations. Hence, only predicted displacements are shown in Fig. 20. It can be seen that significant movements occur only within a distance of about 2·5 times the depth of the excavation. At distances away from the excavated boundary, the vertical movements are approximately of the same order of magnitude as the horizontal movements, whereas closer to the excavated area, the horizontal movements dominate. The observed displacement pattern is quite similar to Osaimi and Clough (1979). The unloading response at a point depended on the state of stress with respect to the individual yield surface at a given time step. The unloading was elastic until the stress

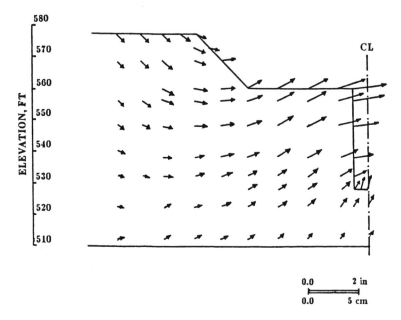

FIG 20.   Predicted displacement vectors.

state approached the yield surface and there upon became elastoplastic. Accordingly, the elastic recovery following the stress release dominated the base heave and lateral movements, which were supplemented by the time dependent dissipation of excess pore pressures.

7 CONCLUSIONS

The anisotropic model provided a realistic effective stress distribution within the soil mass in Welland clay excavation and the stability analysis using these stresses provided a reasonably accurate estimate of the factor of safety of the vertical cut during the post-excavation transient period. The pore pressure response, ground movements and time dependent stability of the cut were successfully predicted by the finite element analysis. The parametric studies showed that the pore pressure boundary conditions along the boundary affected the pore pressure dissipation within the soil mass and governed the stability of the slope. The increased permeability due to opening up of the fissures, a likely reason for the rapid collapse, could not be simulated; however, application of free flow conditions across the excavation surface probably compensated this effect.

REFERENCES

BALASUBRANAMIAN, A.S. (1969). Some factors influencing the stress-strain behavior of clays, Ph.D. Thesis, University of Cambridge, UK.

BALIGH, M. and LEVADOUX, J.N. (1978). Consolidation theory for cyclic loading, J. Geotech. Eng. Div. ASCE, 104(GT4), 415–31.

BANERJEE, P.K. and STIPHO, A.S. (1978). Associated and non-associated constitutive relations for undrained behavior of isotropic constitutive relations for undrained behavior of isotropic soft clays, Int. J. Numer. Anal. Methods Geomech., 2(1), 35–56.

BANERJEE, P.K. and STIPHO, A.S. (1979). An elastoplastic model for undrained behavior of heavily overconsolidated clays, Int. J. Numer. Anal. Methods Geomech., 3(1), 97–103.

BANERJEE, P.K. and YOUSIF, N.B. (1986). A plasticity model for the mechanical behavior of anisotropically consolidated clay, Int. J. Numer. Anal. Methods Geomech., 10, 521–41.

BANERJEE, P.K., STIPHO, A.S. and YOUSIF, N.B. (1984). A theoretical and experimental investigation of the behavior of anisotropically consolidated clay. In Developments in Soil Mechanics and Foundation Engineering, Vol. II, P.K. Banerjee, and R. Butterfield, (Eds), Elsevier Applied Science, London, UK, 1–42.

BANERJEE, P.K., KUMBHOJKAR, A.S. and YOUSIF, N.B. (1988). Finite element analysis of time dependent stability of excavations, *Can. Geotech. J.*, February, 232–40.

BROOKER, E.W. and IRELAND, H.O. (1965). Earth pressure at rest related to stress history, *Can. Geotech. J.*, **2**, 1–15.

CAIRNCROSS, A.M. and JAMES, R.G. (1977). Anisotropy in overconsolidated clays, *Geotechnique*, **27**(1), 31–6.

DiBAGIO, E. and ROTI, J.A. (1972). Earth pressure measurements on a braced slurry-trench wall in soft clay, *Proc. 5th European Conf. on Soil Mechanics and Foundation Engineering*, Vol. 1, pp. 473–84.

DUNCAN, J.M. and DUNLOP, P. (1969). Slopes in stiff fissured clays and shales, *J. Soil Mech. Foundation Eng. Div. ASCE*, **92**(SM2), 467–92.

DYSLI, M. and FONTANA, A. (1982). Deformations around the excavations in clay soil, *Proc. Int. Symp. on Numerical Models in Geomechanics*, Zurich, R. Dunger, G.N. Pande, and J.A. Studer, (Eds), pp. 634–42.

GRAHAM, J., NOONAN, M.L. and LEW, K.V. (1983). Yield stress and stress-strain relationships in a natural plastic clay, *Can. Geotech. J.*, **20**, 502–16.

JAKY (1944). The coefficients of earth pressure at rest, *J. Hung. Arch. Engrs. Soc.*, Budapest, 355–8.

KOUTSOFTAS, D.C. and LADD, C.C. (1985). Design strengths for an offshore clay, *Geotechnique*, **111**(3), 337–55.

KUMBHOJKAR, A.S. (1987). Theoretical and numerical analyses of geotechnical construction problems, Ph.D. Thesis, State University of New York at Buffalo, Buffalo, NY.

KWAN, D. (1971). Observations of the failure of a vertical cut in clay at Welland, Ontario, *Can. Geotech. J.* **8**, 283–98.

LADD, C.D. (1965). Stress strain behavior of anisotropically consolidated clays, *Proc. 6th Int. Conf. on Soil Mechanics and Foundation Engineering*, Montreal, Canada, Vol. I, pp. 282–6.

LAMBE, T.W. (1964). Methods of estimating settlement. *J. Soil Mech. Foundation Eng. Div. ASCE*, **90**(9), 47–70.

LAMBE, J.W. and TURNER, C.K. (1970). Braced excavations, *Proc. ASCE Specialty Conf. on Lateral Stresses in the Ground and Design of Earth-Retaining Structures*, pp. 149–218.

MAYNE, P.W. and KULHAWAY, F.H. (1982). K_0-OCR relationships in soil, *J. Geotech. Eng. Div. ASCE*, **107**(GT6), 851–72.

MITACHI, T. and KITAGO, S. (1976). Change in undrained strength characteristics of a saturated remodeled clay due to swelling, *Soils* and *Foundations*, **16**(1), 45–58.

OSAIMI, A.E. and CLOUGH, G.W. (1979). Pore-pressure dissipation during excavation. *J. Geotech. Eng. Div. ASCE*, **105**(GT4), 481–98.

POPESCU, M.E. (1982). Stability analysis of deep excavations in expansive clays, *Proc. Int. Symp. on Numerical Models in Geomechanics*, Zurich, R. Dunger, G.N. Pande, and J.A. Studer, (Eds), pp. 660–7.

ROSCOE, K.H. and BURLAND, J.B. (1968). On the generalized stress-strain behavior of wet clay. In *Engineering Plasticity*, J.F. Hayman and F.A. Leckie (Eds), Cambridge University Press, Cambridge, UK, pp. 535–609.

SANDHU, R.S. and WILSON, E.L. (1969). Finite element analysis of seepage in elastic media, *J. Eng. Mech. Div. ASCE*, **95**(EM3), 641–52.

SCHOFIELD, A.N. and WROTH, C.P. (1968). *Critical State Soil Mechanics*, McGraw-Hill, London, UK.

SIMPSON, B., CALABRESI, G., SOMMER, H. and WALLAYS, M. (1979). Design parameters for stiff clays, *Proc. 7th European Conf. on Soil Mechanics and Foundation Engineering*, Brighton, UK, Vol. 5, pp. 91–126.

SKEMPTON, A.W. and SOWA, V.A. (1963). The behavior of saturated clays during sampling and testing, *Geotechnique*, **13**(4), 269–90.

STIPHO, A.S. (1978). Theoretical and experimental investigation of the anisotropically consolidated Kaolin, Ph.D. Thesis, University College, Cardiff, UK.

TAVENAS, F, and LEROUEIL, S. (1977). Effects of stresses and time on yielding of clays, *Proc. 9th Int. Conf. on Soil Mechanics and Foundation Engineering*, Tokyo, Japan, Vol. I, pp. 319–26.

TAVENAS, F., GARNEAU, R., BLANCHET, R. and LEROUEIL, S. (1978). The stability of stage constructed embankments on soft clays, *Can. Geotech. J.*, **15**(2), 283–305.

WROTH, C.P. and HOULSBY, G.T. (1985). Soil mechanics — property characterization and analysis procedures, *Proc. 11th Int. Conf. on Soil Mechanics and Foundation Engineering*, San Francisco, Vol. I, pp. 1–55.

YOUSIF, N.B. (1984). Finite element analysis of some time dependent construction problems in geotechnical engineering, Ph.D. Thesis, State University of New York at Buffalo, Buffalo, NY.

ZIENKIEWICZ, O.C. (1971). *The Finite Element Method in Structural and Continuum Mechanics*, McGraw-Hill, New York.

ZIENKIEWICZ, O.C., HUMPHESON, C. and LEWIS, R.W. (1975). Associated and non-associated viscoplasticity and plasticity in soil mechanics, *Geotechnique*, **25**, 671–89.

APPENDIX

Recalling the expressions for the invariants and yield surface

$$\bar{J}_2 = (\tfrac{1}{2}\bar{s}_{ij}\bar{s}_{ij})^{1/2} = \left[\frac{1}{2}\left(s_{ij} - \frac{2}{3}\frac{p}{p_0}s_{ij}^0\right)\left(s_{ij} - \frac{2}{3}\frac{p}{p_0}s_{ij}^0\right)\right]$$

$$\bar{J}_3 = (\tfrac{1}{3}\bar{s}_{ij}\bar{s}_{jk}\bar{s}_{ki})^{1/3}$$

and

$$f(\sigma_{ij}, s_{ij}^0, p_0) = \frac{3}{2g^2(\theta)}\left(\bar{s}_{ij}\bar{s}_{ij} - \frac{1}{9}\frac{p}{p_0}s_{ij}^0 s_{ij}^0\right) - pp_0 + p^2 = 0$$

the gradient vector $\partial f/\partial \boldsymbol{\sigma}$ can be obtained from

$$\frac{\partial f}{\partial \boldsymbol{\sigma}} = \frac{\partial f}{\partial p}\frac{\partial p}{\partial \boldsymbol{\sigma}} + \frac{\partial f}{\partial \bar{J}_2}\frac{\partial \bar{J}_2}{\partial \boldsymbol{\sigma}} + \frac{\partial f}{\partial \bar{\theta}}\frac{\partial \bar{\theta}}{\partial \boldsymbol{\sigma}} \tag{A1}$$

where

$$\frac{\partial \bar{\theta}}{\partial \boldsymbol{\sigma}} = -\frac{\tan 3\bar{\theta}}{\bar{J}_2}\frac{\partial \bar{J}_2}{\partial \boldsymbol{\sigma}} + \frac{\sqrt{3}}{2 \cos 3\bar{\theta}}\frac{3\bar{J}_3^2}{\bar{J}_2^3}\frac{\partial \bar{J}_3^3}{\partial \boldsymbol{\sigma}} \tag{A2}$$

Substituting eqn (A2) in (A1) gives

$$\frac{\partial f}{\partial \boldsymbol{\sigma}} = C_1\frac{\partial p}{\partial \boldsymbol{\sigma}} + C_2\frac{\partial \bar{J}_2}{\partial \boldsymbol{\sigma}} + C_3\frac{\partial \bar{J}_3}{\partial \boldsymbol{\sigma}} \tag{A3}$$

where

$$C_1 = \frac{\partial f}{\partial p} = -\frac{1}{g^2(\bar{\theta})p_0}[2\mathbf{q}^T\bar{\mathbf{s}}^T + \tfrac{1}{6}\mathbf{s}^{0T}\mathbf{s}^0] - p_0 + 2p$$

$$C_2 = \frac{\partial f}{\partial \bar{J}_2} - \frac{\tan 3\bar{\theta}}{\bar{J}_2}\frac{\partial f}{\partial \bar{\theta}}, \qquad \bar{\theta} = \tfrac{1}{3}\sin^{-1}\left[\frac{3\sqrt{3}}{2}\frac{\bar{J}_3^3}{\bar{J}_2^3}\right]$$

$$C_3 = \frac{\sqrt{3}}{2 \cos 3\bar{\theta}}\frac{3\bar{J}_3^2}{\bar{J}_2^3}\frac{\partial f}{\partial \bar{\theta}}$$

$$\frac{\partial f}{\partial \bar{J}_2} = \frac{6\bar{J}_2}{g^2(\bar{\theta})}$$

$$\frac{\partial f}{\partial \bar{\theta}} = -\frac{3 \cos 3\bar{\theta}}{g(\bar{\theta})}\bar{J}_2^2$$

$$g(\bar{\theta}) = \frac{6 \sin \phi}{3 - \sin \phi \sin 3\bar{\theta}}$$

　　The components of each of the vectors on the right-hand side of eqn (A3) can be then easily recovered.

Chapter 4

FINITE ELEMENT SIMULATION OF EMBEDDED RETAINING WALLS

D.M. POTTS

Department of Civil Engineering,
Imperial College of Science, Technology and Medicine,
London, UK

ABSTRACT

This chapter considers the application of the finite element method to simulate the behaviour of embedded retaining walls. The generation of earth pressures, both active and passive, are considered first. In particular the effects of various types of wall movement on both the distribution and magnitude of earth pressure are discussed. Embedded cantilever walls and walls with a prop near to the top of the wall are then considered. Results from finite element analysis are compared with simpler limit equilibrium design approaches and the shortcomings of the latter established. The effects of initial stress conditions within the soil, the stiffness of the wall and the method of construction (excavated or backfilled) are discussed. It is shown that for diaphragm or secant pile walls installed in overconsolidated clays typical design approaches may not be conservative. The chapter ends by describing the application of the finite element approach to complex embedded retaining walls for which simple design methods are not available.

1 INTRODUCTION

Embedded retaining walls, as their name suggests, are partly buried promoting the mobilisation of a resisting force from the soil to help maintain stability. In many cases the soil at higher elevations provides, at least part of, the disturbing force acting on the wall. The simplest example is an embedded cantilever (see Fig. 1(a)) where the retained soil above the excavation level provides the disturbing force and the soil below this level, both in front of and behind the wall, provides the resistance. The wall itself acts as a means of transferring load from the soil above excavation level to be resisted by the soil below. The magnitude and distribution of the loads acting on the wall depend on a complex soil-structure interaction and are highly dependent on the mechanical properties of both the wall and the soil, the initial stresses within the soil and the method of wall construction. For example a situation in which the retained height is formed by backfilling behind the wall is likely to behave in a very different manner to the case where the retained height is formed by excavating soil from in front of the wall.

Further examples of embedded retaining walls are shown in Figs 1(b) and 1(c). Figure 1(b) indicates a propped/anchored embedded cantilever

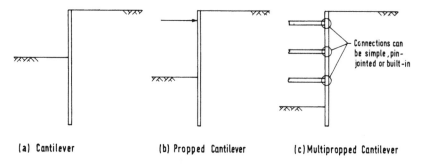

(a) Cantilever   (b) Propped Cantilever   (c) Multipropped Cantilever

FIG. 1.   Embedded retaining walls: (a) cantilever; (b) propped cantilever; (c) multipropped cantilever.

wall in which the prop/anchor acts as a simple support and provides an additional stabilising force. As the height of the retained soil increases then it is often necessary to provide extra support in the form of additional props or anchors. If props are used then these may act as simple supports, they may be pin-jointed to the wall or they may be built-in providing a full or partial moment connection (see Fig. 1(c)). These props or anchors are often installed as construction takes place. For example in the so called 'top down' construction method often used for deep basements, the walls are first cast either using the diaphragm wall or secant pile technique. As excavation proceeds in front of the wall, props, usually in the form of floor slabs, are cast at the appropriate level. Excavation is then continued and the process repeated. The nature of the construction process and the type of connection between prop and wall (i.e. simple, pin-jointed or built-in) affects the magnitude and distribution of the loads induced in both the wall and the props.

During the design of these types of structure it is necessary to determine the structural loads and the likely soil movements such that the effects on adjacent structures may be estimated. These must be determined under working load conditions. Because of the complex nature of the soil-structure interaction involved with this type of structure it is extremely difficult to estimate these design quantities from simple methods. Design guidelines are available for determining the embedment depths for both cantilever and propped/anchored cantilever walls and for determining prop/anchor forces for the simple multipropped walls. However difficulties arise if the props are either pin-jointed or have a full or partial moment connection. There are no simple methods available for predicting ground movements. Many of the design approaches are crude and often based on empirical observation. They may lead to optimistic designs especially in novel situations where little previous experience is available.

The finite element method provides an alternative approach. With this method it is possible to simulate the construction history and obtain predictions of movements and structural loads both under working conditions and at failure. With the recent advances in computer hardware and software development it is becoming cost effective to undertake such analyses as part of the design process for some of the more complex structures. In addition, the method may be used to verify and improve design procedures for the simpler types of embedded structure.

Over the past decade the author has applied the finite element method to many different types of embedded structure ranging from simple cantilever walls to complicated multipropped retaining walls in which the props apply both a lateral and moment restraint to the wall. The conclusions arising from this work and the benefits of using the finite element approach for simulating the behaviour of embedded retaining walls will be summarised in this chapter. The chapter begins with a brief discussion on the mobilisation of earth pressures and in particular the effects of wall deformation. The behaviour of cantilever and propped cantilever walls are then discussed and finally an example of the application of the finite element method to analysing a complicated multipropped retaining wall is described.

2 MOBILISATION OF EARTH PRESSURE

The distribution and intensity of mobilised earth pressure is of major importance in the design of any embedded retaining wall. It affects not only the overall stability of a retaining wall but also the bending moments and any prop/anchor forces that may occur. The limiting effective stress σ'_h for a cohesionless, $c' = 0$, ϕ' soil is often defined as

$$\sigma'_h = K_a \cdot \sigma'_v \quad \text{or} \quad K_p \cdot \sigma'_v \tag{1}$$

where, for a uniform soil deposit, $\sigma'_v = \gamma \cdot z - u$, γ is the bulk unit weight, z is the depth below the soil surface, u the pore water pressure and K_a and K_p are the active and passive earth pressure coefficients, respectively. These coefficients depend on the angle of shearing resistance, ϕ', of the soil and the angle of friction, δ, between soil and wall. It should be noted that for a rough wall, σ'_v in the above equation is not the vertical effective stress immediately adjacent to the wall but rather the free field value. For a uniform deposit of soil, eqn (1) indicates that the active and passive earth pressures increase linearly with depth. Current design practice relies on approximate methods to estimate the active and passive earth pressure coefficients, e.g. limit equilibrium, stress field solutions, limit analysis. For a complete solution the requirements of equilibrium, compatibility, material behaviour and the boundary conditions, both load and displacement, must all be satisfied. It can be shown that all four of the above requirements are only satisfied for a limit analysis solution in which both upper and lower bound calculations lead to the same result. Apart from this special case, all the methods fail to satisfy at least one of the requirements. Implicit in these conventional methods is the assumption that the angle of shearing resistance, ϕ', is fully mobilised. No account is taken of the influence of the mode of wall displacement on the resultant earth pressure, and no indication of the distribution or magnitude of earth pressure prior to ultimate failure is given. The estimation of prop/anchor forces and bending moments in the wall under working load conditions is therefore extremely difficult. Although a great deal of experience has been gained using the approximate methods, the above short comings restrict their usefulness, particularly in solving unusual problems.

An alternative approach is afforded by the finite element method of analysis. All four basic solution requirements are met, albeit in an approximate manner, the accuracy of which depends on the assumed discretization. In addition the geometry and boundary conditions of a specific problem can be accurately modelled. Potts and Fourie (1986) have used this method to study the soil-structure interaction of a rigid wall embedded in an initially horizontal, uniform soil deposit. Three modes of wall deformation were studied, namely horizontal translation, rotation about the wall top and rotation about the toe. The mode of deformation of a real retaining wall is complicated, and depends, in part, on both the flexibility and the type of wall. For embedded cantilever walls, the mode of displacement is essentially one of rotation about the toe. Alternatively, if a prop is installed near the top of the wall, the mode changes to one of rotation about the top. Horizontal translation is more applicable to a gravity type wall.

It must be stressed at the outset that the objective of this work was not to model a real retaining wall situation, but to investigate, at a fundamental level, the effect of the mode of deformation on the generation of earth pressure. A simple elasto-plastic soil model employing a Mohr-Coulomb failure criterion ($c' = 0$, $\phi' = 25°$) as the yield surface was adopted. The finite element mesh used for the study is shown in Fig. 2

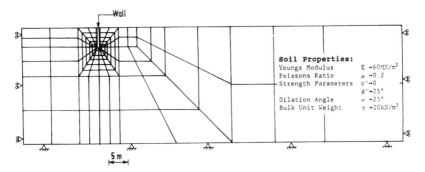

FIG. 2.   Finite element mesh for studying earth pressure mobilisation.

and consists of 154 eight noded isoparametric elements. Also shown in this figure are the soil properties and the assumed boundary conditions. The 5-m deep by 1-m thick embedded wall was modelled as part of the external boundary, and loading was simulated by applying increments of displacement to this part of the boundary. The behaviour of both smooth and rough walls were investigated as well as the effects of soil dilation and the coefficient of earth pressure at rest, K_0. For further details the reader is referred to Potts and Fourie (1986).

As an example of the results obtained from this investigation the development of the active and passive pressure coefficients with increasing wall displacement are shown in Fig. 3. The results are for a rough wall and with an initial coefficient of earth pressure $K_0 = 2$. The equival-

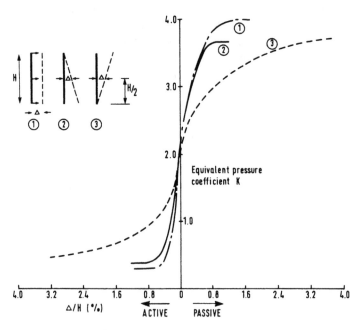

FIG. 3.   Development of earth pressure coefficients with increasing wall displacement (rough wall).

ent coefficient K (active or passive) is defined as $K = 2 \cdot P/(\gamma \cdot H^2)$, where P is the resultant force exerted by the soil on the front (passive) or back (active) of the wall, H is the depth of the wall and γ the bulk unit weight of the soil. Clearly the magnitudes of these displacements are dependent on the elastic stiffness assumed for the soil. However, as the same stiffness was assumed in all three analyses, the results do provide a relative measure of the amount of movement required to mobilise limiting conditions on either side of the wall. Rotation about the base requires significantly more displacement to obtain failure conditions than do the other modes of displacement. For high K_0 values, active and passive conditions can be mobilised at similar displacements (see Fig. 3). This is contrary to the commonly accepted notion that active conditions occur long before passive conditions and invalidates the arguments commonly put forward for imposing restrictions on K_p.

The effect of the mode of wall displacement on the distribution of earth pressure is summarised in Fig. 4. The development of the active and

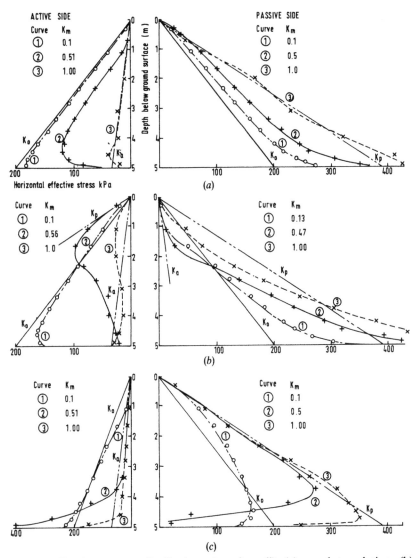

FIG. 4.   Earth pressure distributions (rough wall): (a) equal translation; (b) rotation about top; (c) rotation about toe.

passive pressures for all three modes of deformation are shown for various values of the pressure coefficient K_m where $K_m = (K - K_0)/(K_p - K_0)$ for passive pressures and $K_m = (K_0 - K)/(K_0 - K_a)$ for active pressures. K_m can therefore be seen to be equivalent to a load factor. The lines labelled K_a and K_p shown on these figures correspond to the Caquot and Kerisel (1948) values. For a wall rotating about its top or bottom, the distributions are far from the linear distributions commonly assumed in design and given by eqn (1). For example, for a wall rotating about its top, the soil on the passive side at failure is in an active condition near the top and exceeds the classical passive value lower down. The pressure distributions also change during the mobilisation of a limiting condition. Such vastly different pressure distributions imply that wall bending moments will be highly dependent on the mode of wall deformation.

The above numerical predictions are in agreement with the experimental observations made by Bros (1972), who carried out laboratory model retaining wall tests in which the three modes of displacement were simulated.

From this study it is clear that the development of limiting earth pressures is complex and that many of the simple design approaches may lead to optimistic designs if applied to novel situations. The finite element method on the other hand has considerable advantages.

3 EMBEDDED CANTILEVER WALLS

Embedded cantilever retaining walls may be used in either temporary or permanent works situations. When designing such walls the main criterion is to prevent unserviceability. The embedment depth must be sufficient to ensure overall stability of the retaining wall and the wall itself must be strong enough to resist the maximum applied bending moment. To ensure overall stability, cantilevers are often designed using the fixed-earth support method in which the wall is assumed to rotate about a point near to its toe. The resulting earth pressure distribution is highly idealised as shown in Fig. 5(a). At the point of rotation '0' there is a transformation from active to passive pressures behind the wall and vice versa in front of the wall. Determination of the required embedment depth using these distributions of earth pressures can be particularly tedious (Padfield and Mair, 1984) and a further simplification is usually made. Figure 5(b) shows that the difference between the passive pressure

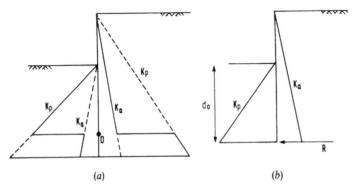

(a) (b)

FIG. 5.   Assumed pressure distributions for design of embedded cantilever
retaining walls.

behind the wall and the active pressure in front of the wall may be
replaced by an equivalent force R, at the toe of the wall. The required
embedment depth is then obtained by taking moments about the toe, and
this depth is increased by an empirical 20% to account for the preceding
assumptions and simplifications. For design, the moment equation is
directly or indirectly used to ensure that restoring moments exceed
overturning moments by a prescribed safety margin. This is often
achieved by the inclusion of a single factor of safety. There are many ways
in which such a factor may be incorporated, for example it may be
applied to the passive pressure (BSI, 1951; NAVFAC, 1982), the net
passive pressure (BSC, 1979), the strength parameters or the net resisting
moment (Burland et al., 1981). In this chapter the latter approach is used,
and the factor of safety is referred to as F_r. This factor of safety is
intended to account for site variability and uncertainties in construction
details.

Current design approaches do not distinguish between cantilever
retaining walls formed by excavation of soil in front of the wall and those
formed by backfilling of material behind the wall. The influence of the
initial soil stresses are also not addressed. Some effects of these parameters have been studied using the finite element method (Fourie and
Potts, 1989) and are discussed below.

The geometry and material properties selected for investigation are
shown in Fig. 6. Such soil properties are appropriate to a stiff clay. The
1-m thick 20-m deep concrete retaining wall is assumed to be linearly
elastic whereas the soil is assumed to be linearly elastic-perfectly plastic,

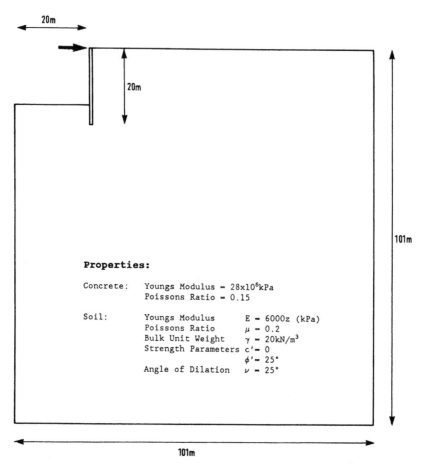

FIG. 6.   Problem geometry and material properties.

with a Mohr–Coulomb yield surface. Drained soil conditions were assumed with zero pore pressures everywhere, and it is therefore the long term stability of a wall in which the water table is at some depth below the toe of the wall that is under investigation. The analyses could have included a more realistic water regime in which differing levels occur in front of and behind the wall. However, this would have increased the number of variables and would have required additional assumptions to be made as to the nature of the pore pressure distribution around the wall. Initial effective stresses of $\sigma_v' = \gamma \cdot z$ and $\sigma_h' = K_0 \cdot \sigma_v'$ were specified in the soil. It was assumed that the retaining wall was installed in the soil without altering the in-situ condition. Values of $K_0 = 2\cdot0$ and $0\cdot5$ were used to represent two possible extremes of in-situ stresses. The adoption of $K_0 = 0\cdot5$ does not necessarily imply that a normally consolidated soil is being modelled, but has been selected to provide an extreme value for comparison with $K_0 = 2\cdot0$. The techniques used for modelling the construction procedure in the finite element analysis are discussed in detail by Potts and Fourie (1984) and will not be repeated here.

A limit equilibrium analysis using the fixed-earth support method as described above, and the previously mentioned soil strength properties indicated that the maximum allowable retained height of soil was approximately $10\cdot54$ m. This resulted in a factor of safety against overall instability of unity assuming full wall friction acted on both the front and back of the wall. To ensure a factor of safety F_r in excess of $2\cdot0$ the retained height should not exceed $8\cdot7$ m. For these calculations, values of active and passive earth pressure coefficients of $K_a = 0\cdot33$ and $K_p = 4\cdot19$ have been adopted. The value for K_a is consistent with those quoted by Caquot and Kerisel (1948), Chen (1975) and Packshaw (1946), whereas the K_p value is consistent with those quoted by Chen (1975), NAVFAC (1982) and Lee and Herrington (1974). If a K_p value of $3\cdot89$ (Caquot and Kerisel, 1948) is adopted, the limiting depth of excavation reduces from $10\cdot54$ m to $10\cdot3$ m.

The variation of the maximum wall displacement with increasing excavation depth predicted by the finite element method is shown in Fig. 7 for K_0 values of $2\cdot0$ and $0\cdot5$. It is apparent that as the depth equivalent to the limit equilibrium factor of safety of unity is approached, the maximum wall displacement increases rapidly. In all cases analysed it was not possible to excavate below the depth of $10\cdot6$ m without resulting in a non-converging solution. The finite element analyses therefore indicate a height of $10\cdot6$ m to be approximately the limiting retained height for drained conditions with no pore pressures present irrespective of the initial state of stress in the soil. This is in agreement with the limit equilibrium analysis used, and justifies the assumptions involved in this approach. In particular the finite element analyses justify the 20% increase in embedment depth arbitrarily assumed in the limit equilibrium approach.

The horizontal displacements of the cantilever retaining wall towards the excavation are shown in Fig. 8 for a retained height of $8\cdot85$ m ($F_r = 2\cdot0$). Predictions for excavated walls with initial stresses corresponding to K_0 values of $2\cdot0$ and $0\cdot5$ are shown together with a prediction for

D.M. POTTS

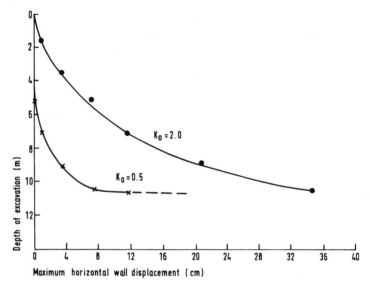

FIG. 7.   Variation of maximum lateral wall movement with increase in depth of
excavation (embedded cantilever).

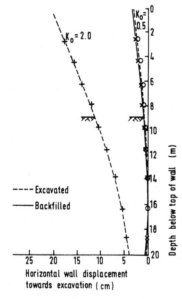

FIG. 8.   Lateral wall displacements for a retained height of 8·85 m (embedded
cantilever).

a backfilled wall. As would be expected, the magnitude of the wall displacement is significantly higher for the excavated wall with the larger value of K_0. An important difference in the predictions shown in Fig. 8 is that for the excavated wall with a high K_0 the base of the wall moves towards the excavation, where as for the excavated wall with $K_0 = 0.5$ and for the backfilled wall little movement occurs.

The predicted wall bending moments are given in Fig. 9 for a depth of

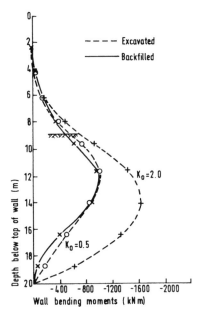

FIG. 9.   Bending moment distributions for a retained height of 8·85m (embedded cantilever).

excavation of 8·85 m and the maximum bending moments are listed in Table 1 for two retained heights, namely 8·85 m $(F_r = 2)$ and 10·6 m $(F_r = 1)$. Predictably, in all cases the maximum bending moment increases as the retained height increases. The difference in the predicted maximum moments at the limiting equilibrium condition is not significant (less than 14%). This is in contrast to differences of in excess of 100% for propped retaining walls as discussed subsequently. At an excavation depth of 8·85 m $(F_r = 2)$ the difference in maximum bending moments is much larger being approximately 60%.

The maximum retaining wall bending moments calculated assuming

TABLE 1
MAXIMUM WALL BENDING MOMENTS FOR CANTILEVER WALLS

Depth of excavation/ retained height (m)	Factor of safety, F_r	Finite element predictions (kNm/m)					Limit equilibrium (kNm/m)	
		Excavated walls				Backfilled wall	Method 1	Method 2
		$K_0=2$	$K_0=1.5$	$K_0=1$	$K_0=0.5$	$K_0=2$		
8·85	$\lesssim 2$	1620	1330	1130	1000	1000	1390	1800
10·6	$\lesssim 1$	2150	2070	1970	1830	1700	2400	2400

earth pressure distributions consistent with the limit equilibrium approach are given in Table 1 for retained heights of 8·85 m ($F_r \approx 2$) and 10·6 m ($F_r \approx 1$). In these calculations full wall friction has been assumed and in line with most design approaches no account has been taken of the moment due to shear stresses acting on the front and back of the wall. For F_r not equal to unity two approaches have been adopted. In the first, method 1, the maximum bending moment has been calculated at limiting equilibrium. For the particular retained height the embedment necessary to just maintain equilibrium ($F_r = 1$) is first determined using unfactored soil parameters. With this geometry and earth pressure distribution the maximum bending moment is calculated. This is the method recommended by Padfield and Mair (1984) and essentially ignores the effect of the extra embedment that is actually present. In the second approach, method 2, the actual embedment depth is used but the earth pressures have been modified by F_r to ensure equilibrium. To be consistent with the definition of F_r only the earth pressures contributing to the net resisting moment have been factored. At limiting conditions when $F_r = 1$ and the retained height is 10·6 m both approaches give the same prediction. Also shown the Table 1 are finite element results for excavated walls with $K_0 = 0·5$, 1·0, 1·5 and 2·0 and for the backfilled wall discussed above.

For a retained height of 10·6 m ($F_r \approx 1$) the limit equilibrium value exceeds the finite element prediction for all situations analysed. This is also true for a retained height of 8·85 m ($F_r \approx 2$) if method 2 is employed. In particular, the maximum moment in the walls formed by backfilling, or by excavation in a low K_0 soil are overpredicted by some 35–45% for $F_r \approx 2$. Method 1 produces lower values than method 2 and for high K_0 soils these are lower than the finite element values. It is evident from the results presented in Table 1 that for all the cantilever walls analysed, whether they be formed by backfilling or by excavation in a low or high K_0 soil, limit equilibrium method 2 provides a conservative estimate of maximum bending moment when compared with the finite element results. Limit equilibrium method 1 produces lower values of maximum bending moment which are in closer agreement with the finite element results.

Although it should be noted that the finite element method is by no means perfect, it does in fact attempt to satisfy equilibrium, compatibility, the assumed boundary conditions and the soil constitutive model, which is more than can be said for any of the approximate limit equilibrium approaches currently in use. It therefore provides a very useful yardstick against which the approximate methods of analysis may be compared. In

addition to providing predictions of wall movements and bending moments the finite element analysis also predicts the displacements occurring in the adjacent soil. This information is of great value if the
behaviour of any adjacent buildings is important.

4 PROPPED RETAINING WALLS

4.1 Introduction
When retained heights increase or restricting soil movements becomes
important, propped embedded cantilever walls are often used. The recent
developments of secant piles and slurry trench methods of in-situ wall
construction have led to the frequent use of such walls for retained
cuttings and cut and cover tunnels in urban environments where land use
is restricted. These walls maintain stability and prevent excessive soil
movement.

Present design procedures involve the use of the free-earth support
method in which the wall is assumed to rotate about the prop to
investigate overall stability. Bending moments are either calculated from
the free- or fixed-earth support approaches. The free-earth approach
leads to higher bending moments and some design procedures recom-
mend the use of moment reduction factors. Current design approaches
provide little information about the distribution and magnitude of soil
movements or about their effects on wall bending moments and lateral
earth pressures. As noted above, for cantilever walls, available design
techniques do not distinguish between walls formed by excavation of soil
in front of the wall and those formed by backfilling of material behind
the wall. The influence of the initial soil stresses and wall flexibility are
often ignored. The effect of these parameters on wall behaviour have been
studied using the finite element approach by Potts and Fourie (1984,
1985). Some of their findings will now be presented.

4.2 Effect of Initial Stresses and Method of Construction
Adopting the geometry and soil properties shown in Fig. 6 but with a
rigid horizontal prop placed at the top of the wall, finite element analyses
have been performed to investigate the effect of initial stresses (i.e. K_0
values) and on the method of construction, backfilled or excavated.

For the geometry and soil properties employed, a stability analysis
using the free-earth support method indicates that the maximum possible
retained height is approximately 15·3 m. This corresponds to a factor of

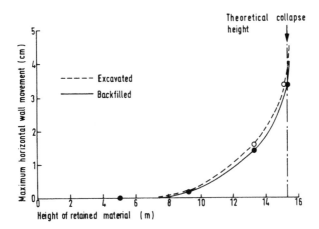

FIG. 10.   Variation of maximum lateral wall movement with increase in height
of retained soil (propped cantilever).

safety of unity and was calculated assuming full wall friction to be acting
on both the front and back of the wall. As for the cantilever wall
discussed in Section 3 values of $K_a = 0.33$ and $K_p = 4.19$ have been
adopted. To ensure a factor of safety $F_r = 2$ the retained height should be
kept less than 13.3 m. The predicted variation of maximum horizontal
wall movement with increase in retained height are given in Fig. 10 for
an excavation analysis employing $K_0 = 0.5$ and for a backfilled analysis
employing $K_0 = 2$ in the foundation. Both predictions indicate that failure
occurs when the retained height approaches 15.3 m and in all analyses it
was found impossible to excavate or construct further material to give a
greater retained height. Thus the finite element analyses indicate that this
is the limiting retained height, which is independent of construction
method or K_0 in-situ, and which is in agreement with the results from
simple limit equilibrium stability analyses.

Horizontal wall movements for a retained height of 13.26 m ($F_r = 2$) are
shown in Fig. 11. Results for both excavated walls with K_0 of 0.5 and 2.0
and for backfilled walls with K_0 of 0.5 and 2.0 in the foundation are
given. It is clear from this figure that while the movements for both
backfilled cases and for the excavated wall with $K_0 = 0.5$ are similar, the
movements for the excavated wall with $K_0 = 2$ are approximately eight
times larger.

The associated bending moments in the wall are given in Fig. 12 and

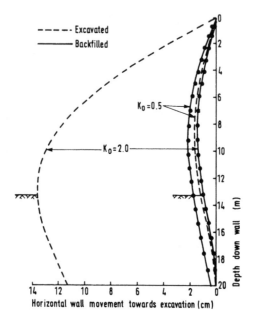

FIG. 11. Comparison of wall displacements for excavated and backfilled walls (propped cantilever).

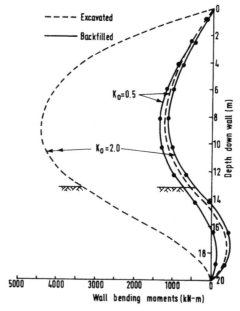

FIG. 12. Comparison of wall bending moments for a retained height of 13·26 m (propped cantilever).

show a similar trend to that of the wall displacements with the maximum bending moment for the excavated case with $K_0 = 2$ being some four times greater than for the other analyses which produce similar predictions. Values of maximum bending moments are tabulated in Table 2 for both excavated and backfilled walls. Additional predictions for excavated walls with K_0 of 1 and 1·5 are given as well as results obtained from free-earth support limit equilibrium type calculations using both methods 1 and 2 (see Section 3). Inspection of this table indicates that the maximum bending moments for the excavated wall with $K_0 = 0·5$ and the backfilled walls are very similar for all three retained heights and are smaller than those from the limit equilibrium calculation. For excavated walls with higher K_0 values the bending moments are larger and substantially exceed the limit equilibrium values. The equivalent prop forces are shown in Table 3 and indicate a similar trend. It may be concluded that for excavated walls in soils with a high initial K_0 value, such as a stiff clay, prop forces and wall bending moments could greatly exceed those predicted by current design methods.

The analyses have been performed assuming zero pore water pressure. While this greatly simplifies the presentation of the results it is unrealistic, since field situations generally involve seepage around the wall. The effects of the pore water is not difficult to include in the finite element calculations, but complicates the interpretation of the results and increases the number of variables. While the presence of pore water pressure will clearly affect the magnitude of predicted movements and stresses it is unlikely to alter the main conclusions.

4.3 Effect of Prop Position

In the above comparisons the prop has been assumed to act at the top of the wall. However, in the majority of field situations the prop is positioned some distance below the top of the wall. As this distance increases then the applicability of the simple active and passive earth pressure distributions assumed in the limit equilibrium type calculations becomes questionable. Fourie and Potts (1988) have employed the finite element approach to investigate this effect. Using the same geometry and soil properties as for the analyses described above, separate calculations were carried out for prop depths of 0 m, 1·5 m, 3·25m, 5 m, 7 m and 9·26 m below ground surface. Initially the walls behaved as unrestrained cantilevers but props were installed immediately the depth of excavation reached the required depth of propping.

The prop forces and maximum bending moments for a depth of

TABLE 2
MAXIMUM WALL BENDING MOMENTS FOR PROPPED CANTILEVERS

Depth of excavation/ retained height (m)	Factor of safety, F_r	Finite element predictions (kNm/m)						Limit equilibrium (kNm/m)	
		Excavated walls				Backfilled walls		Method 1	Method 2
		$K_0=2$	$K_0=1.5$	$K_0=1$	$K_0=0.5$	$K_0=2$	$K_0=0.5$		
9·26	6·1	3020	1990	1040	380	355	540	555	1395
13·26	2·0	4400	3085	1820	1220	1140	1360	1625	2135
15·26	≑1·0	4770	3570	2545	2030	2000	2120	2500	2500

TABLE 3
PROP FORCES FOR PROPPED CANTILEVERS

Depth of excavation/ retained height (m)	Factor of safety, F_r	Finite element predictions (kN/m)						Limit equilibrium (kN/m)	
		Excavated walls				Backfilled walls		Method 1	Method 2
		$K_0=2$	$K_0=1.5$	$K_0=1$	$K_0=0.5$	$K_0=2$	$K_0=0.5$		
9·26	6·1	821	540	271	120	82	113	130	242
13·26	2·0	1013	754	400	237	190	217	270	320
15·26	≑1·0	938	758	461	330	283	296	355	355

excavation of 13·26 m and for K_0 values of 0·5 and 2·0 are given in Table 4. Predictions based on limit equilibrium method 2 (see Section 3) are also given. The results indicate that lowering the position of the prop increases the magnitude of the predicted prop force while the maximum bending moment value decreases. As the depth of the prop position increases beyond 7 m below ground level the finite element analyses predict a change in sign of the maximum bending moment, whereas no change in sign is predicted by the limit equilibrium method. This is attributable to the unrealistic assumptions of earth pressure distribution inherent in the limit equilibrium method. As noted above the limit equilibrium approach underestimates the finite element predictions for $K_0 = 2$.

Figure 13 shows a plot of the absolute value of the maximum bending moment predicted by the finite element method, $|M_{FE}|$ normalised by the limit equilibrium value M_{LE} with the depth of the prop position below ground level D_P. This value has been normalised by the depth of excavation, D_E, which in this case is 13·26 m ($F_r \geqslant 2·0$, see Table 4). It can be seen that up to a prop depth of approximately one-third the excavation depth, a reasonably constant ratio between the finite element and limit equilibrium results exists. It is particularly notable that the constant of proportionality is remarkably similar to the value of K_0 used in the analysis. At prop depths in excess of one-third the excavation depth, the ratio between the maximum bending moments varies erratically.

4.4 Effect of Wall Stiffness

At first the above results seem to be in conflict with those of Rowe (1952) who performed scale model tests on flexible retaining walls. He showed that the limit equilibrium type calculations assuming free-earth support conditions and based on the actual embedment depth (method 2) overestimate the observed bending moments and this leads to the recommendation of moment reduction factors (Terzaghi, 1954).

Rowe's tests were carried out using excavated walls in sand and the K_0 values were less than unity. His results are therefore only comparable with those given above in which low K_0 values were employed. In these analyses the predicted bending moments were less than those given by limit equilibrium calculations.

A direct comparison is complicated by the fact that the above analyses simulated a stiff diaphragm or secant pile wall installed in stiff clay whereas Rowe modelled a more flexible sheet pile wall in sand. Rowe found that the moment reduction factors were dependent on the flexibility

TABLE 4
PROP FORCES AND MAXIMUM BENDING MOMENTS FOR VARIOUS PROP POSITIONS (DEPTH OF EXCAVATION=13·26 m)

Depth of prop below ground surface (m)	Factor of safety, F_r	Maximum bending moment in wall (kNm/m)			Prop forces (kN/m)		
		Finite element		Limit equilibrium	Finite element		Limit equilibrium
		$K_0=2$	$K_0=0.5$	method 2	$K_0=2$	$K_0=0.5$	method 2
0	2·0	4400	1220	2135	1013	237	320
1·5	2·05	4100	1124	1903	1267	245	350
3·25	2·16	3450	915	1610	1185	289	392
5·0	2·33	2550	672	1270	1874	347	446
7·0	2·65	1400	−346	818	2045	420	529
9·26	3·44	−2050	−709	260	1953	465	670

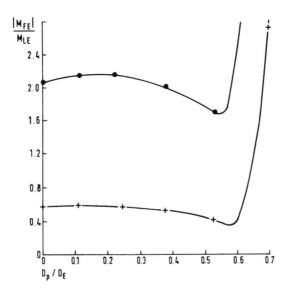

FIG. 13. Variation of bending moment ratio with depth of excavation
($D_E = 13\cdot26$ m, propped cantilever).

of the wall: the stiffer the wall the larger the bending moment. In an attempt to clarify the situation, further finite element analyses were performed varying the stiffness of the wall. These analyses considered an excavated wall and, apart from the wall stiffness, the geometry, material properties and assumptions were all identical to those used above. Four different wall stiffnesses have been investigated corresponding to a 'rigid' wall, a diaphragm or secant pile wall, a sheet pile wall and a 'soft' wall. These walls had bending stiffnesses EI of $2\cdot3 \times 10^9$ kNm^2, $2\cdot3 \times 10^6$ kNm^2, $7\cdot8 \times 10^4$ kNm^2 and $2\cdot3 \times 10^4$ kNm^2, respectively. The diaphragm wall represents a concrete section 1 m thick (as modelled above) and the sheet pile wall models a Larssen 4B section. The rigid and soft wall cases are included as two extreme situations. Two values of K_0, 2 and 0·5, were analysed for each wall stiffness.

The predicted horizontal movements of the walls are given in Fig. 14 for an excavation depth of 13·26 m ($F_r = 2$) and $K_0 = 2$. The mode of deformation of the rigid wall is essentially one of rotation about the top of the wall with the maximum horizontal displacement at the bottom of the wall. In comparison the soft, sheet pile and diaphragm walls have a much greater movement at mid-height.

D.M. POTTS

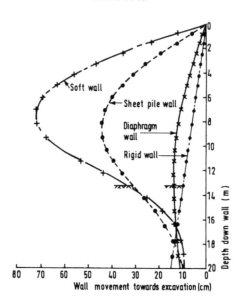

FIG. 14. Effect of wall stiffness on lateral wall movements ($F_r=2$, propped cantilever).

The corresponding horizontal stresses acting on the back of the wall are given in Fig. 15. The effect of wall flexibility on the distribution of earth pressure is clearly evident. For the rigid wall a parabolic distribution is predicted with a maximum value at mid-height of the wall. The results for both the soft and the sheet pile wall indicate pressures that are slightly lower than the 'classical' active distribution over the middle of the wall with a dramatic increase near the toe of the wall and a bulge of high pressure at the top of the wall. For the diaphragm wall the pressure distribution is essentially in between those for the rigid and sheet pile walls: however, it does show higher pressures, approaching a passive condition, near the top of the wall. The corresponding predictions for $K_0=0.5$ show similar trends and therefore are not presented.

As a result of the differences in the pressure distributions on the back of the wall the bending moments are also very different. These are shown in Fig. 16 for an excavation depth of 13·26 m. The maximum values of the bending moment for excavation depths of 13·26 m ($F_r=2$) and 15·26 m ($F_r=1$) are given in Table 5. Values for both $K_0=0.5$ and $K_0=2$ are given along with values obtained from a free-earth support limit equilibrium calculation based on the actual embedment depth (method

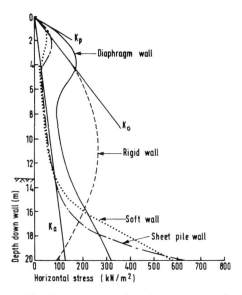

FIG. 15.   Effect of wall stiffness on lateral earth pressures on back of wall ($F_r = 2$, propped cantilever).

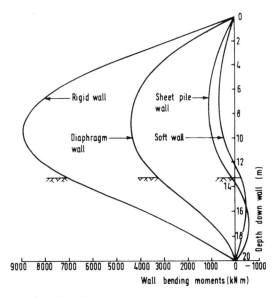

FIG. 16.   Effect of wall stiffness on wall bending moments ($F_r = 2$, propped cantilever).

TABLE 5
EFFECT OF WALL STIFFNESS ON MAXIMUM WALL BENDING MOMENT

			Maximum bending moments: (kNm/m)				
K_0	Excavation depth (m)	F_r	Finite element predictions				Limit equilibrium Method 2
			Rigid wall	Diaphragm wall	Sheet pile wall	Soft wall	
2·0	13·26	2·0	8900	4400	1160	700	2135
2·0	15·26	1·0	7850	4770	1500	1025	2510
0·5	13·26	2·0	1670	1220	816	570	2135
0·5	15·26	1·0	2600	2030	1212	840	2510

2). For the low K_0 value the predicted bending moments are all less than the limit equilibrium values except for the rigid wall at an excavation depth of 15·26 m $(F_r = 1)$ where the predicted value slightly exceeds the limit equilibrium value. However, for the analyses employing $K_0 = 2$ the maximum bending moments are higher and for the rigid and diaphragm wall greatly exceed the limit equilibrium values. The influence of K_0 on the bending moments is larger the stiffer the wall.

Predicted prop forces for excavation depths of 13·26 m $(F_r = 2)$ and 15·26 m $(F_r = 1)$ are presented in Table 6. Values for K_0 of 0·5 and 2·0 are given along with values obtained from a limit equilibrium calculation (method 2). For $K_0 = 0·5$ the prop forces are all lower than the limit

TABLE 6
EFFECT OF WALL STIFFNESS ON PROP FORCE

			Prop force (kN/m)				
K_0	Excavation depth (m)	F_r	Finite element predictions				Limit equilibrium Method 2
			Rigid wall	Diaphragm wall	Sheet Pile wall	Soft wall	
2·0	13·26	2·0	1483	1013	317	239	320
2·0	15·26	1·0	1304	938	340	276	360
0·5	13·26	2·0	259	237	245	194	320
0·5	15·26	1·0	348	330	335	268	360

equilibrium values. However, for the rigid and diaphragm walls with $K_0 = 2$ much higher prop forces are predicted, For example for the diaphragm wall the prop forces are approximately three times the limit equilibrium values. Prop forces increase with an increase in both K_0 and the stiffness of the wall.

Rowe (1952) carried out his tests on sheet pile walls in loose and dense sands (with angles of shearing resistance in excess of 30°). He presented the effects of wall flexibility in a diagram of bending moment reduction factor M/M_{LE} (where M is the maximum observed bending moment and M_{LE} is the maximum bending moment from a limit equilibrium calculation (method 2)) against the logarithm of the wall flexibility number ρ ($= H^4/EI$). Figure 17 shows such a diagram (ρ has the units m^4/(kNm^2 per metre)) and the shaded zones represent Rowe's results for dense and loose sands. Rowe's flexibility number does not include the stiffness of the ground. Intuitively it is reasonable to expect that the stiffness of the ground as well as that of the wall and K_0 must influence both the bending moments and prop forces. Since the present numerical results are for one stiffness distribution, the influence of soil stiffness cannot be assessed and it could be misleading to plot these results on the same graph as Rowe's data which presumably relate to two stiffness distributions (corresponding to loose and dense sands). Further work is required to assess the significance of the soil stiffness.

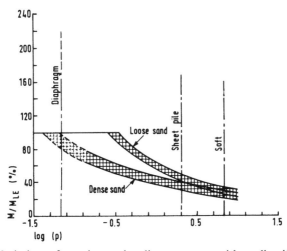

FIG. 17. Variation of maximum bending moment with wall stiffness (after Rowe, 1952).

Notwithstanding this and noting that the present numerical results are for material with $\phi' = 25°$ it is of interest to compare them with Rowe's experimental results. The open points in Fig. 18 show the computed moment reductions for walls supporting soil for which $K_0 = 0.5$. It can be seen that the theoretical results are similar in form to the experimental results and for flexible walls are in good quantitative agreement.

The theoretical moment reduction factors for $K_0 = 2$ lie above those for $K_0 = 0.5$ although for flexible walls the differences are much less than for stiffer walls. It is important to note that for $K_0 = 2$ the arbitrary cut-off at $M/M_{LE} = 1$ is not valid and for diaphragm walls the bending moments can be much larger than those given by the simple limit equilibrium calculation. Rowe did not perform tests with walls in a soil with a high K_0 value.

Inspection of Table 6 indicates a similar trend for the prop forces, namely for stiff walls and a high K_0 value the prop forces exceed the limit equilibrium value. These results are consistent with those of Rowe who found that the prop force increased with wall stiffness and that for his experiments in sand (low K_0 values) the magnitude of the prop force was less than the limit equilibrium value.

Results of further finite element calculations are combined with those presented above to give the summary plots shown in Figs 19 and 20.

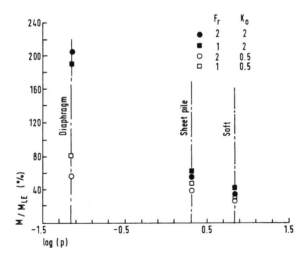

FIG. 18. Variation of maximum bending moment with wall stiffness (finite element predictions).

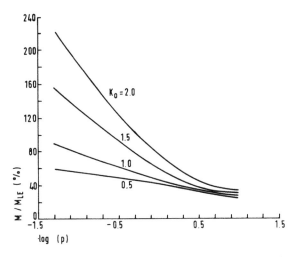

FIG. 19. Variation of maximum bending moment with wall stiffness ($F_r = 2$, propped cantilever).

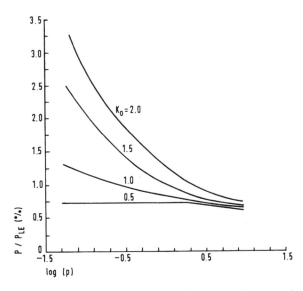

FIG. 20. Variation of prop force with wall stiffness ($F_r = 2$, propped cantilever).

These indicate the variation of M/M_{LE} and P/P_{LE} with log ρ, respectively. P_{LE} is the prop force from a limit equilibrium calculation (method 2). Both figures refer to a wall propped at its top with an excavation depth of 13·26 m ($F_r = 2$). These may be useful for preliminary design. However, in cases where bending moments or prop forces are a critical factor in design it is advisable to carry out a sensitivity study using the finite element method for the particular structure along the lines described by Potts and Burland (1983) and Hubbard et al. (1984).

5 MULTIPROPPED RETAINING WALLS

In the above sections of this chapter application of the finite element method to idealised embedded wall situations has been discussed. Results of such analyses have been compared with the simple calculations currently used in design. In this respect finite element analysis has been used to theoretically calibrate the simpler approaches and highlight conditions for which such calculations may or may not be appropriate. In many practical situations several rows of anchors/props are often employed and props may provide both a lateral and moment restraint. Soil conditions are usually not uniform and complex construction procedures are often involved. For many of these situations design procedures do not exist and it is for these cases that the finite element approach has the greatest potential. Examples of the use of the finite element method in the design of cut and cover tunnels can be found in Hubbard et al. (1984) and Potts and Knights (1985). The author has also used this approach for the design of several multipropped deep basements in the London area.

To illustrate the use of the approach in a complex field situation the preliminary design of the George Green tunnel (GGT) (Potts and Knights, 1985), will be taken as an example. This is a cut and cover tunnel which is to be built under an existing trunk road on the outskirts of London. It is essential that traffic flow be maintained along the trunk road and adjoining side roads while construction is proceeding. The density and nature of surface structures imposed a severe constraint on the alignment of the tunnel. In particular, the close proximity of the London Transport (LT) underground station at Wanstead and the accompanying LT running tunnels imposes severe design and construction restrictions on the GGT with respect to surface settlement and sequencing of construction. As a consequence of these restrictions it was felt that either a diaphragm or bored pile wall construction followed by casting of the roof slab before the major part of the excavation takes place will be necessary. Such an approach will minimise soil movements and, with careful sequencing of the wall and roof construction, will keeep disruption of traffic flows to a minimum. Once the wall and roof slabs are in place, full traffic flow at the surface can continue while tunnel construction is carried out beneath the existing road level.

To assess the feasibility of such a solution a preliminary design was initiated prior to the final design being carried out. In this respect it was necessary to assess both the movements of and the structural loads in the embedded walls and the road and roof slabs. It was also necessary to predict the likely movements of the soil in and around the excavation. This was particularly important as the twin LT running tunnels cross obliquely under the GGT within a distance of 4 m at the nearest point (see Fig. 21). The close proximity of the LT tunnels also constrains the wall embedment that can be achieved in some places.

The proposed GGT is clearly a complex soil-structure interaction problem and it quickly became apparent that conventional design approaches were unlikely to be applicable. It should be noted that the road and roof slabs provide both lateral and moment restraints to the walls. Consequently, it was decided to adopt a finite element approach, which allows the movements of, and stresses in both the structure and the soil

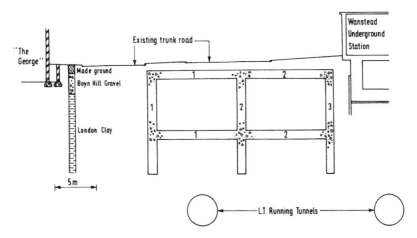

FIG. 21.   Sketch showing typical cross-section of George Green Tunnel.

to be calculated from the same analyses. In addition, it is possible to model the construction sequence.

For the preliminary design of the GGT several cross-sections of the tunnel were analysed, but the results of only one section are discussed here.

The geometry of the tunnel and soil conditions are shown in Fig. 21 and correspond to the main part of the tunnel. The tunnel consists of three embedded walls, to be installed prior to any construction. Although the geometry of the tunnel is symmetric about a vertical plane through the centre wall, the construction sequence was not symmetric. Consequently, it was necessary to consider the complete section. The assumed sequence of construction and that simulated in the analysis were as noted below.

(a) Construct all embedded walls.
(b) Excavate soil between walls 1 and 2 (see Fig. 21) to a depth of 3 m below original ground level.
(c) Construct roof slab 1.
(d) Excavate soil between walls 1 and 2 to a depth of 9 m below original ground level.
(e) Construct road slab number 1.
(f) Backfill 1 m above the left-hand portal (roof slab 1) and apply a surcharge loading to the surface of the fill of $10.5 \, \text{kN/m}^2$ to represent traffic loading.
(g) Excavate soil between walls 2 and 3 to a depth of 3 m below original ground level.
(h) Construct roof slab 2.
(i) Excavate soil between walls 2 and 3 to a depth of 9 m below original ground level.
(j) Construct road slab 2.
(k) Replace 1 m of soil above the right-hand portal (roof slab 2) and apply traffic surcharge of $10.5 \, \text{kN/m}^2$.
(l) Dissipate excess pore water pressure.

It was assumed that the roof and road slabs were cast into the embedded walls, forming a joint that could transmit both thrusts, shear forces and bending moments. During the construction stages of the analysis, the London Clay was assumed to be undrained and the Boyn Hill Gravel and made ground was assumed to behave in a drained manner. For the final stage (1) of the anaysis all the soil was assumed to be drained and the excess pore water pressures developed in the London

Clay during the construction process were dissipated. As only the excess pore water pressures were dissipated, it was implicitly assumed that the final pore water pressure distribution in the soil returns to that prevailing prior to construction (see Fig. 22). Due to underdrainage the initial pore water pressure distribution is not hydrostatic.

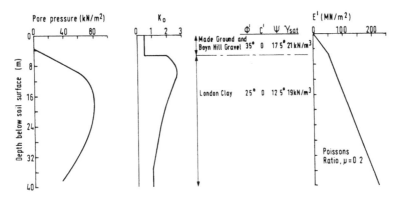

FIG. 22.   Soil conditions: George Green Tunnel.

The soil parameters were obtained from the site investigation data and are given in Fig. 22. The concrete that forms the walls, road and roof slabs were modelled as isotropic elastic with Young's modulus of $28 \times 10^6 \, kN/m^2$, Poisson's ratio of 0·15 and bulk unit weight of $24 \, kN/m^3$. In the modelling of the undrained behaviour of the London Clay the bulk modulus of the pore water was set to $10^7 \, kN/m^2$.

As an example of the results from the analysis horizontal deformation of, and bending moments in the three diaphragm walls for various stages of construction are given in Fig. 23. For the outside walls (1 and 3) the maximum bending moment occurs after completion of undrained construction, whereas for wall 2 the bending moments are greater after construction of roof slab 2. In all cases the maximum bending moments are reduced during the dissipation of excess pore water pressures. It should be noted that at all times wall 2 is subjected to appreciable bending moments, a result that might not have been expected if the construction sequence had been ignored. The analysis also provided predictions of soil displacements and of the movements of, and stresses within the road and roof slabs.

During the preliminary design several cross-sections of the tunnel were

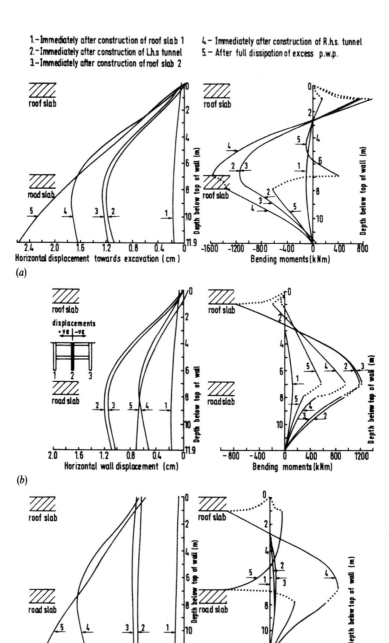

FIG. 23. Horizontal displacements and bending moments in tunnel walls, George Green Tunnel: (a) wall 1; (b) wall 2; (c) wall 3.

considered. Analyses varying such input parameters as the material properties, soil conditions, type of wall to slab connection (pin-jointed or moment), depth of wall embedment and construction procedure were performed. The results of these analyses showed that the concept of a cut and cover tunnel was feasible and were used to establish the final design. Further finite element analyses were then performed for the final design. In particular analyses were carried out in which the LT running tunnels were modelled with respect to the GGT. This was done to enable heave predictions of the LT tunnels to be observed.

6   CONCLUDING REMARKS

In this chapter application of the finite element method to analysing embedded retaining walls has been considered. For cantilever and singly propped retaining walls these analyses have been used as a yardstick against which the simpler limit equilibrium type design calculations have been compared. It has been shown that for cantilever walls the simpler methods of analysis predict embedment depths for stability in agreement with the more sophisticated finite element analyses. They also predict conservative values of maximum bending moment. To some extent the finite element analyses justify the approximations made in the simpler approaches. The same conclusions are also valid for walls singly propped near their top with the exception of stiff walls constructed by excavation in soils with a high initial K_0. For such cases the finite element analyses indicate that the simpler approaches may substantially underestimate the bending moments. Diaphragm and bored pile walls installed in overconsolidated soil fall into this category. For all cases analysed by the finite element approach the earth pressure distribution on the wall differed from that assumed in the simpler calculations. This is particularly true at working load conditions and results in different bending moment distributions. As well as providing information on stability and of stresses in the wall the finite element analysis also provides estimates of both wall and soil movements.

For the more complex situations where there are several rows of props/anchors or where a single row of props/anchors act nearer to the excavation level than the top of the wall and/or where the props supply both a lateral and moment restraint, no simple design procedures exist. It is for these situations that the finite element approach has its greatest potential. The example of the preliminary design of the George Green tunnel shows how such an approach can be applied to complex situations. It is possible to simulate complicated construction sequences and a single analysis provides information on both stresses in and deformation of the structure and of soil movements.

The use of such an approach for the design of the more complex structures is now economically viable and is rapidly becoming standard practice. Costs of the analyses are likely to fall in the future as advances are made in both computer hardward and software.

REFERENCES

BRITISH STANDARDS INSTITUTION (1951). *Earth Retaining Structures CP2: 1951* (currently under revision).

BRITISH STEEL CORPORATION (1986). *Piling Handbook*, 5th edn.

BROS, B. (1972). The influence of model retaining wall displacements on active and passive earth pressure in sand, *Proc. 5th European Conf. on Soil Mech. Found. Eng.*, Madrid, Vol. 1, pp. 241–9.

BURLAND, J.B., POTTS, D.M. and WALSH, N.M. (1981). The overall stability of free and propped embedded cantilever retaining walls, *Ground Engng* **14**(5), 28–37.

CAQUOT, A. and KERISEL, J. (1948). *Tables for the Calculation of Passive Pressure, Active Pressure and Bearing Capacity of Foundations*, Gauthier-Villars, Paris.

CHEN, W.F. (1975). Limit analysis and soil plasticity, *Developments in Geotechnical Engineering*, Vol. 7, Elsevier, Amsterdam.

FOURIE, A.B. and POTTS, D.M. (1988). The estimation of design bending moments for retaining walls using finite elements, *6th Int. Conf. on Numer. Methods in Geomech.*, Innsbruck, A.A. Balkema (Ed.) pp. 1101–7.

FOURIE, A.B. and POTTS, D.M. (1989). Comparison of finite element and limiting equilibrium analyses for an embedded cantilever retaining wall, *Geotechnique*, **39**(2) 175–88.

HUBBARD, H.W., POTTS, D.M., MILLER, D. and BURLAND, J.B. (1984). Design of the retaining walls for the M25 cut and cover tunnel at Bell Common, *Geotechnique* **34**(4), 495–512.

LEE, I.K. and HERINGTON, J.R. (1974). *Soil Mechanics New Horizons*, I.K. Lee (Ed.), Newnes-Butterworths, London.

NAVFAC DM-7 (1982). *Design Manual: Soil Mechanics, Foundations and Earth Structures*, US Department of the Navy, Washington, DC.

PACKSHAW, E. (1946). Earth pressure and earth resistance, *J. Inst. Civ. Engrs.*, **25**, 233–56.

PADFIELD, C.J. and MAIR, R.J. (1984). The design of propped and cantilever walls constructed in stiff clay, CIRIA Project 317, London, Construction Industry Research and Information Association.

POTTS, D.M. and BURLAND, J.B. (1983). A numerical investigation of the retaining walls of the Bell Common Tunnel, Supplementary report 783, Crowthorne, Transport and Road Research Laboratory.

POTTS, D.M. and FOURIE, A.B. (1984). The behaviour of a propped retaining wall: results of a numerical experiment, *Geotechnique* **34**(3), 383–404.

POTTS, D.M. and FOURIE, A.B. (1985). The effect of wall stiffness on the behaviour of a propped retaining wall, *Geotechnique* **35**(3), 347–52.

POTTS, D.M. and FOURIE, A.B. (1986). A numerical study of the effects of wall deformation on earth pressures, *Int. J. Numer. Anal. Methods Geomech.*, **10**(3) 383–404.

POTTS, D.M. and KNIGHTS, M.C. (1985). Finite element techniques for preliminary assessment of a cut and cover tunnel, *Tunnelling '85*, The Institute of Mining and Metallurgy, pp. 83–92.

ROWE, P.W. (1952). Anchored sheet pile walls, *Proc. Inst. Civ. Engrs.*, Part 1, **1**, 27–70.

TERZAGHI, K. (1954). Anchored bulkheads, *Trans. Am. Soc. Civ. Engrs.* **119**, 1243–81.

Chapter 5

FINITE DEFORMATION ANALYSIS
OF SOIL PENETRATION PROBLEMS

Manoj B. Chopra, Gary F. Dargush and Prasanta K. Banerjee

*Department of Civil Engineering, State University of New York at Buffalo,
Buffalo, NY 14260, USA*

ABSTRACT

A consistent finite element formulation has been developed for time
dependent problems involving finite deformations in a two-phase medium such as soils. An updated Lagrangian approach has been adopted
as large plastic strains and finite rotations occur in such problems. The
soil is considered to be either an elastic or an elastoplastic, critical state
material. Nominal stress measures are introduced in the formulation and
all terms are retained in deriving the governing integral equations. This
leads to corrections for equilibrium, yield and applied load in addition to
the geometric stiffening terms. Some simple numerical examples are
included to validate the proposed formulation, followed by a detailed
analysis of a class of penetration problems into soil involving piles and
sampling tubes. Particular emphasis is placed on the physical interpretation of the results from the study of these problems.

1 INTRODUCTION

Classical finite element algorithms used in analysing the behavior of soils
assume that small strains occur in the soil due to the applied loads.
However, this assumption is no longer valid for problems involving the
penetration of cylindrical objects such as piles and sampling tubes, into
the soil. The excessive movement of the medium, particularly around the

169

boundaries of the pile or sampler, during the embedment process causes substantial alterations in the geometry of the solution domain. As a result, strains are no longer linearly related to displacement gradients in such regions, and the equilibrium equations must be modified to take into account these changes in geometry. Of course, irrecoverable plastic deformation is also prevalent.

In the case of the sampling tube, the effects of the sides of the tube are known to cause remolding of the soil around the sampler and hence alter the strength and deformation characteristics of the material inside. In order to realistically simulate the distortional behavior and stress changes around the sampling tube, an approach considering large deformations and the elastoplastic nature of soils is necessary. So far, analytical investigations of this sampling disturbance problem are very sparse in the literature. Alonso *et al.* (1981) presented an axisymmetric finite element algorithm to numerically analyse the effects of sampling disturbance. Their approach was adapted from an algorithm of Zienkiewicz (1977), which has been used to solve extrusion and other forming processes. However, Alonso *et al.* (1981) did not consider the soil as a multi-phase material but simply modeled its behavior as a Drucker–Prager solid. Even though the discretization of the soil domain and the simulation of tube advancement were quite crude and the results unrealistic in a quantitative sense, it should be noted that this study is a good first step to a more rigorous analysis. Other investigators (Kallestenius, 1958, 1963; Lang, 1967) also studied the effects of disturbance due to sampler installation. Actually, Hvorslev (1949) was the first to recognize the importance of sampler geometry. His study combined with further investigations (Broms, 1980; Begemann 1961, 1965, 1971) have led to optimization of testing equipment and a code of practice for the installation of samplers.

For piles, it is well known that the ultimate bearing capacity increases with time after driving. The high excess pore water pressures induced as a result of the driving process and the subsequent dissipation of this build-up of pressures is said to be the main cause of this increase in bearing capacity (e.g., Cummings *et al.*, 1950; Seed and Reese, 1955; Butterfield and Banerjee, 1970). However, most of the previous work in pile penetration problems is based upon the simulation of the driving process as simply an expansion of a cylindrical cavity from a zero radius to a finite radius (Soderberg, 1962; Butterfield and Banerjee, 1970; Randolph *et al.*, 1979). More recently, Banerjee and Fathallah (1979) studied the process in more detail by using a large deformation Eulerian formulation for insertion, followed by a small strain consolidation analysis.

In this chapter, a consistent finite element formulation is developed for finite deformation, with the soil considered as a two-phase, critical state, elastoplastic material. Since large incremental plastic strains and finite rotations take place during the penetration process, an updated Lagrangian (UL) approach is adopted. Within this algorithm, the pile or tube advancement is simulated with a gradual splitting of elements at the tip of the penetrating object. The resulting methodology permits realistic analysis of this entire class of geotechnical problems.

In the next section, the modified cam-clay model employed in the present work is discussed. Next, the focus shifts to the development of the governing equations for a two-phase consolidating body undergoing finite deformation. During this development, a generic incremental elastoplastic material model is assumed. Temporal and spatial discretization is then introduced in the updated Lagrangian integral equations to produce a Galerkin finite element formulation. In order to validate the proposed algorithms, several simple numerical examples are investigated. Afterwards, the sampler and pile penetration problems are examined in detail, with emphasis placed on the physical interpretation of the results. The chapter concludes with a summary of major findings. Standard indicial notation is used throughout. Summations are therefore implied by repeated indices. In particular, Latin indices i, j, k, l vary from 1 to the number of dimensions of the medium.

2   MODIFIED CAM-CLAY MODEL

In the formulation by Roscoe and Burland (1968), the state of stress and strain are both represented in terms of the corresponding invariants. The stress invariants are p and q which represent mean effective and deviatoric stress, respectively. The strain invariants are ε_v, the volumetric strain and ε_d, the deviatoric strain. In terms of principal effective stresses and strains, the invariants are given by

$$p = \tfrac{1}{3}(\sigma_1' + \sigma_2' + \sigma_3') \tag{1a}$$

$$q = \frac{1}{\sqrt{2}}[(\sigma_1' - \sigma_2')^2 + (\sigma_2' - \sigma_3')^2 + (\sigma_3' - \sigma_1')^2]^{1/2} \tag{1b}$$

$$\varepsilon_v = \varepsilon_1 + \varepsilon_2 + \varepsilon_3 \tag{1c}$$

$$\varepsilon_d = \tfrac{2}{3}[(\varepsilon_1 - \varepsilon_2)^2 + (\varepsilon_2 - \varepsilon_3)^2 + (\varepsilon_3 - \varepsilon_1)^2]^{1/2} \tag{1d}$$

The assumptions made by the modified cam-clay model in deriving the elastoplastic stress–strain relation based on energy principles are listed below:

(i) The critical state angle (ϕ) or the slope of critical state line (M) along with λ and κ are fundamental soil parameters. λ is the slope of the normally consolidated line (NCL) and κ is the slope of the elastic rebound line in void ratio-mean effective stress space ($e - \ln p$ space).

(ii) Stress and strain are coaxial allowing the use of the associated flow rule.

(iii) Work done during elastic deformation is conserved whereas a dissipation of energy occurs during plastic straining.

(iv) Changes in the void ratio are related to the total volumetric strain by the relation

$$\dot{\varepsilon}_v = \frac{\Delta e}{1 + e_0} \tag{2}$$

where e_0 is the initial void ratio and Δe is the change in the current void ratio.

The equation of the yield surface takes the form

$$\frac{q^2}{M^2} - pp_0 + p^2 = 0 \tag{3}$$

where p_0 is the maximum mean effective pressure to which the soil was ever subjected. In the context of the incremental plasticity theory, p_0 becomes the hardening parameter defining subsequent locations of the yield surface.

The incremental stress–strain relationship is now written in a form suitable for numerical implementation as

$$\dot{\sigma}'_{ij} = D^{ep}_{ijkl} \dot{\varepsilon}_{kl} \tag{4}$$

where

$$D^{ep}_{ijkl} = \left[D^e_{ijkl} - \frac{D^e_{ijpq}(\partial F/\partial \sigma'_{pq})(\partial F/\sigma'_{mn})D^e_{mnkl}}{H^e + H^p} \right]$$

where D^e_{ijkl} represents the elastic constitutive tensor. The yield function given by the modified cam-clay model is expressed as

$$F(\sigma'_{ij}, p_0) = \frac{3}{2g^2(\theta)} S_{ij} S_{ij} + p^2 - pp_0 = 0 \tag{5a}$$

where

$$S_{ij} = \sigma'_{ij} - \delta_{ij} p \tag{5b}$$

$$g(\theta) = \frac{2Kg(\pi/6)}{(1+K) - (1-K)\ \sin 3\theta} \tag{5c}$$

$$K = \frac{3 - \sin \phi}{3 + \sin \phi}, \qquad g(\pi/6) = \frac{6\ \sin \phi}{3 - \sin \phi} = M \tag{5d,e}$$

and θ is the Lode angle. The hardening moduli, H^e and H^p corresponding to the elastic and plastic moduli, respectively, are defined by

$$H^e = \left(\frac{\partial F}{\partial \sigma'_{ij}}\right) D^e_{ijkl} \left(\frac{\partial F}{\partial \sigma'_{kl}}\right) \tag{6a}$$

$$H^p = \left(\frac{1 + e_0}{\lambda - \kappa}\right) pp_0 \left(\frac{\partial F}{\partial \sigma'_{ii}}\right) \tag{6b}$$

The response of the material at any stage of loading can be predicted from eqn (4) with the knowledge of the in-situ state of effective stress σ'_{ij} and the past maximum stress p_0. Those equations can then be numerically integrated to obtain the complete stress–strain response. It should be noted that for soils, the Young's modulus E becomes a function of the state of stress and is given by

$$E = \frac{3p(1 + e_0)(1 - 2v)}{\kappa} \tag{7}$$

where p is the current value of mean stress.

3   FINITE DEFORMATION THEORY

Having completed the definition of the material model, in this section, geometrically nonlinear behavior is examined. The fundamental concepts underlying finite deformation theory are discussed at length in textbooks by Green and Adkins (1960), Fung (1965) and Washizu (1975), and consequently will not be considered here. Instead, the presentation is directed toward obtaining an integral representation that can serve as the basis for a finite element formulation for finite deformation elastoplastic soil consolidation.

Hibbitt *et al.* (1970) presented a total Lagrangian (TL) formulation which corrected the deficiencies of some earlier formulations in the finite element method. This work properly accounted for finite rotations and loading variations as effects of large deformation. However, the stiffness matrix is quite complicated with the TL approach, since all deformation must be referred to an initial configuration. A more attractive algorithm is the so-called updated Lagrangian (UL) formulation which shifts the reference configuration to coincide with the current state at the end of each increment. Details on the UL approach can be found in the classic paper by McMeeking and Rice (1975). More recently, UL formulations have been developed by Banerjee and Fathallah (1979) for undrained pile penetration, while Wifi (1982) and Chandra and Mukherjee (1984) examined metal forming.

For a fluid infiltrated porous body under quasistatic loading, these formulations must be augmented to include an expression for the conservation of fluid mass. The original paper in this area was by Sandhu and Wilson (1969), who presented a finite element method for the small deformation consolidation of poroelastic media. Additional work on this topic includes the investigations by Hwang *et al.* (1971) and Smith and Hobbs (1976). The extension to finite deformational response was accomplished by Carter *et al.* (1977). Later, material nonlinearity was added. Prevost (1981, 1982) and Zienkiewicz and Shiomi (1984) all present detailed formulations. (The two most recent of these include inertia effects as well.) However, none of these researchers detailed the contributions of initial load stiffness, nor are corrections made for violation of the plastic yield surface and overall equilibrium. Without these corrections, errors can accumulate and soon invalidate the analysis, particularly when very large deformation is involved.

The present formulation utilizes a UL approach based primarily upon the work of Hill (1959) and McMeeking and Rice (1975). As will be seen, equilibrium is written in terms of unsymmetric first Piola–Kirchhoff stress, and all terms are retained in deriving the governing integral equations.

In order to begin this development, consider a two-phase body initially occupying a volume V^0 with surface S^0. Material points are defined in this undeformed reference state by the cartesian coordinates X_i. Meanwhile, in the current deformed configuration, the body resides in the volume V, bounded by the surface S. The material points are now located at x_i, where

$$x_i = X_i + u_i \tag{8}$$

with u_i representing the total displacement vector.

Under quasistatic conditions, the conservation of linear momentum can be written (e.g., Hill, 1959) as

$$\frac{\partial T_{ji}}{\partial X_j} + \rho^0 b_i = 0 \tag{9}$$

where T_{ji} is the total first Piola–Kirchhoff stress tensor, ρ^0 is the reference mass density, and b_i is the body force per unit mass. Equation (9) is written at the current time, but with reference to the initial configuration. Thus, actually b_i represents the current body force at x_j per unit reference mass at X_j.

Since the equilibrium equations (9) are valid at all points X_j in V^0, then

$$\int_{V^0} \left[\frac{\partial T_{ji}}{\partial X_j} + \rho^0 b_i \right] \delta v_i \, dV = 0 \tag{10}$$

in which δv_i is an arbitrary virtual velocity. Performing integration by parts on the first term yields the principle of virtual velocities

$$\int_{V^0} T_{ji} \frac{\partial \delta v_i}{\partial X_j} \, dV - \int_{S^0} T_{ji} n_j^0 \, \delta v_i \, dS - \int_{V^0} \rho^0 b_i \delta v_i \, dV = 0 \tag{11}$$

Then, taking the derivative of (11) produces the incremental form

$$\int_{V^0} dT_{ji} \frac{\partial \delta v_i}{\partial X_j} \, dV = \int_{S^0} dT_{ji} n_j^0 \delta v_i \, dS + \int_{V^0} \rho^0 \, db_i \delta v_i \, dV \tag{12}$$

This is an appropriate starting point for a finite element formulation. However, the unsymmetric incremental stress dT_{ji} is not a suitable measure for use in a constitutive relationship, since it is not objective. Instead, the co-rotational Jaumann rate of Kirchhoff stress $\overset{\triangledown}{t}_{ji}$ or the Jaumann rate of Cauchy stress $\overset{\triangledown}{\sigma}_{ji}$ are introduced to provide a frame indifferent stress rate. Then,

$$\overset{\triangledown}{t}_{ji} = D^{ep}_{jikl} d_{kl} - \delta_{ji} \, du_p \tag{13a}$$

or

$$\overset{\triangledown}{\sigma}_{ji} = D^{ep}_{jikl} d_{kl} - \delta_{ji} \, du_p \tag{13b}$$

where

$$d_{kl} = \frac{1}{2}\left(\frac{\partial\,du_k}{\partial x_l} + \frac{\partial\,du_l}{\partial x_k}\right) \tag{14}$$

is the deformation rate tensor, D_{ijkl}^{ep} is the usual elastoplastic constitutive tensor as defined in the previous section, and du_p is the incremental excess pore pressure. The Jaumann rates in (13) are defined by

$$\overset{\ast}{t}_{ji} = d\tau_{ji} - \omega_{jk}\tau_{ki} + \tau_{jk}\omega_{ki} \tag{15a}$$

and

$$\overset{\ast}{\sigma}_{ji} = \overset{\ast}{t}_{ji} - \sigma_{ji}\,d_{kk} \tag{15b}$$

in terms of the skew-symmetric spin tensor

$$\omega_{jk} = \frac{1}{2}\left(\frac{\partial\,du_j}{\partial x_k} - \frac{\partial\,du_k}{\partial x_j}\right) \tag{16}$$

In (13), it is assumed that the additive decomposition of the strains, into elastic and plastic components, is valid even under finite deformation (Nemat-Nasser, 1982).

Next, the Kirchhoff stress is related to the first Piola–Kirchhoff stress via

$$T_{ji} = \frac{\partial X_j}{\partial x_k}\tau_{ki} \tag{17}$$

Therefore, in rate form, this becomes

$$dT_{ji} = \frac{\partial X_j}{\partial x_k}\left(d\tau_{ki} - \tau_{li}\frac{\partial\,du_k}{\partial x_l}\right) \tag{18}$$

and thus from (15),

$$dT_{ji} = \frac{\partial X_j}{\partial x_k}\left(\overset{\ast}{t}_{ki} + \omega_{kl}\tau_{li} - \tau_{kl}\omega_{li} - \frac{\partial\,du_k}{\partial x_l}\tau_{li}\right) \tag{19}$$

Introducing (19), along with (13), into equation (12) produces

$$\int_{V^0}\left[D_{jikl}^{ep}\,d_{kl} + \omega_{jk}\tau_{ki} - \tau_{jk}\omega_{ki} - \frac{\partial\,du_j}{\partial x_k}\tau_{ki} - \delta_{ji}\,du_p\right]\frac{\partial\delta v_i}{\partial x_j}\,dV$$

$$= \int_{S^0} dT_{ji}n_j^0\,\delta v_i\,dS + \int_{V^0}\rho^0\,db_i\delta v_i\,dV \tag{20}$$

Some further manipulation is still needed on the right-hand side, since known components of incremental tractions are generally not specified in terms of dT_{ji}, but rather $d\sigma_{ji}$, where σ_{ji} represent the real symmetric total Cauchy stresses. These are related to T_{ji} via

$$T_{ji}=\frac{\rho^0}{\rho}\frac{\partial X_j}{\partial x_k}\sigma_{ki} \tag{21}$$

in which ρ is the current density. Then taking the derivative of (21),

$$dT_{ji}=\frac{\rho^0}{\rho}\frac{\partial X_j}{\partial x_k}\left(d\sigma_{ki}+\sigma_{ki}\frac{\partial\,du_l}{\partial x_l}-\frac{\partial\,du_k}{\partial x_l}\sigma_{li}\right) \tag{22}$$

Using (14), (16) and (22), eqn (20) can be written

$$\int_{V^0}[D^{ep}_{jikl}\,d_{kl}-d_{jk}\tau_{ki}-\tau_{jk}\omega_{ki}-\delta_{ji}\,du_p]\frac{\partial\delta v_i}{\partial x_j}\,dV$$

$$=\int_{S^0}\frac{\rho^0}{\rho}\frac{\partial X_j}{\partial x_k}[d\sigma_{ki}+\sigma_{ki}\,d_{ll}-(d_{kl}+\omega_{kl})\sigma_{li}]n^0_j\,\delta v_i\,dS \tag{23}$$

$$+\int_{V^0}\rho^0\,db_i\delta v_i\,dV$$

It should be noted that the final volume integral may also require a reformulation for certain problems in which the body force is dependent upon the deformation. Examples include gravitational and centrifugal loads.

In an updated Lagrangian formulation, which is adopted in the present work, the reference configuration for the *nth* load increment is established as the final configuration from the previous increment. Thus,

$$X^n_j=x^{n-1}_j \tag{24}$$

In this new reference configuration, eqn (23) can be approximated as

$$\int_{V^{n-1}}[D^{ep,n-1}_{jikl}\,d^n_{kl}-\sigma^{n-1}_{ki}\,d^n_{jk}+\sigma^{n-1}_{jk}\,\omega^n_{ki}-\delta_{ij}\,du^n_p]\frac{\partial\delta v_i}{\partial X^n_j}\,dV$$

$$=\int_{S^{n-1}}[d\sigma^n_{ji}+\sigma^{n-1}_{ji}\,d^{n-1}_{kk}-\sigma^{n-1}_{ki}(d^{n-1}_{jk}+\omega^{n-1}_{jk})]n^{n-1}_j\,\delta v_i\,dS$$

$$=\int_{V^{n-1}}[\rho^{n-1}\,db^n_i]\delta v_i\,dV \tag{25}$$

where it is assumed that

$$d_{jk}^n = \frac{1}{2}\left(\frac{\partial\, du_j^n}{\partial X_k^n} + \frac{\partial\, du_k^n}{\partial X_j^n}\right) \tag{26a}$$

$$\omega_{jk}^n = \frac{1}{2}\left(\frac{\partial\, du_j^n}{\partial X_k^n} + \frac{\partial\, du_k^n}{\partial X_j^n}\right) \tag{26b}$$

and σ_{ji}^{n-1}, d_{jk}^{n-1}, ω_{jk}^{n-1}, $D_{jikl}^{ep,\,n-1}$, n_j^{n-1} and ρ^{n-1} are all known quantities obtained from the previous increment. Notice that the surface load correction terms are written completely in terms of known quantities, and thus lag one increment behind the solution. This is done to maintain symmetry of the resulting finite element stiffness matrix. Additionally, in the above, the incremental quantities are written explicitly as follows:

$$du_i^n = u_i^n - u_i^{n-1} \tag{27a}$$

$$du_p^n = u_p^n - u_p^{n-1} \tag{27b}$$

$$d\sigma_{ji}^n = \sigma_{ji}^n - \sigma_{ji}^{n-1} \tag{27c}$$

$$db_i^n = b_i^n - b_i^{n-1} \tag{27d}$$

The approximations introduced in forming (25) from (23) somewhat limit the size of the load increment that is permissible. In order to take larger increments, a more sophisticated iterative process could be employed utilizing (23) directly. However, such an approach is not warranted here, since with material nonlinearity present, the load step must be small in any case to capture the proper response.

For two-phase media, eqn (25) must be supplemented with a statement of conservation of mass. This can be developed rigorously from a Lagrangian viewpoint, by introducing mass flux quantities analogous to T_{ji} and t_{ji}. Instead, the following simplified approach is adopted. Assuming incompressible constituents and no mass sources, the divergence of the pore fluid velocity must balance the time rate of change of solid dilatation. Thus,

$$\frac{\partial v_i}{\partial X_i} - \frac{\partial^2 u_i}{\partial t \partial X_i} = 0 \tag{28}$$

where v_i is the velocity of the fluid. This is related via Darcy's Law to the pore pressure. That is,

$$v_i = -\frac{k}{\gamma}\frac{\partial u_p}{\partial X_i} \tag{29}$$

with k as the isotropic permeability and γ as the unit weight of the pore fluid. Then, introducing (29) into (28), multiplying by a virtual pressure rate of δq and integrating over the reference volume, produces the identity

$$\int_{V^0}\left[-\frac{\partial}{\partial X_i}\left(\frac{k}{\gamma}\frac{\partial u_p}{\partial X_i}\right)-\frac{\partial^2 u_i}{\partial t\partial X_i}\right]\delta q\,\mathrm{d}V=0 \tag{30}$$

Integration by parts and application of the divergence theorem yields

$$\int_{V^0}\left[\frac{k}{\gamma}\frac{\partial u_p}{\partial X_i}+\frac{\partial u_i}{\partial t}\right]\frac{\partial\delta q}{\partial X_i}\,\mathrm{d}V=\int_{S^0}\left[\frac{k}{\gamma}\frac{\partial u_p}{\partial X_i}n_i+\frac{\partial u_i}{\partial t}n_i\right]\delta q\,\mathrm{d}S \tag{31}$$

In the current updated Lagrangian formulation, this becomes

$$\int_{V^{n-1}}\left[\frac{k}{\gamma}\frac{\partial u_p^n}{\partial X_i^n}+\frac{\partial u_i^n}{\partial t}\right]\frac{\partial\delta q}{\partial X_i^n}\,\mathrm{d}V=\int_{S^{n-1}}\left[\left(\frac{k}{\gamma}\frac{\partial u_p^n}{\partial X_i^n}+\frac{\partial u_i^n}{\partial t}\right)n_i^{n-1}\right]\delta q\,\mathrm{d}S \tag{32}$$

Equation (32), along with (25), provides the basis for the finite element formulation which is presented in the following section.

4 FINITE ELEMENT FORMULATION

The load increments that were introduced, through the superscripts n and $n-1$, in the previous development have been considered simply as pseudo-time steps. This is generally satisfactory in elastoplastic analysis of solids. However, in a two-phase medium, even under quasistatic conditions, time dependence arises from the process of pore fluid diffusion through the solid skeleton. This dissipative effect occurs due to the presence of the volumetric strain rate term in the mass balance equation. As a result, in a consolidation formulation, load increments must be associated with real time.

In the present finite element formulation, the time derivative in (30) is approximated by an implicit finite difference operator. Thus, let

$$\theta\frac{\partial u_i^n}{\partial t}+(1-\theta)\frac{\partial u_i^{n-1}}{\partial t}=\frac{u_i^n-u_i^{n-1}}{\Delta t} \tag{33}$$

where θ is a parameter and Δt is the time step. As discussed, for example, by Zienkiewicz (1977) this scheme is unconditionally stable for $\theta\geqslant 0\cdot5$. Consequently, $\theta=0\cdot55$ is used for all of the analyses reported herein.

Equation (32) can be written at both the beginning and end of the nth time increment. Multiplying the former by $(1-\theta)$, the latter by θ and then summing the results produces

$$\int_{V^{n-1}} \left[\frac{k}{\gamma} \left\{ \theta \frac{\partial u_p^n}{\partial X_i^n} + (1-\theta) \frac{\partial u_p^{n-1}}{\partial X_i^n} \right\} + \left\{ \theta \frac{\partial u_i^n}{\partial t} + (1-\theta) \frac{\partial u_i^{n-1}}{\partial t} \right\} \right] \frac{\partial \delta q}{\partial X_i^n} \, dV$$

$$= \int_{S^{n-1}} \left[\frac{k}{\gamma} \left\{ \theta \frac{\partial u_p^n}{\partial X_i^n} + (1-\theta) \frac{\partial u_p^{n-1}}{\partial X_i^n} \right\} + \left\{ \theta \frac{\partial u_i^n}{\partial t} + (1-\theta) \frac{\partial u_i^{n-1}}{\partial t} \right\} \right] n_i^{n-1} \delta q \, dS \tag{34}$$

After substituting (33) and utilizing the relationships in (27), this can be simplified as

$$\int_{V^{n-1}} \left[\frac{k}{\gamma} \theta \Delta t \frac{\partial}{\partial X_i^n} \frac{du_p^n}{} + du_i^n \right] \frac{\partial \delta q}{\partial X_i^n} \, dV$$

$$= \int_{S^{n-1}} \left[\frac{k}{\gamma} \Delta t \left(\theta \frac{\partial \, du_p^n}{\partial X_i^n} + \frac{\partial u_p^{n-1}}{\partial X_i^n} \right) + du_i^n \right] n_i^{n-1} \delta q \, dS \tag{35}$$

$$- \int_{V^{n-1}} \left[\frac{k}{\gamma} \Delta t \frac{\partial u_p^{n-1}}{\partial X_i^n} \right] \frac{\partial \delta q}{\partial X_i^n} \, dV$$

Spatial discretization of this equation, along with (25), is now required to complete the finite element formulation. In the present work, six- and eight-noded quadratic elements are employed for the geometric representation of two-dimensional and axisymmetric bodies (Zienkiewicz, 1977). Within each element,

$$X_i^n = N_w X_{iw}^n \tag{36}$$

in which N_w are quadratic shape functions and X_{iw}^n are nodal coordinates. The summation in (36) is from 1 to Ω, the total number of nodes in the element. Similarly, for the incremental displacements

$$du_i^n = N_w \, du_{iw}^n \tag{37a}$$

with du_{iw}^n representing the nodal values. On the other hand, the incremental pore pressures are described by linear shape functions, such that

$$du_p^n = M_y \, du_{py}^n \tag{37b}$$

where M_y are linear shape functions with summation over y from 1 to Γ, the total number of vertex nodes in the element. Then, by using

$$\delta v_i = N_w \delta v_{iw} \tag{38a}$$

and

$$\delta q = M_\gamma \delta q_\gamma \qquad (38b)$$

a Galerkin formulation is obtained. After eliminating the arbitrary virtual nodal quantities, the resulting discretized matrix equations can be written symbolically as

$$\begin{bmatrix} K_{ij}^n & K_{ip}^n \\ K_{pj}^n & K_{pp}^n \end{bmatrix} \begin{Bmatrix} du_j^n \\ du_p^n \end{Bmatrix} = \begin{Bmatrix} df_i^n \\ df_p^n \end{Bmatrix} \qquad (39)$$

where K_{pj}^n is the transpose of K_{jp}^n. Of course, the incremental displacements and pore pressures are actually grouped on a nodal basis in order to preserve the bandedness of the global 'stiffness' matrix. Both linear and nonlinear stiffness contributions are included in K_{ij}^n. In eqn (39), notice that the entire matrix is symmetric if both K_{ij}^n and K_{pp}^n are symmetric. The symmetry of the latter is evident from (35), however, K_{ij}^n is more complicated. Here, symmetry is maintained by shifting a few volumetric nonlinear terms in K_{ij}^n to the right-hand side. Meanwhile, in addition to the incremental applied loads $d\sigma_{ij}^n$, the right-hand side vector df_i^n in (39) also includes the equilibrium and yield corrections. Other terms from the surface integral in (25) besides $d\sigma_{ij}^n$, are included in df_i^n as load corrections.

At each step, K^n and df^n are formed, and (39) is solved using a frontal method. Total displacements and pore pressures are then accumulated via

$$u_j^n = u_j^{n-1} + du_j^n \qquad (40a)$$

$$u_p^n = u_p^{n-1} + du_p^n \qquad (40b)$$

Finally, stresses and strains are determined, and the geometry is updated for the next time step. As mentioned above, any equilibrium or yield surface violations are imposed as corrections to the subsequent step.

It should be noted that at small times two-phase media is essentially incompressible since the fluid is unable to escape through the pores. However, because the present formulation includes pore pressure as an additional primary variable, no special treatment, beyond that discussed above, is required. In particular, there is no need for penalty function methods nor reduced integration in any of the consolidation analyses discussed in the next section. These techniques were employed, however, for all undrained analyses, which utilize a formulation involving only displacements as primary variables.

5 NUMERICAL EXAMPLES

5.1 Introduction

The completeness and accuracy of the finite deformation FEM formulation is validated through a number of examples in this section. Results obtained from the developed computer program, which is a modified version of CRISP (Gunn and Britto, 1984), are compared to analytical results for problems of simple extension with large displacements and the problem of simple shear with large rotations. Next the one-dimensional elastic consolidation behavior under small and large strains is investigated. Geotechnical applications, including the undrained triaxial compression of a clay sample and the cavity-expansion problem, are simulated in a finite deformation environment and the solutions are compared with available results. Finally, the problems of penetration of a sampling tube and a pile into the ground are extensively studied in order to gain insight into the behavior of the surrounding soil as a result of the insertion.

5.2 Finite Elastic Extension of a Bar Under Plane Strain

This example illustrates the effect of the geometric stiffening terms in the FEM formulation resulting in the development of geometric nonlinearity even if the material is assumed to behave in a linear, elastic manner. The effects of the choice of an objective stress measure are not experienced in this problem since no rotations take place.

Two cases of objective stress strain relations (eqn (13)) were considered: (i) the Jaumann–Cauchy stress measure which pertains only to incompressible materials and (ii) the Jaumann–Kirchhoff stress measure which is valid for both compressible and incompressible material behavior. It was found that a simple updated Lagrangian analysis with continuously updated geometry gives results similar to the Jaumann–Cauchy stress approach since no rotational contributions are encountered. Analytical solutions for plane strain extension were developed by Osias (1973) by carrying out a direct integration of the constitutive equations. Osias and Swedlow (1974) present some analytical results for the Jaumann–Cauchy case which were compared with the present FEM analysis.

It was observed that the present algorithm shows very good correspondence with the analytical results for the Jaumann–Cauchy stress case. However, the Jaumann–Kirchhoff stress measure more correctly accounts for the compressibility of the material. As the Poisson's ratio approaches 0·5, the two cases coincide as both become incompressible.

5.3   Finite Simple Shear

Numerical results obtained from the present analysis were compared with analytical solutions presented by Osias and Swedlow (1974). The nonzero stress components for plane strain simple shear analysis are given by (Osias and Swedlow, 1974)

$$\tau_{xy} = G \sin \zeta \tag{41a}$$

$$\sigma_x = G(1 - \cos \zeta) \tag{41b}$$

$$\sigma_y = G(\cos \zeta - 1) \tag{41c}$$

where G is the elastic shear modulus. It may be noted that the two normal stresses are equal in magnitude and opposite in direction.

Shear stress values are predicted to an accuracy of within 2% in the $0° \leqslant \gamma \leqslant 45°$, while the error in the normal stress values builds up to about 5% at $\gamma = 45°$, where γ is the angle of shear ($\tan^{-1} \zeta$).

Unlike the finite extension problem, simply updating the geometry and load at every stage is no longer sufficient to produce correct results. The choice of an objective stress measure becomes relevant in this problem as large rotations are involved. However, the two types of Jaumann stress rates, namely Jaumann–Cauchy and Jaumann–Kirchhoff, yield the same result as the compressibility of the material is irrelevant.

5.4   One-dimensional Elastic Consolidation

The uniaxial consolidation behavior is another example of the effect of nonlinear compressibility and geometric stiffening on an otherwise elastic material skeleton. The finite element mesh for this problem is shown in Fig. 1. It consists of 16 LST and the boundary conditions were specified as shown. The H/B ratio was initially established as 5 and the traction q_z is applied instantaneously at the top surface. Thereafter, the load is held constant and consolidation is allowed to take place with drainage permitted only at the top surface. Two different levels of nondimensional loads (q_z/E) were applied to gauge the response at low and high load levels.

Analytical solutions for the small deformation case for this well known problem of soil mechanics are documented by Lambe and Whitman (1969). The results obtained for the average degree of consolidation as a function of a dimensionless time factor T from both the analytical solution and the finite deformation analysis, are shown in Fig. 2. In this problem

$$T = C_v t / L_0^2 \tag{42}$$

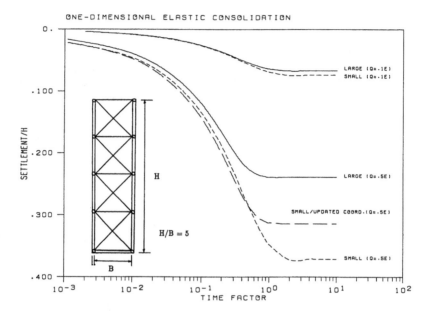

FIG. 1. Settlement of the top surface as a function of time factor T. Inset: FEM mesh ($v = 0.3$).

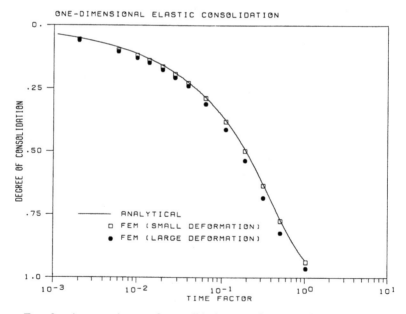

FIG. 2. Average degree of consolidation as a function of time factor T.

where C_v is the coefficient of consolidation and L_0 is the initial diffusion length. The small deformation FEM results show extremely good agreement with the analytical solution. Meanwhile, the finite deformation FEM analysis predicts a higher average degree of consolidation for each level of time compared to the small deformation analysis. This is in agreement with the theoretical results reported by Gibson et al. (1967). Physically, the increase can be attributed to a reduction in the diffusion length as the specimen consolidates and compresses. This, in turn, increases the rate at which further consolidation may take place.

Figure 1 also illustrates the settlement u_z at the top surface as a function of the dimensionless time factor T, for two different q_z/E ratios. It can be seen from the flat portion of each curve that consolidation has been permitted to occur for sufficient time that the ultimate drained settlement response has been achieved. The curves show that the finite deformation analysis predicts lower settlements than the small deformation case. The difference is seen to be more pronounced at a larger load level, i.e., at q_z/E of 0·5. The curves labeled 'LARGE' depict the settlement for a large deformation, Jaumann–Kirchhoff analysis. The intermediate curve, labeled 'SMALL-UPDATED COORD.', corresponds to a small deformation FEM analysis within an updated Lagrangian framework, wherein the geometry is constantly updated and load corrections are applied at the end of each increment. It should be noted that for problems such as this one where no rotations are involved, the Jaumann–Cauchy stress analysis and the small strain analysis with updated coordinates coincide and are used interchangeably.

These results can be explained in the light of the changes in the nominal stress measure due to the compressibility of the material undergoing consolidation. In a large deformation analysis, as a material is compressed, the compressibility subsequently reduces and the resulting settlements are lower.

5.5 Expansion of a Cylindrical Cavity and Subsequent Consolidation

The expansion of a cylindrical cavity in an infinite medium is a well known problem in applied mechanics and is of particular interest to geotechnical engineers. Several solutions to the problem have been proposed in the literature for a number of ideal materials (Butterfield and Banerjee, 1970; Soderberg, 1962). The process of installation of a pile into clayey soils has been modeled as an undrained expansion of a cylindrical cavity. Earlier, estimates of pile shaft capacity utilized the undrained shear strength of soils prior to pile installation, occasionally with a factor introduced to account for disturbances caused by the driving process. However, it was observed that at depths not close to either end of the pile, the soil is displaced predominantly in an outward radial direction. Thus, the process of pile installation was assumed to be adequately simulated by the solutions of a cavity expansion from a zero radius to the radius of the pile.

The geometry of this problem is considered to be axially symmetric as well as under plane strain conditions. The excess pore pressure built up during the driving process is assumed to dissipate by outward radial flow. The consolidation process that follows the pile installation is responsible for the increase in the bearing capacity of the pile with time. As the excess pore pressure around the pile dissipates, the void ratio decreases and hence, the strength goes up.

Randolph *et al.* (1979) carried out a finite element analysis of this problem using the work hardening, elastoplastic modified cam-clay model for soils proposed by Roscoe and Burland (1968). They carried out an exhaustive study of several facets of this problem including the effect of different OCR and of the sensitivity of the clay, on the stress changes in the soil surrounding the pile. It was one of the earliest efforts to incorporate the path dependent elastoplastic behavior of the soil for solving this problem. They were able to set up simple rules for the magnitudes of excess pore pressures and residual stresses at the end of the installation and also at the end of the final consolidation. A similar effort using ordinary elastoplastic von Mises soil skeleton was described earlier by Butterfield and Banerjee (1970).

A small deformation FEM analyses using the present algorithm was conducted on the Boston Blue clay discussed by Randolph *et al.* (1979). In addition, a finite deformation analysis using both Jaumann stress measures was also carried out. The modified cam-clay (MCC) critical state theory was used to model the soil behavior. The effective stresses and excess pore pressures were monitored at every stage of the loading and consolidation phases. For the undrained loading stage, very small increments were used and a modified FEM formulation using reduced integration (Zienkiewicz, 1977) was found to yield better results. However, for the consolidation phase, larger time steps were used.

5.5.1   Soil Parameters and In-situ Stress Conditions

The soil parameters were selected to be those corresponding to Boston Blue clay quoted by Randolph *et al.* (1979). Only a normally consolidated specimen was studied (i.e., OCR was 1). The other properties are listed below:

$$\lambda = 0\cdot15, \qquad \kappa = 0\cdot03, \qquad e_{cs} = 1\cdot744$$

$$M = 1\cdot2, \qquad v = 0\cdot3, \qquad K_0 = 0\cdot55$$

$$G/C_u = 74, \qquad G/\sigma'_z(0) = 25, \qquad G/p(0) = 36$$

in which $K_0 = \sigma'_r(0)/\sigma'_z(0)$. These parameters yield an in-situ stress state:

$$C_u(0) = 29\cdot45 \text{ kN/m}^2, \qquad \sigma'_z(0) = 87\cdot20 \text{ kN/m}^2, \qquad \sigma'_r(0) = 47\cdot96 \text{ kN/m}^2$$

where $C_u(0)$ is the initial undrained shear strength. Also, $G = 2180 \text{ kN/m}^2$ and the preconsolidation pressure (past maximum) $p'_c = 78\cdot53 \text{ kN/m}^2$.

5.5.2   Analysis of the Cavity Expansion Phase

The modeling of the initial phase of the expansion of the cavity is restricted by the present algorithm to begin with a nonzero initial radius. This is required to avoid the occurrence of infinite circumferential strain. Hence, the cavity must begin with a finite initial radius and then be expanded to its final value. But this restriction does not involve any inconsistencies as the ultimate response is obtainable to a sufficiently accurate level by doubling the radius of a cavity with critical radius a_0 (Randolph et al., 1979). The changes in stress and strain behavior around a cavity of radius r_0 are obtainable if a_0 is chosen as $r_0/\sqrt{3}$.

The axisymmetric FEM mesh used in analysing this problem is shown in Fig. 3. It consists of 39 linear strain quadrilaterals (LSQ) elements with 3 degrees of freedom per corner node, in order to monitor the pore pressure changes. The initial radius of the cavity was chosen to be $1\cdot0$ and gradually expanded to a final value of $2\cdot0$. The boundary conditions were specified in a manner suitable for plane strain conditions along the length of the cylinder as shown in Fig. 3. The outer radius of the cylindrical soil sample was found to be adequate at a value of $35\cdot0$, since no appreciable changes were seen to take place beyond this point.

The results of this first phase of the analysis are shown in Figs 4–6. Figure 4 shows the variation of total radial stress σ_r and excess pore pressure u_p at the inner surface of the cavity, as a function of the cavity radius a. It may be noted that σ_r and u_p have approached their limiting states by the time the cavity is doubled in radius. The total stress values do not include any ambient pore pressures in the soil before the commencement of the expansion. Both the small and the finite deformation results are shown in this figure.

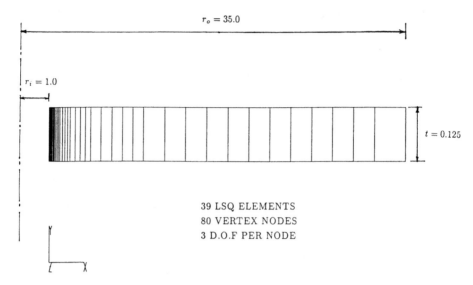

FIG. 3. Axisymmetric FEM mesh for the cavity expansion and consolidation problem.

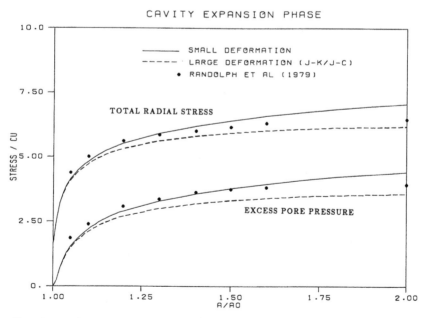

FIG. 4. Variation of total radial stress and excess pore pressure close to the inner surface of the cavity (at $r = 1 \cdot 15 r_0$) as a function of cavity radius a.

STRESSES AT THE END OF CAVITY EXPANSION

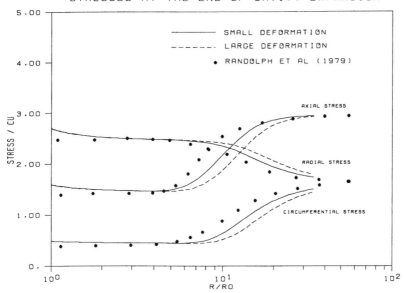

FIG. 5. Effective stress distributions in the soil around the cavity along the radial distance *r* at the end of expansion.

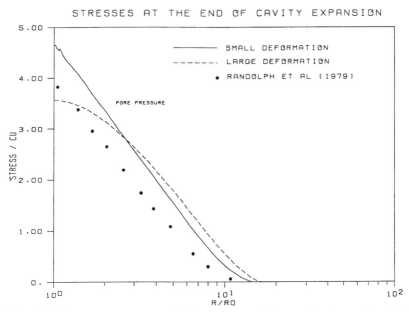

FIG. 6. Excess pore pressure distribution around the cavity along the radial distance *r* at the end of expansion.

The finite deformation results for σ_r and u_p using either the Jaumann–Kirchhoff stress measure (J-K) or the Jaumann–Cauchy (J-C) measure for incompressible materials, are consistently lower than the corresponding small deformation results owing to geometric stiffening. Also, since only purely radial movement of soil occurs, both J-K and J-C results are identical in nature since a simple updating of the geometry in a small strain analysis would also provide these results. A comparison with results obtained by Randolph et al. (1979) shows that there is a small difference in the two sets of small deformation results. The authors feel that this may be attributed to the choice of the FEM mesh and the types of elements used. This analysis uses a nonuniform mesh with a concentration of elements towards the cavity gradually expanding to larger elements as the radial distance away from the cavity increases. This mesh was found to be more accurate and efficient than a uniform mesh with larger elements close to the cavity surface.

Figure 5 shows the distribution of effective stresses along the radial distance away from the cavity, at the end of the cavity expansion phase. The axial and circumferential stresses reduce from their in-situ values and the radial stress goes up. Both small and finite deformation results are plotted and it is seen that there is no effect of the finite deformation analysis close to the cavity surface. This is due to the fact that the material is in a critical state up to a radial distance (r/r_0) of about 5. Beyond this distance, the difference due to a finite deformation analysis begins to emerge. This difference may be attributed to the updated geometry and geometric striffness in the finite deformation algorithm. However, at a large distance away from the cavity surface, the effective stresses converge to the in-situ values, in both the cases. These results agree well with those previously reported by Randolph et al. (1979) for changes in stresses close to the cavity surface.

Figure 6 shows the distribution of excess pore pressure along the radial distance. The pore pressure values close to the cavity differ between the small and finite deformation analyses. Since the material is in a critical state, no changes in effective stress take place. Therefore, the pore pressure must account for the difference in total stress.

5.5.3   Analysis of the Consolidation Phase Following Cavity Expansion
After the expansion of the cavity to double its original radius, the soil around the expanded cavity is allowed to consolidate by dissipating the excess pore pressure. Theoretical (Butterfield and Banerjee, 1970; Randolph et al., 1979) and field measurements (Bjerrum and Johannssen,

1961; Lo and Stermac, 1965) of the excess pore pressure around driven piles, indicate that the pressure dissipation takes place in a radial fashion over most of the pile shaft. This fact is incorporated in the present analysis by assuming only radial flow of pore water and also radial motion of the solid particles. Since the pile is relatively rigid and impermeable, the inner surface of the cavity will be assumed to be rigid and impermeable as well.

The rate of consolidation depends upon the coefficient of consolidation C_v along with certain other parameters. For the poroelastic case, a nondimensional time factor of the nature

$$T = C_v t / r_0^2 \qquad (43)$$

is considered suitable to describe the consolidation behavior. However, since C_v is no longer a constant for the modified cam-clay model, a different nondimensional time variable T^* has been used (Randolph *et al.*, 1979), where

$$T^* = \frac{k}{\gamma} \frac{C_u(0)t}{r_0^2} \qquad (44)$$

and $C_u(0)$ is the initial undrained shear strength.

The variation of the express pore pressure with time is plotted in Fig. 7, at a distance close to the cavity surface. Figure 8 shows the variation of the total and effective radial stress with time. Since the pore pressure decreases due to dissipation, the effective stress must increase. However, the total stresses also decrease, though not very much. This may be attributed to the change in the volume taking place as the pore water diffuses radially. The small difference between *J-K* and *J-C* stress measures is due to the small strain deformation in this phase. However, the difference is much less than the case of one-dimensional consolidation because in this case the material does not undergo as much compression and the effects of compressibility are lower.

Finally, once the consolidation is allowed to be completed, the residual stresses in the soil around the cavity are plotted in Fig. 9. It can be seen that cavity expansion alters the stress state in the soil significantly, up to about 20 times the initial cavity radius. The resulting radial effective stress close to the cavity is about 5 $C_u(0)$ whereas it was 1·7 $C_u(0)$ in its in-situ state. Similarly, the axial stress increases to 3·4 $C_u(0)$ from an in-situ value of 2·9 $C_u(0)$ and the circumferential stress goes up to 3·6 $C_u(0)$ from 1·7 $C_u(0)$.

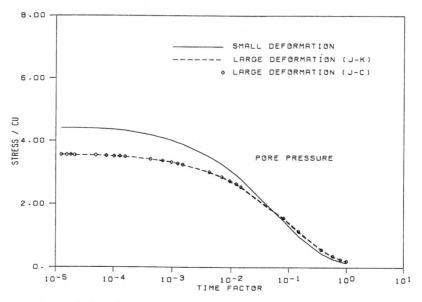

FIG. 7. Variation of excess pore pressure with time factor T during consolidation.

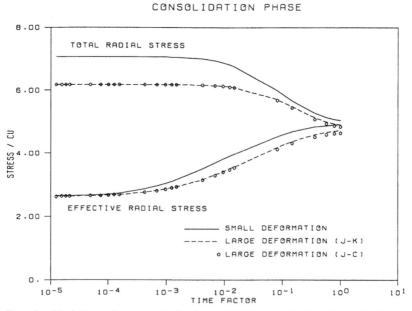

FIG. 8. Variation of total and effective radial stress with time factor T during consolidation.

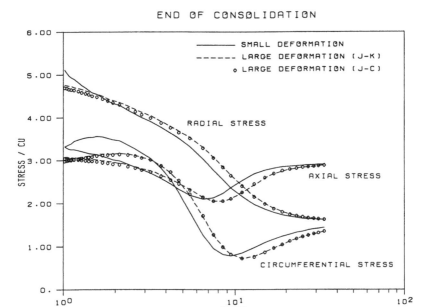

FIG. 9. Variation of stresses with radial distance r at the end of consolidation.

5.6 Insertion of Sampling Tube into Soil

In this example, the mechanical disturbance that occurs during the process of soil sampling is investigated using the large deformation finite element algorithm. This forms an attempt to establish the value of such a numerical tool for practical applications.

The sampling tube has typical dimensions, 6·0 in (15·2 cm) length and 1·5 in (3·8 cm) internal diameter, and has three different thicknesses based upon the norms prescribed for the area ratios (A_r):

$$A_r = \frac{D_e^2 - D_i^2}{D_i^2} \tag{45}$$

where D_e = external diameter and D_i = internal diameter. In addition, there are three different tip angles based upon the three thicknesses. Three test simulations are carried out using the large deformation algorithm and Table 1 lists the relevant parameters for each test. Skin friction at the soil-sampler interface is neglected, thus assuming a smooth surface, both inside and outside the tube. In the future, frictional effects may be included in order to study the insertion of a rough sampler.

TABLE 1

DETAILS OF THE TESTS FOR SAMPLING TUBE INSERTION[a]

Test no.	Tube thickness (in)	Tip angle (degrees)	A_r (%)	Depth (in)
LSI	0·03	1·72	8·2	6·0
LS2	0·05	2·86	13·8	6·0
LS3	0·08	4·57	22·5	6·0

[a]Radius of the tube = 0·75 in. All tests assume smooth samplers.

However, element distortions and other nonlinear effects would make that problem considerably more difficult, highlighting most limitations of large deformation FE algorithms.

5.6.1 Simulation of the Sampling Tube Penetration

The FEM axisymmetric mesh used for this analysis is shown in Fig. 10. It consists of 95 linear strain elements and 119 vertex nodes. Gradation is provided in the mesh to capture the behavior close to the tube in a more accurate manner. Figure 11 is an enlarged view of that area. The eventual location of the sampling tube within the soil mass is indicated by the bold line *ag*. As can be seen from that diagram, a total of twelve elements represent the soil inside the tube, which will eventually be used for laboratory testing.

The simulation of the penetration process is illustrated in Fig. 12. The analysis begins with the sharp tip of the sampler about to enter the soil at node *a*, as shown in Fig. 12(a). Then, during the first load sequence, the tip penetrates to node *b*. The adjoining elements are separated by introducing *a'* and gradually incrementing the enforced radial displacements until the state portrayed in Fig. 12(b) is attained. Sufficiently small steps are required to ensure that error does not accumulate. The parameter *D* represents the depth of penetration. Figure 12(c) shows the position of the sampler at the end of the second load sequence, while the final location is provided in Fig 12(d). Notice that a series of vertex nodes (*a'b'c'd'e'f'*) have been introduced, requiring considerable housekeeping for the stiffness, load and displacement matrices at each step. Additionally, in the updated Lagrangian scheme, geometric load corrections must be computed for all of the boundaries, including these newly formed surfaces.

The analysis was carried out both as a fast consolidation analysis with

SAMPLING TUBE INSERTION - FEM MESH

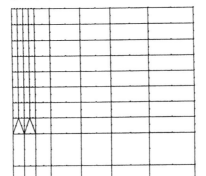

FIG. 10.   Axisymmetric FEM mesh for the sampling tube penetration problem.

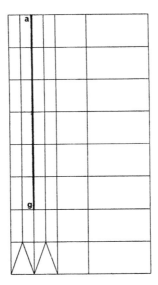

FIG. 11.   Enlarged view of the FEM mesh close to the tube used for results evaluation.

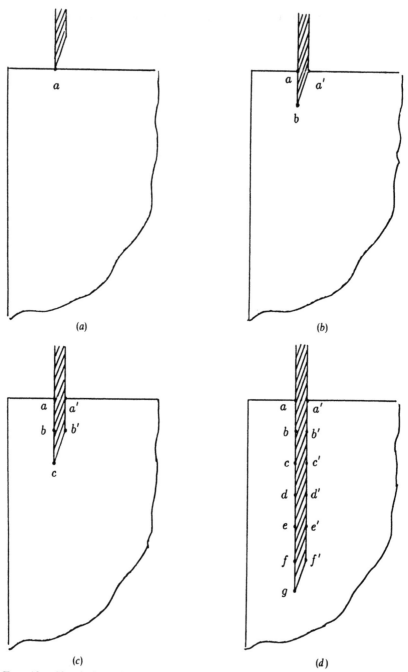

FIG. 12. Simulation of the penetration process for different depths of penetration, (a) $D=0\cdot0$ in, (b) $D=1\cdot0$ in, (c) $D=2\cdot0$ in and (d) $D=6\cdot0$ in.

a rate of penetration of 18 in (45·7 cm)/h and as an undrained (instantaneous) analysis. It was found that consolidation effects during the insertion stage are insignificant and that the two methods yield very similar results. Hence, the undrained analysis results are reported, with a bulk compressibility of the pore fluid approximately fifty times that of the soil skeleton. As mentioned previously, reduced integration is employed for the undrained problems.

5.6.2 The Medium

The soil medium used for these tests is a K_0-consolidated Kaolin clay. The modified cam-clay elastoplasticity model is used for the analysis and the relevant soil parameters are listed below.

$$OCR = 1·0, \qquad v = 0·3, \qquad \lambda = 0·14$$

$$\kappa = 0·05, \qquad e_0 = 1·13, \qquad \phi = 26·5° \ (M_c = 1·05)$$

The in-situ stress state is

$$\sigma'_r = 10·24 \text{ psi}, \qquad \sigma'_z = 16·0 \text{ psi}, \qquad p = p_0 = 12·6 \text{ psi}$$

5.6.3 Results of the Numerical Analysis

Although the primary concern is with events taking place within the sampling tube, it is also of interest to observe how these events relate to changes occurring outside the sample and the mechanism of transfer between the two regions. However, the entire discretized soil domain is not necessary for plotting the results of the sampler penetration as only a zone of limited extent in the vicinity of the tube feels the effect. Consequently, the section, shown in Fig. 11, is used for the contour plots of the pore water pressure and its deviatoric stress variations in and around the sample. The plots are given in a nondimensional form, normalized with respect to the initial preconsolidation pressure (p_0). Results for samplers of three thickness values (i.e., $T = 0·03$, 0·05, 0·08 in) are plotted side-by-side to reflect the influence of the area ratio (A_r) on the extent of disturbance during sampling.

It was found that negative pore pressures develop in the soil sample and the degree of this alteration in pore pressure conditions depends upon the thickness of the sampler. This is a result of the outward movement of the soil around the sample creating a tensile condition in the soil. Figure 13 illustrates the variation of the pore pressure with the depth of embedment averaged over the two elements at the bottom of the laboratory sample. These elements are selected in order to reflect the area of maximum disturbance at the conclusion of sampling and eventual

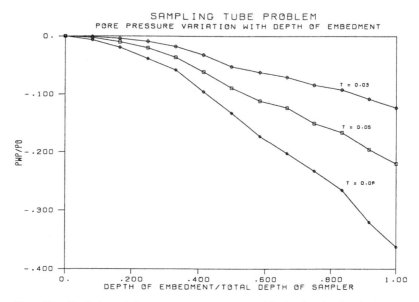

FIG. 13.   Variation of the excess pore pressure with the depth of embedment.

withdrawal of the tube. It is for this reason that the soil sample must be trimmed at the two edges and a sample from the mid-portion selected for testing.

Figure 14 (a)–(c) shows the pore pressure contours for the three thicknesses at the completion of the penetration process ($D = 6$ in (15·2 cm)). These plots indicate that the maximum pore pressures develop around the tip. In addition, there is very little effect of the insertion for the thinnest ($A_r = 8·2\%$) and intermediate size ($A_r = 13·8\%$) tubes on the soil sample inside. For the thickest case ($A_r = 22·5\%$), however, considerable negative pore pressure develops inside. Figure 14 indicates that the change in pore pressure at the bottom of the sample at the end of penetration is about $-0·4p_0$. Meanwhile, the behavior outside the tube, particularly close to the tip, is quite complex. Figure 15 (a)–(c) illustrates the behavioral pattern of the deviatoric stress with the depth of penetration for the thickest sampler.

It is evident from Fig. 14 that most of the changes are confined to the outside of the samplers for the first two cases ($A_r \leqslant 13·8\%$) and therefore the samplers of these area ratios are potentially capable of carrying out their desired task of extracting an undisturbed sample, although no attempt is made here to simulate this actual extraction process.

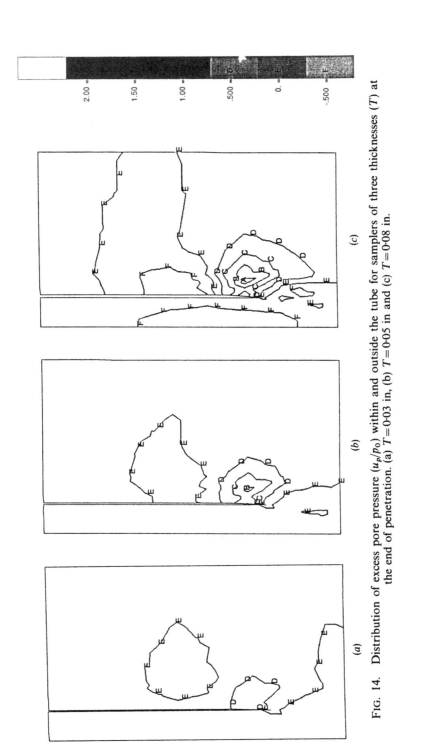

FIG. 14. Distribution of excess pore pressure (u_p/p_0) within and outside the tube for samplers of three thicknesses (T) at the end of penetration. (a) $T = 0\cdot03$ in, (b) $T = 0\cdot05$ in and (c) $T = 0\cdot08$ in.

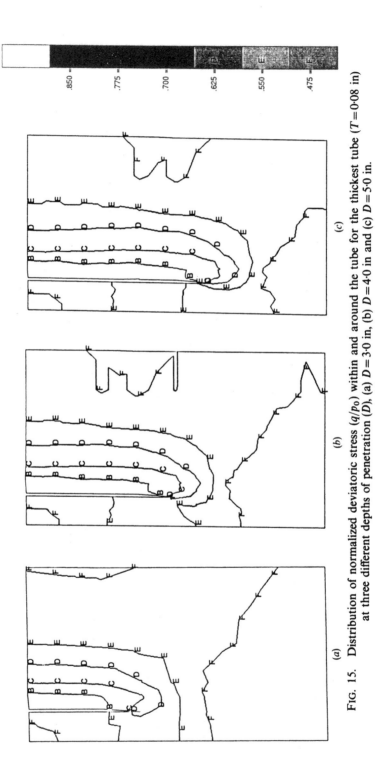

FIG. 15. Distribution of normalized deviatoric stress (q/p_0) within and around the tube for the thickest tube ($T = 0.08$ in) at three different depths of penetration (D), (a) $D = 3.0$ in, (b) $D = 4.0$ in and (c) $D = 5.0$ in.

However, for the thicker tube ($T = 0.08$ in (2 mm)), some noticeable alterations in the state of the sample begin to appear. This is in good agreement with, and forms an interesting confirmation of Terzaghi's judgement (e.g., Terzaghi and Peck, 1967) that an area ratio of less than 20% is necessary to keep the disturbances in the sample to a minimum. Ratios in excess of 20% may cause severe disturbances resulting in changes in the physical properties of the sample.

A similar study of the problem of sampling tube penetration was also conducted by Karim (1984). However, the finite element formulation used by Karim had several inconsistencies. In addition, unrealistic dimensions of the sampling tube as well as inadequate depths of penetration render any comparison with that analysis irrelevant.

5.7 Pile Penetration and Consolidation Problem

The experiments conducted by Fathallah (1978) and described by Banerjee et al. (1982) on the study of the pile driving and consolidation process, form the basis of this section. However, lessons learned from carrying out the experimental verification are extended to the study of the model pile under real (or in-situ) ground conditions. Besides providing insight into the physics of the problem, these examples are attempts to establish the practical utility of the present numerical algorithm. Of course, the results obtained by using any numerical tool are only as good as the assumptions made in simulating the actual experimental or in-situ conditions. For the analysis, material properties and boundary conditions were selected based upon the best data available, exclusive of the experiment itself. There was no attempt to fit the numerical model to the experimental results, since the latter were obtained from an isolated test.

5.7.1   Case I: Pile Driving Under Experimental Conditions:

Problem description. The entire loading time history of the experiment was simulated in five stages. These stages are tabulated in Table 2 and are illustrated in Fig. 16. Stage 1 involves the unloading of a tank containing the soil by a uniform stress of 16 psi. This unloading was carried out in order to obtain more accurate measures of the changes in effective stresses and total stresses in the soil. Since this stage was done in a very short period of time (15 min), most of the applied stress was absorbed by the pore pressure creating a state of suction (-16 psi) in the sample.

This was followed (Stage 2) by a period of consolidation for 70 min.

TABLE 2

DIFFERENT STAGES IN THE PILE PENETRATION-CONSOLIDATION EXPERIMENT

Test stage	Description	Time	
		Δt	$t(h)$
1	Removal of surcharge	15 min	0·25
2	Initial consolidation	70 min	1·42
3	Pile driving	20 min	1·75
4	Replacement of surcharge	15 min	2·00
5	Final consolidation	60 days	1442·0

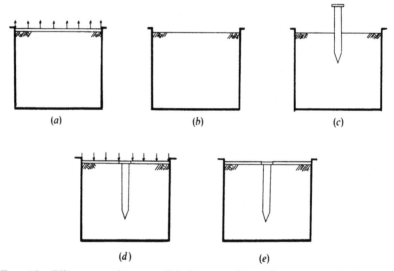

FIG. 16. Pile penetration-consolidation experimental stages; (a) Stage 1: Unloading of surcharge, (b) Stage 2: Initial consolidation, (c) Stage 3: Pile driving, (d) Stage 4: Reloading of surcharge, (e) Stage 5: Final consolidation.

This period enabled the technicians to prepare the instrumentation for the model pile to be driven into this medium.

Stage 3 consisted of a 20-min period during which the pile driving operation was carried out. Upon the completion of driving, Stage 4 was to reload the top surface of the medium by replacing the top lid of the tank and applying an overburden pressure of 16·0 psi. This was done in about 15 min. Finally, the soil, along with the embedded pile, was allowed to consolidate for a period of 60 days (Stage 5).

During each of these stages, measurements were recorded by means of four piezometers (P1–P4) located 6·0 in (15·2 cm) away from the side of the tank (i.e., 18 in (45·7 cm) from the centre of the pile) as shown in Fig. 17. In addition, three pore water pressure and total pressure cells (C1–C3) located on the pile shaft furnished data during the driving, reloading and subsequent consolidation stages.

The dimensions of the tank and the model pile along with the experimental instrumentation are illustrated in Fig. 17. The soil used for

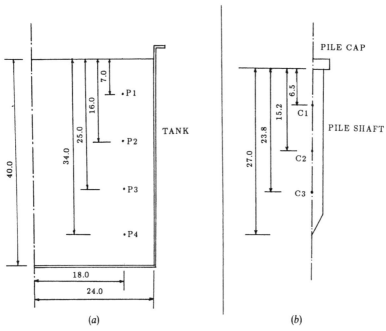

(a)   (b)

FIG. 17. Experimental instrumentation for the tank and model pile, (a) The tank showing the piezometer locations (P1–P4). (b) The model pile with the cell locations (C1–C3).

the experiment was Kaolin clay, for which the parameters of the modified cam-clay model are as follows:

$$\lambda = 0·14, \qquad \kappa = 0·05, \qquad e_0 = 1·13, \qquad \phi = 26.5°, \qquad v = 0·3$$

The permeability of the soil sample was found to be

$$\frac{k}{\gamma} = 0·0014 \text{ in}^2/\text{lb-h}$$

The in-situ state of stress was

$$\sigma'_r = 10 \cdot 24 \text{ psi}$$

$$\sigma'_z = 16 \cdot 0 \text{ psi}$$

Additionally, the soil was normally consolidated with a past maximum consolidation pressure of $12 \cdot 16$ psi ($C_u = 3 \cdot 8$ psi).

FEM simulation. The axisymmetric FEM mesh used to analyze this experiment is shown in Fig. 18. It consists of 78 LSQ elements, with 98 vertex nodes. This relatively coarse mesh was found to yield reasonable results for this problem.

In order to simulate the experimental conditions to the best possible

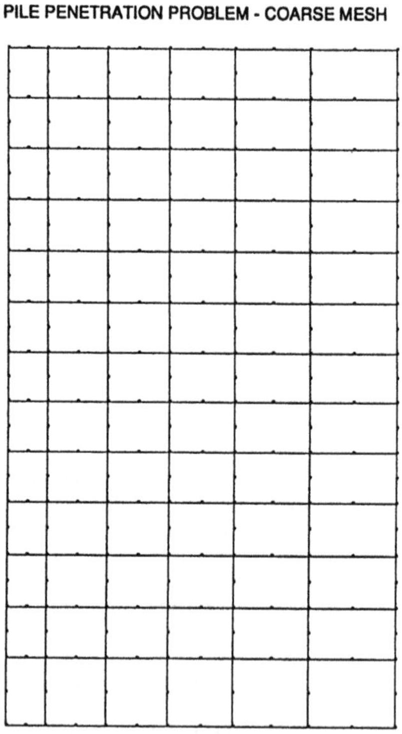

PILE PENETRATION PROBLEM - COARSE MESH

FIG. 18. Axisymmetric FEM mesh used for analysing the pile driving-consolidation experiment.

degree, the boundary conditions used for this FEM analysis are as follows:

(i) Top surface of the tank was sealed and is assumed impermeable.

(ii) Sides of the tank are assumed to be smooth. (This may not be strictly true, but alternatives were not feasible). Also, the sides of the tank contributed to the drainage and are taken as permeable boundaries.

(iii) Bottom of the tank is assumed to be smooth and permeable.

(iv) The pile surface is considered smooth with the pile-soil interface nodes assumed to be free in the vertical direction but restrained in the radial direction.

The entire pile driving process is simulated in nine sub-stages, depicting the passage of the pile through each layer of elements on its way to a total depth of embedment of 27·0 in (68·6 cm). Each increment block models the entry of the pile tip into the layer by imposing radial displacements from an initial cavity of 0·1 r_p to the final pile radius ($r_p = 1·25$ in (3·2 cm)). The number of increments of radial displacements required for each block were found to vary with depth indicating that as the severity of the problem increases, smaller increments of load need to be imposed. The number of increments required for each stage of this experiment are tabulated in Table 3. The entire analysis was carried out

TABLE 3

NUMBER OF INCREMENTS USED FOR EACH STAGE OF THE FEM
SIMULATION OF THE PILE PENETRATION-CONSOLIDATION EXPERIMENT

Test stage	Description	Number of increments
1	Removal of 16 psi surcharge	1
2	Initial consolidation	50
3	Pile driving	9011
4	Replacement of 16 psi surcharge	1
5	Final consolidation for 60 days	300

as a large deformation analysis. It should be noted that the inclusion of equilibrium and yield correction is extremely important in this problem, which involves significant large deformation and extensive material nonlinearity. In fact, with the present numerical approach severe element distortion prevented the use of a more refined model in the vicinity of the pile. For more detailed and more accurate information in that area, remeshing algorithms must be introduced.

Results of the analysis. This section reviews the results obtained from the FEM algorithm and compares them to those measured during the course of the experiment. The measured excess pore pressure at the four piezometer locations (P1–P4) and the corresponding results obtained from the FEM analysis are cited in Table 4 for each of the loading stages. At the end of Stage 1 (removal of 16 psi overburden surcharge), the predicted FEM results maintain a negative pore pressure of -16 psi at all four piezometers as compared to recordings ranging from -10.6 psi to -11.9 psi. This increase in pore water pressure may be attributed to the presence of air pockets in the soil that cannot be accounted for in the numerical analysis which assumes fully saturated soil conditions. In addition, certain amount of slippage at the tank surface may also alter the state of the soil medium during this initial unloading. Right at the beginning, it becomes evident that the results from the FEM analysis must be viewed subjectively in the context of uncertainties in the modeling of the experimental set up.

Continuing with the comparisons in Table 4, the pore pressure measurements at the beginning of pile driving (Stage 2) are about -10.0 psi averaged over the piezometers as compared to -14.0 psi obtained from the FEM analysis. This increase is a result of the initial lag period allowing some consolidation to occur. The difference between the measured and FE results reflects the initial imbalance as the changes and the trend across the piezometers is very comparable.

The next column of information depicts quite a large difference between the two measurements. As is evident from the data reported, the pile driving process results in substantial increase of pore pressure at all four piezometers, ranging from 3.9 psi at P4 to 7.3 psi at P3. Based upon reasonable assumptions for the coefficient of consolidation for Kaolin clay and the time involved in driving, this experimental change in pore pressure cannot be attributed to diffusion processes. However, the FEM analysis shows that most of the pore pressure changes occur close to the pile. This fact will become clearer in the contour plots showing these changes throughout the soil tank. No significant changes in pore pressure can be seen to occur at the remote piezometer locations as a result of pile driving. The difference in results may once again be attributed to the effects of the tank wall which is only 6 in (15.2 cm) from the side or to

TABLE 4
MEASURED AND FE RESULTS FOR PORE WATER PRESSURE AT PIEZOMETER LOCATIONS P1–P4

Piezometer	Depth below surface (in)	Stage 1: After removal of surcharge		Stage 2: At the start of pile driving		Stage 3: At the end of pile driving		Stage 2→3: Change in pore pressure		Stage 4: After reapplication of surcharge	
		Measured	FEM	Measured	FEM	Measured	FEM	Measured	FEM	Measured	FEM
P1	7·0	−11·7	−16·0	−9·1	−14·08	−5·0	−13·92	4·1	0·16	6·1	2·07
P2	16·0	−11·3	−16·0	−9·8	−13·65	−4·5	−13·64	5·3	0·01	6·1	2·35
P3	25·0	−10·9	−16·0	−10·1	−13·53	−2·8	−13·41	7·3	0·12	7·8	2·84
P4	34·0	−10·6	−16·0	−9·6	−14·79	−5·7	−14·33	3·9	0·46	4·1	1·63

other uncertain behavior during the experiment. The analysis, of course, assumes idealized conditions. Finally, the last column of the table shows the pore pressure measurements at the end of Stage 4 (reapplication of 16 psi overburden pressure). These predictions average to about 2·3 psi which is lower than the 6·0 psi average measured from the experiment. All of the pressures dissipate to zero at the end of the 60 day period of consolidation.

Table 5 gives the measurements of the total radial stress and the pore pressure at the three cell locations (C1–C3) at the pile-soil interface. The experimental results at cell location C2 were not included due to its failure during driving. The increase in pore pressure at the end of Stage 4 were reported as 13·3 psi at C1 and 14·6 psi at C3. The FE results for these locations provide an increase of 13·7 psi and 13·6 psi respectively. These numbers are in good agreement with the experimental recordings, in spite of the differences in the values at the early stages. The total residual radial stress at the end of Stage 5 (after the final consolidation is complete) show that the experimental measurements average to about 24·8 psi whereas the predictions average to about 24·9 psi over the three cells. These show the excellent correlation between the data for total stress close to the pile and may be attributed to the fact that the equilibrium conditions ensure a strict balance of total stresses for both types of tests.

The changes in pore water pressure near the pile surface during all the stages are shown in Fig. 19. These are plotted by the increase of the current pore pressure values at the three cell locations followed by the decay with time during consolidation. In addition, the variation of total radial stresses with time is plotted in Fig. 20. Further, Fig. 21 shows the variation of the total radial stress with the depth of penetration of the pile at the pile-soil interface measured from the moment a cell (C1, C2 or C3) enters the soil until the total embedment of the pile. It should be noted that some judgement was exercised in extrapolating these values between the initial and final positions of the penetrating cell, since the analysis results are continuous while the test measurements become available only when the cell has sufficiently penetrated the soil. The average increase in total radial stress at the surface due to pile driving was found to be 13·7 psi. Also plotted on this figure are the measured values from the experiment. The agreement is quite satisfactory though the FE analysis predicts higher total stresses close to the pile than those measured.

Figure 22(a, b) shows the contours of the distribution of effective radial

TABLE 5
MEASURED AND FE RESULTS FOR TOTAL RADIAL STRESSES AND PORE WATER PRESSURES AT THE PILE-SOIL INTERFACE (AT DIFFERENT CELL LOCATIONS C1–C3)

Readings	Cell locations	Stage 2: Beginning of pile driving		Stage 3: End of pile driving		Stage 2→3: Change due to driving		Stage 4: After reapplication of surcharge		Stage 5: Residual σ'_r and $\Delta\sigma'_r$ after consolidation	
		Measured	FEM	Measured	FEM	Measured	FEM	Measured	FEM	Measured	FEM
Total radial stress (psi)	Top	*	−3·7	15·0	15·5	15·0	19·2	26·1	31·4	22·7, 11·6	24·7, 14·5
	Middle	*	−4·0	17·0	17·0	17·1	21·0	27·7	32·9	24·7, 13·1	24·5, 14·3
	Bottom	*	−3·9	19·1	18·0	19·1	21·9	29·8	33·6	27·2, 16·5	25·6, 15·3
Pore water pressure (psi)	Top	−9·1	−14·8	2·3	−2·3	11·4	12·5	13·3	13·7	—	—
	Middle	*	−15·1	*	0·7	*	15·8	*	16·8	—	—
	Bottom	−10·1	−15·2	3·9	−2·6	14·0	12·6	14·6	13·6	—	—

*data not available

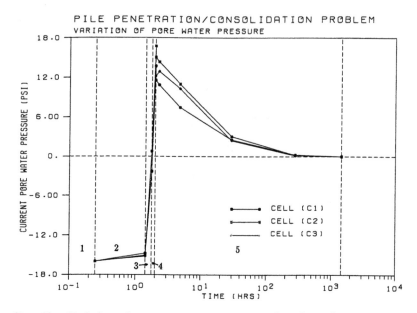

FIG. 19. Variation of excess pore pressure near the pile surface (at the cell locations) with time during all the stages.

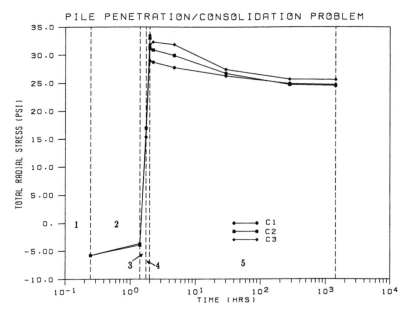

FIG. 20. Variation of total radial stress near the pile surface (at the cell locations) with time during all stages.

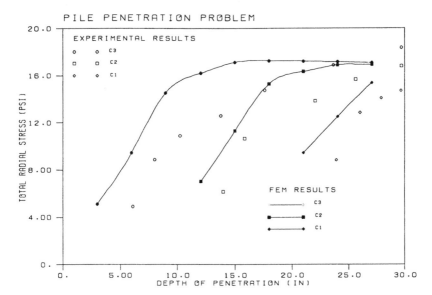

FIG. 21. Variation of total radial stress with depth of penetration of the pile.

stress at the beginning and end of the pile driving. Further, Fig. 22(c) illustrates, to the same scale, the distribution of residual effective stresses at the end of the final consolidation period.

Other numerical simulations of this experiment have been attempted by Fathallah (1978) and Karim (1984), which have yielded similar results to those from the present analysis. In some cases, particularly away from the pile, a better correlation with the experimental results is shown in these previous works. This may be attributed to the fact that they used several distinct FEM analyses to simulate the various stages of the experiment. This enabled them to judicially average the quantities at the beginning of each stage and reduce the differences between the numerical and experimental results. In the present analysis, however, the entire experiment is modeled as a continuous process with no external interference, utilizing a single, large deformation FEM formulation.

5.7.2 (b) Case 2: Pile Driving Under Real Ground Conditions

Since some of the conditions imposed during the course of the experiment were not justifiable as practical under real conditions, a need was felt to carry out a parallel analysis with the simulation of more realistic conditions. However, for convenience as well as comparison, the soil properties, dimensions of the soil medium and the pile selected were

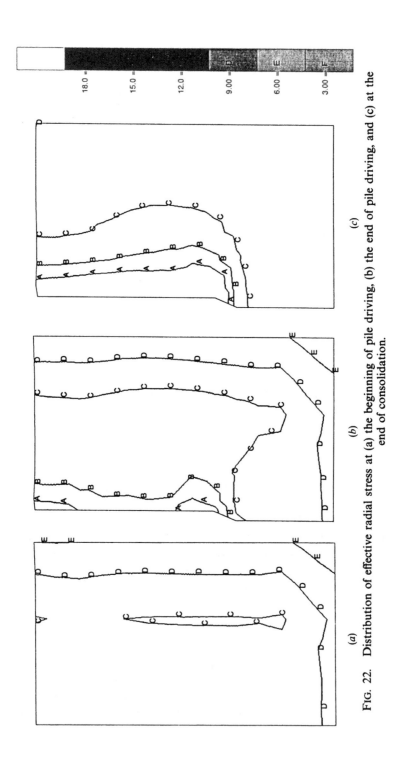

FIG. 22.   Distribution of effective radial stress at (a) the beginning of pile driving, (b) the end of pile driving, and (c) at the end of consolidation.

identical to the experiment. Changes were made in the types of loading stages and the boundary conditions for the in-situ case.

Problem description. Out of the five stages carried out in the experimental simulation, only two stages are required to be conducted in this case. These are:

Stage 1:   pile driving for 20 min;
Stage 2:   consolidation of the soil with the embedded pile for 60 days.

The in-situ stress state is again assumed to be

$$\sigma'_2 = 16 \cdot 0 \, \text{psi}, \qquad \sigma'_r = 10 \cdot 24 \, \text{psi}, \qquad p_0 = 12 \cdot 16 \, \text{psi}$$

However, the pile driving operation begins from a state of zero excess pore water pressure in the soil. Subsequently, the pore pressure built up due to the driving is allowed to dissipate during the consolidation stage (Stage 2).

FE simulation. The FEM mesh used for the experimental verification runs is used again for this case. Additionally, a finer mesh is prepared in order to investigate the influence of smaller elements close to the pile. The finer mesh consists of 119 LSQ and 37 LST elements, as shown in Fig. 23. Considerable refinement is carried out close to the pile in the hope of capturing the behavior more closely.

In order to carry out a fair comparison between the coarse and the fine mesh results, the pile tip is made to penetrate through two layers of elements (each 1·5 in (3·8 cm) in depth) in nine sub-stages. Thus the imposed radial displacement pattern is identical to the coarse mesh, except that more elements are available around the pile for studying the results. Once again, the pile is assumed to be smooth and restrained in the radial direction.

However, the boundary conditions used here reflect the real ground conditions and vary considerably from the experiment. The tank bottom and side from the experimental setup are no longer considered to exist in an infinite space and are replaced by smooth permeable boundaries. The top surface, however, is considered to be permeable to reflect the ground conditions. Large deformation effects are included for all the stages of the problem.

Results of the analysis. An overall idea of the distribution of several variables is presented through the use of contour plots over the entire

PILE PENETRATION PROBLEM - FINE MESH

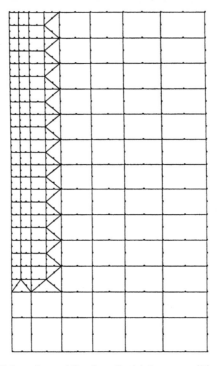

FIG. 23.　Fine FEM mesh used for the pile driving-consolidation problem under real conditions.

domain. Figure 24 shows the distribution of the pore pressure throughout the medium as a result of the driving process for both the coarse and the fine meshes. The overall nature of the curves is very similar except close to the tip of the pile, where the fine mesh shows large variations. This may be attributed to the existence of a singularity at a re-entrant corner. The coarse predictions are smoother due to larger elements close to the tip reflecting a more averaged behavior. In addition, the distribution of effective radial stress in the soil tank at the end of driving and at the end of final consolidation are also presented in Figs 25 and 26, respectively. Once again, the pattern of both sets of the contour results appears to be very similar throughout the medium except for a zone close to the tip. This indicates the convergence in the context of the discretization patterns employed in the FEM analysis.

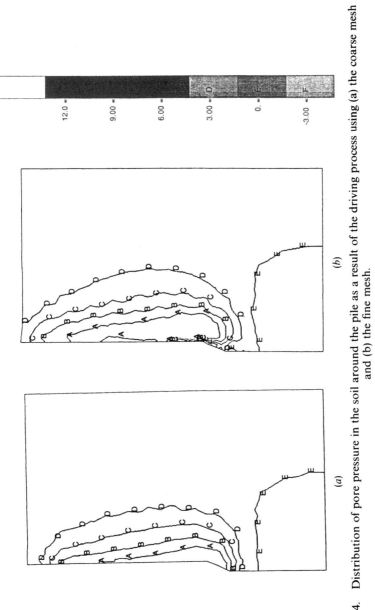

FIG. 24. Distribution of pore pressure in the soil around the pile as a result of the driving process using (a) the coarse mesh and (b) the fine mesh.

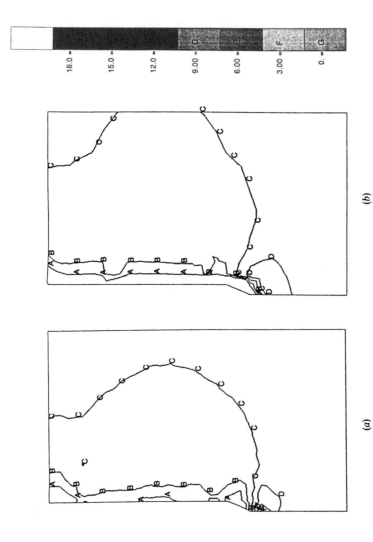

(b)

(a)

FIG. 25.   Distribution of effective radial stress in the soil at the end of driving using (a) the coarse mesh and (b) the fine mesh.

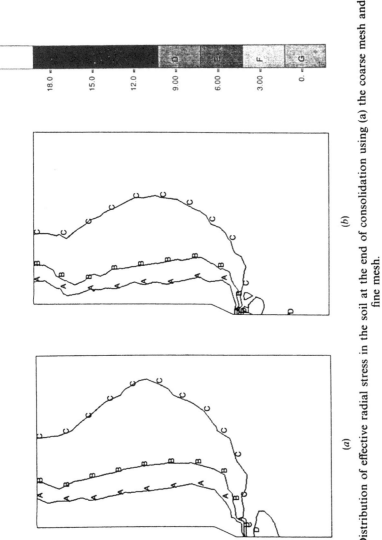

FIG. 26.   Distribution of effective radial stress in the soil at the end of consolidation using (a) the coarse mesh and (b) the fine mesh.

From the observations above, it can be seen that the soil is changed around the pile up to a certain distance as a result of both the driving and the subsequent consolidation process. The pore water pressure builds up around the pile to an average value of $3\cdot4$ C_u at the end of driving. This is in good agreement with the value of $3\cdot7$ C_u reported by Banerjee *et al.* (1982) from the experimental observations. In addition, the total radial stresses close to the pile also increase from their in-situ value of $2\cdot7$ C_u to $5\cdot6$ C_u. The experimental results from Banerjee *et al.* (1982) indicate an average increase of total radial stresses of $4\cdot5$ C_u due to the driving process. At the completion of consolidation, the radial stresses end up at a value of $6\cdot5$ C_u as compared to $6\cdot3$ C_u from the experiment. As a result of this increase in radial stress, the undrained shear strength of the soil increases considerably as reported by a number of investigators of this problem.

6   CONCLUSIONS

An updated Lagrangian finite element formulation has been presented in this chapter for problems involving finite deformations. A class of problems involving objects such as samplers and piles, penetrating a soil medium were analysed in detail to illustrate the practical validity of such a formulation.

The finite element formulation was developed for a two-phase material such as soil and used consistent stress measures redefined in order to account for continuous changes in geometry. Equilibrium equations were written using the first Piola–Kirchhoff stress measure and all the terms appearing in the subsequent integral equations were retained. In addition, the objective nature of the Jaumann rate of Kirchhoff stresses was utilized in casting the constitutive equation for the material. Also included in the formulation are terms for equilibrium and yield correction as well as a load correction term on the surface.

Material behavior was assumed to be either elastic or elastoplastic. A linear hypoelastic stress–strain law was used to model the purely elastic material whereas a critical state model (namely the modified cam-clay model) was used for the elastoplastic behavior of cohesive soils. Several illustrative numerical examples of both types were solved to establish the validity and correctness of the present algorithm. Finally, a comprehensive study of the two penetration problems was embarked upon.

In the case of the sampling tube insertion into soil as a part of the sampling process, a number of samplers of different thicknesses but the same inner radius were modeled. Each of these were penetrated up to identical depths into the same soil and the effects on the sample inside the tube were analysed. It was found, as expected, that the degree of alterations in the state of the soil sample depends upon the thickness of the tubes and hence, the area ratios of the samplers. Only the thickest sampler ($A_r = 22\%$) was observed to have caused significant disturbance to the sample. The disturbing effects of the other two thicknesses ($A_r \leqslant 14\%$) was found to be negligible. This appears to be in good agreement with Terzaghi's assessment that a sampler with $A_r < 20\%$ is required in order to keep the disturbance to the sample at a minimum. The other penetration problem studied was the penetration of a pile into soil. Initially, an effort was made to simulate the experiment described by Banerjee et al. (1982). However, it was found that right from the initial stages of the simulation, the numerical results were different from the experiment. This may be accounted for due to inadequate modeling of the boundary conditions. Slippage of soil at the sides, existence of pockets of air in a semi-saturated soil tank or other inherent inhomogeneities in the experiment cannot be taken account of in the present numerical analysis. Therefore, it was evident that the results of a numerical tool are only as good as the assumptions made while simulating the experimental set-up.

But the results obtained displayed a consistent pattern. Plots of the distribution of the excess pore pressures showed that most of the changes are localized around the pile, particularly close to the pile tip. No significant changes in pore pressure were observed at the remote piezometer locations. Close to the pile, both the total radial stress and the pore pressure changes agreed well with the experimental results. The residual effective radial stress values after consolidation were found to be in very good agreement at the pile–soil interface indicating that there was insignificant effect of the initial discrepancies, at locations away from the edges of the soil tank. It may be mentioned here, that some earlier numerical results to this experiment presented by Fathallah (1978) and Karim (1984) have indicated better correlation with the experimental observations. However, they obtained these results by using several unrelated FEM analyses for each stage and therefore were able to use average values of the quantities around the pile and the soil at the beginning of each analysis, thereby eliminating any continuation of difference in the results at the end of each stage.

Finally, the experimental set-up was discarded and the same pile was analysed under more practical, in-situ conditions. The properties as well as the penetration process were kept identical to the experimental set-up. However, a finer mesh was developed to establish the convergence of the results obtained from utilizing the coarse mesh for this case. It was found that the fine mesh predicted a higher value of changes in both the effective radial stress and the pore pressure close to the pile, when compared to the coarse mesh. This was attributed to the fact that the fine mesh was able to capture the response more accurately due to more elements around locations of high gradients. The final residual effective stress values were higher than that in-situ by an average of $2\cdot7$ C_u at the end of the driving and about $4\cdot0$ C_u at the end of final consolidation. This indicates an increase in the undrained shear strength of the soil around a driven pile, a beneficial factor reported by several previous investigators.

The entire finite deformation FE algorithm was introduced into an existing FE Fortran program called CRISP (Gunn and Britto, 1984) and is capable of analysing general problems of three-dimensional, two-dimensional plane strain or axisymmetry.

REFERENCES

ALONSO, E., ONATE, E. and CARANOVAS, J. (1981). An investigation into sampling disturbance, *Proc. 10th Int. Conf. on Soil Mechanics and Foundation Engineering*, Stockholm, Vol. 2, pp. 419–23.

BANERJEE, P.K. and FATHALLAH, R.C. (1979). An Eulerian formulation of the finite element method for predicting the stresses and pore water pressures around a driven pile, *Proc. 3rd Int. Conf. on Numerical Methods in Geomechanics*, Aachen, pp. 1053–60.

BANERJEE, P.K., DAVIES, T.G. and FATHALLAH, R.C. (1982). Behavior of axially loaded driven piles. In *Developments in Soil Mechanics and Foundation Engineering-1*, P.K. Banerjee and R. Butterfield, (Eds), Elsevier Applied cience, London, Chap 1.

BEGEMANN, H.K. (1961). A new method for taking of samples of great length, *Proc. 5th Int. Conf. on Soil Mechanics and Foundation Engineering*, Paris, France, Vol. 1, pp. 437–40.

BEGEMANN, H.K. (1965). The new apparatus for taking a continuous soil sample, *LAM Mededelingen*, 10(4), 85–105.

BEGEMANN, H.K. (1971). Soil sampler for taking undisturbed sample 66 mm in diameter and with a maximum length of 17 m, *Proc. Specialty Session on Quality in Soil Sampling, 4th Asian Conf. Int. Soc. on Soil Mechanics and Foundation Engineering*, Bangkok, pp. 54–7.

BJERRUM, L. and JOHANNESSEN, I. (1961). Pore pressures resulting from driven piles in soft clay, *Proc. Conf. on Pore Pressure and Suction in Soils*, London, pp. 108–11.

BROMS, B. (1980). Soil sampling in Europe: state-of-the-art, *J. Geotech. Eng. Div. ASCE*, **106**(SM1), 65–98.

BUTTERFIELD, R. and BANERJEE, P.K. (1970). The effects of pore water pressures on the ultimate bearing capacity of driven piles, *Proc. 2nd South East Asian Regional Conference on Soil Mechanics and Foundation Engineering*, Bangkok, pp. 385–94.

CARTER, J.P., SMALL, J.C. and BOOKER, J.R. (1977). A theory of finite elastic consolidation, *Int. J. Solids Struct.*, **13**, 467–78.

CHANDRA, A. and MUKHERJEE, S. (1984). A finite element analysis of metal-forming problems with an elastic-viscoplastic material model, *Int. J. Numer. Methods Engrg*, **20**, 1613–28.

CUMMINGS, E.A., KERHOFF, G.O. and PEK, R.B. (1950). Effects of driving piles in soft clay, *Trans. Am. Soc. Civ. Engrs*, **115**, 275.

FATHALLAH, R.C. (1978). Theoretical and experimental investigation of the behavior of axially loaded single piles driven in saturated clays, Ph.D. Thesis, University College, Cardiff, UK.

FUNG, Y.C. (1965). *Foundations of Solid Mechanics*, Prentice Hall, Englewood Cliffs, NJ.

GIBSON, R.E., ENGLAND, C.L. and HUSSEY, J.J.L. (1967). The theory of one-dimensional consolidation of saturated clays, I, Finite non-linear consolidation of thin homogeneous layers, *Geotechnique*, **17**, 261–273.

GREEN, A.E. and ADKINS, J.E. (1960). *Large Elastic Deformation and Nonlinear Continuum Mechanics*, Oxford University Press, London.

GUNN, M.J. and BRITTO, A.M. (1984). *CRISP-User's and Programmer's Guide*, Engineering Department, Cambridge University, UK.

HIBBIT, H.D., MARCAL, P.V. and RICE, J.R. (1970). A finite element formulation for problems of large strain and large displacement, *Int. J. Solids Structures*, **6**, 1069–86.

HILL, R. (1959). Some basic principles in the mechanics of solids without natural time, *J. Mech. Phys. Solids*, **7**, 209–25.

HVORSLEV, M.J. (1949). Subsurface exploration and sampling of soils for civil engineering purposes, Report on a Research Project of ASCE, US Army Engineer Experiment Station, Vicksburg, Miss., p. 521.

HWANG, C.T., MORGENSTERN, N.R. and MURRAY, D.W. (1971). On solutions of plane strain consolidation problems by finite element methods, *Can. Geotech. J.*, **8**, 109–18.

KARIM, U.F. (1984). Large deformation analysis of penetration problems involving piles and sampling tubes in soils, Ph.D. Thesis, State University of New York at Buffallo, Buffalo, NY.

KALLESTENIUS, T. (1958). Mechanical disturbances in clay samples taken with piston samplers, *Proc. Swedish Geotechnical Institute*, No. 16, Stockholm, p. 75.

KALLESTENIUS, T. (1963). Studies on clay samples taken with standard piston sampler, *Proc. Swedish Geotechnical Institute*, No. 21, p. 207.

LAMBE, T.W. and WHITMAN, R.V. (1969). *Soil Mechanics*, John Wiley, New York.

LANG, J.G (1967). Longitudinal variations of soil distribution within tube samplers, *Proc. 5th Australian–New Zealand Conference on Soil Mechanics and Foundation Engineering*, Auckland, Vol. 1, pp. 39–42.

LO, K.Y. and STERMAC, A.G. (1965). Induced pore pressures during pile driving operations, *Proc. 6th Int. Conf. on Soil Mechanics and Foundation Engineering*, Montreal, Canada, Vol. 2, p. 285.

MCMEEKING, R.M. and RICE, J.R. (1975). Finite-element formulations for problems of large elastic-plastic deformation, *Int. J. Solids Structures*, 11, 601–16.

NEMAT-NASSER, S. (1982). On finite deformation elasto-plasticity, *Int. J. Solids Struct.*, 18, 857–72.

OSIAS, J.R. (1973). Finite deformation of elasto-plastic solids, NASA Cr-2199.

OSIAS, J.R. and SWEDLOW, J.L. (1974). Finite elasto-plastic deformation–I. Theory and numerical examples, *Int. J. Solids Structures*, 10, 321–39.

PREVOST, J.H. (1981). Consolidation of an elastic porous media, *Prod. ASME*, EM1, 169–86.

PREVOST, J.H. (1982). Nonlinear transient phenomena in saturated porous media, *Comp. Meth. Appl. Mech. Engrg*, 20, 3–18.

RANDOLPH, M.F., CARTER, J.P. and WROTH, C.P. (1979). Driven piles in clay–the effects of installation and subsequent consolidation, *Geotechnique*, 29(4), 361–93.

ROSCOE, K.H. and BURLAND, J.B. (1968). On the generalized stress–strain behavior of 'wet' clays. In *Engineering Plasticity*, J.V. Hayman and F.A. Leckie (Eds), Cambridge University Press, UK, pp. 535–609.

SANDHU, R.S. and WILSON, E.L. (1969). Finite element analysis of seepage in elastic media, *J. Engrg Mech. Div. ASCE*, 95(EM3), 641–52.

SEED, H.B. and REESE, L.C. (1955). The action of soft clay along friction piles, *Proc. Am. Soc. Civ. Engrs*, Vol. 81, Paper 842.

SMITH, I.M. and HOBBS, R. (1976). Biot analysis of consolidation beneath embankments, *Geotechnique*, 26, 149–71.

SODERBERG, L.O. (1962). Consolidation theory applied to foundation pile time effects, *Geotechnique*, 12, 217–25.

TERZAGHI, K. and PECK, R.B. (1967). *Soil Mechanics in Engineering Practice*, John Wiley, New York.

WASHIZU, K. (1975). *Variation Methods in Elasticity and Plasticity*, Pergamon Press, Oxford, UK.

WIFI, A.S. (1982). Finite element correction matrices in metal forming analysis (with application to hydrostatic bulging of a circular sheet), *Int. J. Mech. Sci.*, 24(7), 393–406.

ZIENKIEWICZ, O.C. (1977). *The Finite Element Method*, 3rd edn., McGraw-Hill, London.

ZIENKIEWICZ, O.C. and SHIOMI, T. (1984). Dynamic behavior of saturated porous media; the generalized Biot formulation and its numerical solution, *Int. J. Numer. Methods Engrg*, 8, 71–96.

Chapter 6

ANALYSIS OF THE DYNAMICS OF PILE DRIVING

M.F. RANDOLPH

Department of Civil and Environmental Engineering,
The University of Western Australia

ABSTRACT

Analysis of the dynamic response of piles during driving is generally achieved by treating the pile as an elastic bar along which the stress-waves travel axially. Numerical solutions of the one-dimensional wave equation, with simple spring and dashpot soil models distributed along the pile, have been in common use over the last thirty years. However increased use of field monitoring of stress-waves during pile driving has provided the impetus for a number of recent advances, both in numerical techniques and in modelling of the soil response. This chapter outlines these advances, with particular emphasis on improved soil models which take due account of the inertial resistance of the soil continuum. A detailed description of one-dimensional wave propagation is included, and the basis for calculating the dynamic pile capacity from stress-wave data is outlined. The chapter concludes with a discussion of inherent limitations in predicting the static capacity of piles from dynamic measurements.

NOTATION

a_0	dimensionless frequency $= \omega r_0 / V_s$
A	cross-sectional area of pile
c	wave speed in pile
c_u	undrained shear strength of soil

C	damping constant
d	diameter of pile
E	Young's modulus of pile
E_c	Young's modulus of cushion
f	function
F	force in pile
F_d	force in pile associated with downward travelling wave
F_u	force in pile associated with upward travelling wave
g	function
G	shear modulus of soil
i	square root of minus one
I_r	rigidity index $= G/c_u$
j_c	CASE damping constant
J	damping constant (Smith (1960))
k	stiffness of cushion or capblock
K	spring stiffness
K_0	modified Bessel function of order zero
K_1	modified Bessel function of order one
l	pile length
m_a	mass of anvil
m_r	mass of ram
n	exponent in non-linear viscous damping relationship
N_q	bearing capacity factor
p_a	atmospheric pressure $= 100\,\text{kPa}$
q_b	limiting end-bearing pressure
Q	quake
Q_b	point resistance offered by soil
r_0	radius of pile
R	soil resistance
R_d	dynamic soil resistance
R_s	static soil resistance
S_1	in phase stiffness coefficient
S_2	out of phase stiffness coefficient
t	time
t_m	rise time for stress-wave
T	soil resistance at given node
T_S	total soil resistance along pile shaft
v	particle velocity in pile
v_d	particle velocity associated with downward travelling wave
v_i	impact velocity

v_0	reference velocity (1 m/s) in non-linear viscous damping relationship
v_u	particle velocity associated with upward travelling wave
V_s	shear wave velocity in soil
w	displacement
z	depth
Z	pile impedance $= EA/c$
α	viscous parameter
β	viscous parameter
δ	loss angle for viscous soil response
Δ	increment of
ε_z	axial strain in pile
ζ	parameter in static response along pile shaft
μ	parameter in analytical solution of hammer impact
ν	Poisson's ratio for soil
π	mathematical constant
ρ	density of pile
ρ_s	saturated density of soil
ρ_w	density of water
σ'_v	vertical effective stress
σ_z	axial stress in pile
τ	shear stress
ω	angular frequency

Subscripts

b	base of pile
d	dynamic
o	original value
p	pile
r	reflected or return
s	shaft of pile, static or soil, depending on context
t	transmitted

1 INTRODUCTION

Since the early 1970s, there has been increasing use of numerical analysis in the planning and construction control of pile driving. On the planning side, the engineer needs to ensure the drivability of the required pile with a given hammer, and also to assess the likelihood of damage to the pile due to excessive driving stresses. During construction, instrumentation may be used to capture dynamic force and acceleration data at the pile head. The data enable calculation of the energy transmitted by the hammer (and hence an assessment of the efficiency of the driving system) and also provide a means of estimating the current resistance of the soil to penetration of the pile. Field data also provide a means of assessing the integrity of a pile, either during a normal driving operation, or by the application of relatively light blows after construction (particularly for cast-in-situ piles).

The fastest developing area of piling engineering is undoubtedly the capturing and interpretation of dynamic 'stress-wave' data during pile driving. Although such measurements, and an analysis for interpreting them, were achieved as long ago as the 1930s (Fox, 1932; Glanville et al., 1938), widespread use of the technique has had to await modern advances in field instrumentation and, more importantly, powerful microprocessors that permit real time analysis of the data.

There has been extensive publication in connection with the analysis of stress-wave propagation in driven piles. The early work of Smith (1960), and computational approaches that evolved from his work, have been summarised by Coyle et al. (1977). Since then, there has been a series of speciality conferences on the application of stress-wave theory to piles: two in Stockholm (1980 and 1984) and one in Ottawa (1988). The state-of-the-art paper by Goble et al. (1980) in the first of these conferences provides a useful guide to the different facets of the subjects. Two extensive lectures given at the second conference (Fischer, 1984; Rausche, 1984) give detailed accounts of the theory of one-dimensional stress-wave propagation and application of that theory to interpretation of stress-wave data. One further notable publication arose from the International Symposium on Penetrability and Drivability of Piles, held in San Francisco in 1985. This symposium was organised by a Technical Committee of the ISSMFE, and includes a number of National Reports from member countries.

This chapter summarises some of the more recent advances in modelling the dynamic interaction between pile and soil, and highlights areas where there are still major shortcomings. A full dynamic analysis of pile driving entails a two-dimensional (axisymmetric) or three-dimensional model of the hammer, pile and soil system. In principle, such an analysis may be achieved by means of the finite element method. However, this approach is limited by the high level of computational resources required, and by limitations in constitutive relations for the soil. The vast majority of pile driving analyses are conducted using a simplified one-dimensional model of the pile. The pile is treated as an elastic rod with only axial stress-wave propagation considered. The soil response is represented by spring/dashpot/mass elements distributed at discrete points along the length of the pile.

There is a powerful research role for finite element studies that model the full soil continuum, and that is to improve existing simplified models of pile-soil interaction, and to highlight shortcomings in the one-dimensional approaches. For example, finite element computations reported by Smith and his co-workers (Smith and Chow, 1982; Smith et al., 1986) and by Randolph and Simons (Simons, 1985; Simons and Randolph, 1985; Randolph and Simons, 1986) have emphasised major limitations in the widely used spring/dashpot model of Smith (1960).

Such studies have led recently to a much improved understanding of the dynamic interaction between pile and soil during driving, and to the development of improved 'one-dimensional' models of the soil response. This is an aspect of pile driving analysis which is currently receiving particular attention, as may be seen from the most recent speciality conference on stress-wave theory. The chapter outlines the basis for soil models based on elastodynamic theory of the continuum, and emphasieses the importance of modelling the inertial resistance of the soil correctly.

A detailed discussion of wave propagation is included at the start of the chapter, since this forms the basis of computer codes for pile driving analysis and of methods for estimating the soil resistance directly from stress-wave measurements in the field.

2 SOLUTION OF THE ONE-DIMENSIONAL WAVE EQUATION

Wave propagation in the pile, treated as an elastic bar where only axial motion is considered, is governed by the differential equation (e.g. Timoshenko and Goodier, 1970)

$$\frac{\partial^2 w}{\partial t^2} = c^2 \frac{\partial^2 w}{\partial z^2} \tag{1}$$

where w is the axial displacement at a position, z, and time t. The parameter, c, is the wave speed, given by

$$c = \sqrt{E/\rho} \tag{2}$$

where E is the Young's modulus and ρ the density of the pile material.

In early numerical approaches, eqn (1) was approximated by finite difference operators in position and in time, and solved using an explicit time integration with a sufficiently small time step to provide stability. Explicit time integration is inherently unstable; most programs now use implicit time integration, such as the Newmark scheme (Bathe and Wilson, 1976), with parameters chosen to ensure stability unconditionally. Even using implicit time integration, the time step has to be chosen sufficiently small to provide an accurate solution. In most cases, the required time step is not much different in either explicit or implicit time integration.

An alternative method of approximating eqn (1) is by means of one-dimensional finite elements (Smith, 1985; Chow et al. 1988). Such an approach, together with an implicit time integration scheme, requires the formulation of a complete stiffness matrix for the pile. However, the narrow bandwidth leads to relatively low computational effort for solution of the set of equations at each time step.

The most recent method that has been adopted for the solution of eqn (1) is based on characteristic solutions of the form

$$w = f(z - ct) + g(x + ct) \tag{3}$$

where f and g are unspecified functions which represent downward (increasing z) and upward travelling waves, respectively. Taking downward displacement and compressive strain and stress as positive, eqn (3) leads to the following expressions for the axial strain, ε_z, stress, σ_z, and force, F in the pile:

$$\varepsilon_z = -\frac{\partial w}{\partial z} = -(f' + g') \tag{4}$$

$$\sigma_z = E\varepsilon_z = -E(f' + g') \tag{5}$$

$$F = A\sigma_z = -EA(f' + g') \tag{6}$$

where the prime denotes the derivative of the function with respect to its argument, and A is the cross-sectional area of the pile.

The particle velocity, v, at any position and time is given by

$$v = \frac{\partial w}{\partial t} = -c(f' - g') \tag{7}$$

The velocity and force can each be considered as made up of two components, one due to the downward travelling wave (represented by the function f) and one due to the upward travelling wave (represented by the function g). Using subscripts d and u for these two components, the velocity is

$$v = v_d + v_u = -cf' + cg' \qquad (8)$$

The force F is similarly expressed as

$$F = F_d + F_u = -EAf' + (-EAg') \qquad (9)$$

Comparing eqns (8) and (9), it may be seen that

$$F = F_d + F_u = Zv_d + (-Zv_u) = Z(v_d - v_u) \qquad (10)$$

where $Z = EA/c$ is referred to as the pile impedance. [Note, some authors have referred to the pile impedance as $Z = E/c$ relating axial stress and velocity rather than force and velocity. The more common definition of pile impedance as $Z = EA/c$ will be adopted here.]

The relationships given above may be used to model the passage of waves down and up piles of varying cross-section, allowing for interaction with the surrounding soil. It is helpful to consider the pile as made up of a number of elements, each of length Δz, with any soil resistance concentrated at the nodes (see Fig. 1). Numerical implementation of the characteristic solutions involves tracing the passage of the downward and upward travelling waves from one node to the next. The time increment, Δt, is chosen such that each wave travels across one element in the time increment. Thus

$$\Delta t = \Delta z/c \qquad (11)$$

If the material of the pile changes down its length, then the element size, Δz, must be changed to satisfy eqn (11).

At each node, continuity of velocity and equilibrium of force must be satisfied. These conditions enable the magnitude of the transmitted and reflected waves to be calculated for a given magnitude of wave arriving at the node in question. This is illustrated below, considering the effects of changes in cross-section (more precisely, changes in impedance, Z) of the pile.

2.1 Changes in Impedance

Consider a downward travelling wave of velocity $v = v_d = v_i$ arriving at a point in the pile where the impedance changes (due to changes in either cross-section or material properties or both) from Z_1 (in the region of the incident wave) to Z_2 (in the region of the transmitted wave). The incident

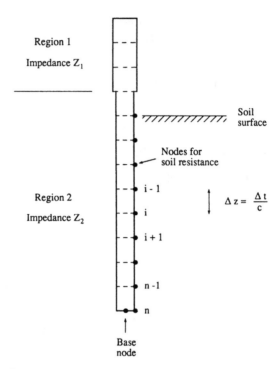

FIG. 1.   Idealisation of pile as elastic rod with soil interaction at discrete nodes.

wave will give rise to a reflected wave with a velocity $v_u = v_r$ in region 1, and a transmitted wave with a velocity $v_d = v_t$ in region 2. Assuming that there is no incident upward travelling wave from region 2 (which may be treated in an analogous way), the particle velocity and force at the boundary of the two regions just after arrival of the downward wave are given by

$$v = (v_d + v_u)_1 = v_i + v_r$$
$$= (v_d + v_u)_2 = v_t \tag{12}$$

$$F = Z_1(v_d - v_u)_1 = Z_1 v_i - Z_1 v_r$$
$$= Z_2(v_d - v_u)_2 = Z_2 v_t \tag{13}$$

From these sets of equations, it may be shown that

$$v_r = \frac{Z_1 - Z_2}{Z_1 + Z_2} v_i \tag{14}$$

and

$$v_t = \frac{2Z_1}{Z_1 + Z_2} v_i \tag{15}$$

It is useful to note also that the transmitted force, F_t, is given by

$$F_t = Z_2 v_t = \frac{2Z_2 Z_1}{Z_1 + Z_2} v_i = \frac{2Z_2}{Z_1 + Z_2} F_i \tag{16}$$

where F_i is the incident force from region 1.

Changes in pile impedance can lead to increases or decreases in the force transmitted down the pile, depending on the relative magnitudes of the impedance in regions 1 and 2. For a uniform pile with no external soil resistance, an impact of magnitude F_0 will give rise to a return wave of the same magnitude, but reversed in sign (that is, a tensile wave). If there are changes in impedance down the length of the pile, the magnitude of the return wave will be reduced. For example, for a pile consisting of two sections, of impedance Z_1 and Z_2, an impact force of F_0 will give rise to a return wave (at time $2l/c$ later, where l is the pile length) of

$$F_u = -\frac{4Z_1 Z_2}{(Z_1 + Z_2)^2} F_0 \tag{17}$$

Since the geometric mean of two numbers is always less than the arithmetic mean, the magnitude of F_u will always be less than F_0. This has important implications in the application of dynamic formulae to predicting the bearing capacity of composite piles. This point is discussed in more detail later.

2.2 Interaction With Soil

As shown schematically in Fig. 1, the soil resistance can be considered as lumped at the pile nodes. At any node, i, the soil resistance may be taken as T_i, the value of which will depend on the local soil displacement and velocity (see later). Taking T_i as positive when acting upwards on the pile (that is, with the soil resisting downward motion of the pile), the soil resistance will lead to upward and downward waves of magnitude

$$\Delta F_u = -\Delta F_d = T_i/2 \tag{18}$$

These waves will lead to modification of the waves propagating up and down the pile.

The procedure for calculating new values of wave velocities at each node is shown schematically in Fig. 2. Thus, consider the downward and upward waves at nodes $i-1$ and $i+1$, at time t. The new downward travelling wave fractionally below node i at time $t+\Delta t$ is given by

$$(v_d)_i[t+\Delta t]=(v_d)_{i-1}[t]-T_i[t+\Delta t]/(2Z) \qquad (19)$$

While the new upward travelling wave fractionally above node i is

$$(v_u)_i[t+\Delta t]=(v_u)_{i+1}[t]-T_i[t+\Delta t]/(2Z) \qquad (20)$$

The particle velocity at the node is

$$v_i=(v_d)_i+(v_u)_i+T_i/(2Z) \qquad (21)$$

where all the quantities refer to time $t+\Delta t$. This equation is still consistent with eqn (8), since the quantities v_d and v_u refer to downward and upward travelling waves which are respectively just below and just above the node. In a similar manner, the axial force in the pile at time

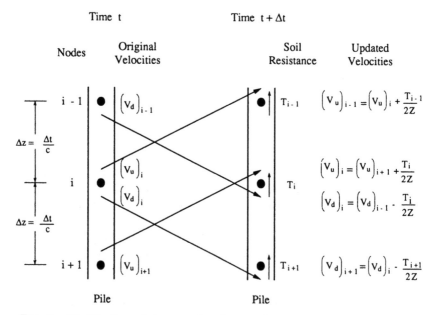

FIG. 2. Modification of downward and upward waves due to soil interaction (after Middendorp and van Weele, 1986).

$t + \Delta t$ is

$$F_i = Z[(v_d)_i - (v_u)_i] \pm T_i/2 \qquad (22)$$

where the '+' sign operates just above the node and the '−' sign operates just below the node.

As will be seen later, the value of T_i is generally a function of the local velocity as well as the displacement. Equation (21) is therefore recursive. For a linear relationship between velocity and resistance (as occurs with a simple dashpot), a set of simultaneous equations results which may be solved to give T_i explicitly. For a non-linear relationship, it is generally simplest to adopt an iterative approach.

At the base of the pile, the downward travelling wave will be reflected, with the magnitude of the reflected wave dependent on the point resistance, Q_b, offered by the soil. The axial force in the pile must balance the point resistance, which leads to an expression for the reflected (upward travelling) wave velocity of

$$(v_u)_n[t + \Delta t] = (v_d)_{n-1}[t] - Q_b[t + \Delta t]/Z \qquad (23)$$

The tip velocity is

$$v_n = 2v_u + Q_b/Z = 2v_d - Q_b/Z \qquad (24)$$

where all quantities refer to time $t + \Delta t$. As for the soil resistance along the shaft, allowance must be made for any dependence of the point resistance on the pile velocity, iterating where such dependence is non-linear.

For a force F_d arriving at the pile tip, eqn (23) implies a reflected force of

$$F_u = -Zv_u = Q_b - F_d \qquad (25)$$

The magnitude of the reflected wave thus varies from $-F_d$, where the tip resistance is zero, to F_d, where the base velocity is zero and the base resistance is twice the magnitude of the incident force (see eqn (24)).

2.3 Solution Procedure

The solution procedure involves looping through each pile node at every time step, updating the velocity components, internal pile force and soil resistance. The pile displacements are updated according to the pile velocity at the previous time step:

$$w_i[t + \Delta t] = w_i[t] + \Delta t\, v_i[t] \qquad (26)$$

At the top of the pile, the force (or velocity) may be specified explicitly, or the driving hammer, cushion and anvil may be modelled directly, by specifying 'pile segments' of the appropriate geometry and material parameters. Each blow is then initiated by specifying an initial velocity for the first pile segment (the ram of the hammer).

It is generally found that a node spacing of about 1–2 pile diameters leads to an adequate solution. Where the hammer is being modelled, it may be necessary to reduce the size of the elements. However, as Middendorp and van Weele (1986) have pointed out, the calculated force–time response near the top of the pile is largely unaffected by detailed modelling of the precise hammer geometry, provided the overall length and mass of the hammer are approximately correct. Thus, one or two elements to represent the hammer will generally prove adequate. Similarly, the cushion may be modelled by a single element, generally with a length rather smaller than that of the pile elements, owing to the lower wave speed in the cushion (see eqn (11)).

Overall, there are a number of advantages to the use of the characteristic solutions of the wave equation in pile driving analysis, rather than a finite difference or finite element approximation. The method has the simplicity of explicit time integration (avoiding the need to assemble and solve a global stiffness matrix for the pile) and yet is completely stable numerically. Wave propagation within the pile is modelled exactly, with only the soil resistance being 'lumped' at nodes. The time increment is directly proportional to the length of the pile elements, and will generally be rather larger than is necessary for accurate solution using finite element or finite difference approaches. Thus, Chow et al. (1988) describe a finite element approach for one-dimensional wave equation analysis, and present an example where a steel H-pile was discretised into 1·5 m long elements. It was found that the time step required for an accurate solution varied from 0·005 ms for a velocity imposed boundary condition at the pile head, to 0·1 ms for a force boundary condition. A characteristic solution would require a time step of about 0·3 ms (assuming a wave speed of 5000 m/s) and would provide a superior solution owing to the exact modelling of wave propagation in the pile.

3 PILE-SOIL INTERACTION

3.1 Traditional Approaches
Accurate prediction of the performance of piles during driving requires modelling of the dynamic response of the soil around (and, for open

ended pipe piles, inside) the pile, both along the shaft and at the base. Following traditional approaches for the analysis of machine foundations, the soil response can generally be represented by a combination of a spring and dashpot. However, it is also necessary to consider limiting values of soil resistance where, along the shaft, the pile will slip past the soil and, at the tip, the pile will penetrate the soil plastically.

In the original work of Smith (1960), which still forms the basis of many commercially available pile driving programs, the soil response was modelled conceptually as a spring and plastic slider, in parallel with a dashpot (see Fig. 3). For such a model, the soil resistance may be written as

$$R = R_s + R_d = Kw + Cv \tag{27}$$

where the subscripts s and d refer to static and dynamic resistances respectively, subject to $R_s \leqslant R_{max}$ (the limit of the plastic slider). The parameters K and C represent the spring stiffness and dashpot constant, respectively. Although not strictly consistent with the model shown in Fig. 3, Smith (1960) suggested for simplicity that this expression could be

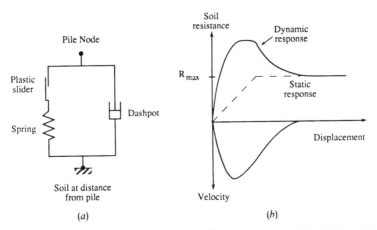

FIG. 3.   Traditional spring and dashpot soil model (after Smith, 1960). (a) Soil model; (b) typical response.

replaced by

$$R = R_s(1 + Jv) = Kw(1 + Jv) \tag{28}$$

with the dimensions of the damping coefficient J now being the inverse of velocity, rather than the dimensions of C in eqn (27) which are force per velocity. Another form of eqn (27) that is commonly utilised is

$$R = Kw + j_c Zv \tag{29}$$

where j_c is referred to as the Case damping coefficient (Goble et al., 1980). By the introduction of the pile impedance, Z, the damping coefficient j_c is rendered dimensionless. The logic behind taking the dynamic resistance of the soil as proportional to the pile impedance is discussed later.

In Smith's original work, the dashpot was introduced to allow for viscous (or material) damping, and no consideration was given to radiation (or inertial) damping due to the axisymmetric geometry. The viscous enhancement of the soil resistance was taken as a linear function of the velocity, although this assumption has since been questioned (see below). While the importance of radiation damping is now accepted, most commercially available programs for pile driving analysis still lump all damping effects into the parameters J or j_c.

In each of the expressions (27)–(29), common practice is to express the stiffness K in terms of the pile displacement to mobilise R_{max}. This displacement is referred to as the quake, Q, from which the stiffness K may be inferred as

$$K = R_{max}/Q \tag{30}$$

The value of R_{max} along the pile shaft and at the pile base must be assessed from the soil conditions, and will clearly range widely for different types of soil. In contrast, the values of the phenomenological parameters Q, J and j_c have generally been taken to lie in a relatively narrow band. With certain exceptions, values of quake along the pile shaft and at the pile base are generally taken to lie in the range 1–5 mm, with 3 mm a commonly assumed value (independent of pile diameter). Values of the Smith damping constant J are generally taken in the range 0·1–0·2 s/m along the pile shaft, and 0·1 s/m (sand)–0·5 s/m (clay) at the pile base. The value of the Case damping coefficient j_c is taken in the range 0·05–0·2 for sand and up to 0·6–1·1 for clay (Rausche et al., 1985).

Laboratory experiments reported by Gibson and Coyle (1968) (triaxial tests) and by Litkouhi and Poskitt (1980) (penetration tests in clay) show that a non-linear variation of resistance with velocity is more appropriate than the linear relationships given above. The non-linear relationship may be expressed as

$$R = R_s[1 + J'(v/v_0)^n] \tag{31}$$

where the quantity $v_0 = 1$ m/s is introduced in order to avoid confusion over the units of the modified damping coefficient J' (now dimensionless). Both sets of workers recommended a value of $n = 0.2$, regardless of soil type. In the penetration tests into clay, Litkouhi and Poskitt (1980) give typical values of J' ranging from 0.5 to 2.5 with an average of 1.5 for side resistance, and about half those values for point resistance. It should be noted that modern recommendations favour the use of lower damping values at the pile base than along the shaft, in contrast with the original recommendations of Smith (1960).

The traditional approach for modelling dynamic pile-soil interaction has proved relatively robust and simple. However, there are major limitations:

(1) No attempt is made in the model to distinguish between radiation damping due to the inertia of the soil (which will always be present), from viscous damping (the magnitude of which may be expected to vary more strongly with soil type).
(2) The parameters in the model have been arrived at empirically, and there is no logical relationship between these parameters and conventional soil properties such as modulus and damping ratio.

These limitations may be overcome simply, by recourse to elastodynamic theory. This is discussed in the following sections, treating conditions along the pile shaft and at the pile base separately.

3.2 Soil Model Along the Pile Shaft (External)
Figure 4 shows a slice of the pile and soil after deformation of magnitude, w, due to a force per unit length of pile, T. The force is in equilibrium

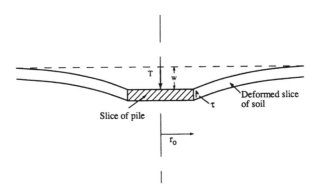

FIG. 4.   Vibration of thin slice of pile and soil (after Novak et al., 1978).

with shear stress, τ, mobilised at the pile-soil interface, such that $T = 2\pi r_0 \tau$, with r_0 being the radius of the pile. A soil model for use in pile driving analysis requires specification of:

(1) a rule for calculating, prior to slip, the shear stress at the pile wall for given local displacement and velocity (and possibly acceleration) of the pile;
(2) a value of limiting friction at which the pile will start to slip past the soil;
(3) an allowance for viscous enhancement of the limiting friction, due to the *relative* velocity between pile and soil.

The soil model is essentially similar to load transfer models used in analysis of static axial loading of piles (Coyle and Reese, 1966; Randolph, 1986), but with viscous and inertial damping effects allowed for additionally, owing to the high strain rates associated with pile driving. It is helpful to summarise the basis of static load transfer curves before embarking on a discussion of a model for dynamic load transfer.

The basis for static load transfer curves has been discussed in detail by Kraft *et al.*, (1981), who make use of the elastic solutions for axially loaded piles proposed by Randolph and Wroth (1978). In that solution, local values of shear stress, τ, at the pile-soil interface were related to the local displacement, w, by

$$\tau = \frac{G}{\zeta r_0} w \tag{32}$$

where r_0 is the radius of the pile, and G is the shear modulus of the soil at that horizon. In homogeneous soil, the parameter, ζ, is given in terms of pile length, l, radius, r_0, and Poisson's ratio for the soil, v, as

$$\zeta = \ln[2 \cdot 5(1 - v)l/r_0] \tag{33}$$

with typical values lying in the range 3–4·5. Allowance can be made for radially varying soil stiffness due to a non-linear stress–strain response. The particular case of a hyperbolic stress–strain relationship has been considered by Randolph (1977) and Kraft *et al.* (1981).

For a value of $\zeta = 4$, which is commonly adopted, the static load transfer stiffness (ratio of force per unit pile length to displacement) is

$$K_s = \frac{2\pi r_0 \tau}{w} = 2\pi G/\zeta = 1 \cdot 6G \tag{34}$$

The displacement required to mobilise the full static skin friction, τ_s, is given by

$$w_{\text{slip}} = \zeta r_0 \frac{\tau_s}{G} \tag{35}$$

This leads to displacements which are typically 1–2% of the pile radius (0·5–1% of the pile diameter) to mobilise peak skin friction.

The relationships (32)–(35) are based on the assumption that the horizontal slice of soil is fixed at some distance, r_m, representing the maximum radius of influence of the pile, with the parameter, ζ, being equal to $\ln(r_m/r_0)$ (Randolph and Wroth, 1978). For static loading, it is necessary to introduce such a limiting radius in order to arrive at a physically meaningful stiffness. Under dynamic conditions, no such assumption is necessary – in fact it would be inappropriate as it would eliminate energy being radiated into the far field.

The work of Baranov (1967) has been adapted by Novak et al. (1978) in studies of the vibration of pile foundations, to arrive at expressions for the dynamic load transfer stiffness. Under dynamic conditions, the applied shear stress and resulting displacement will no longer be in phase. The phase shift arises from both material and inertial damping. At low strains, it is customary to assume hysteretic material damping in the soil (taken here to mean damping which is independent of frequency, as opposed to viscous damping which is frequency or velocity dependent). This can be represented by a complex shear modulus of the form

$$G^* = G(1 + i \tan \delta) \tag{36}$$

where δ is referred to as the loss angle. For most soils, $\tan \delta$ will lie in the range 5–15%.

Novak et al. (1978) give the response of a pile element such as that shown in Fig. 4, subjected to harmonic motion with circular frequency, ω. The shear wave velocity of the soil is given by

$$V_s = \sqrt{G/\rho_s} \tag{37}$$

where ρ_s is the saturated density of the soil. A dimensionless frequency a_0 may be introduced, given by

$$a_0 = \omega \frac{r_0}{V_s} \tag{38}$$

The force, T, on the pile element is then related to the displacement, w, by

$$T = K^* w = 2\pi G^* a_0 * \frac{K_1(a_0^*)}{K_0(a_0^*)} \tag{39}$$

where K_0 and K_1 are modified Bessel functions of order zero and one, respectively, and the quantity a_0^* is a (complex) dimensionless frequency given by

$$a_0^* = \frac{a_0}{\sqrt{1+i\ \tan\delta}} i \qquad (40)$$

It is more convenient to express eqn (39) in the form

$$T = 2\pi r_0 \tau = G[S_1 w + a_0 S_2 v] \qquad (41)$$

The stiffness coefficients S_1 and S_2 are functions of the non-dimensionalised frequency, a_0, and also of the damping quantity $\tan\delta$, as shown in Fig. 5. (Note, for convenience, the coefficients are plotted as S_1/π and S_2/π.)

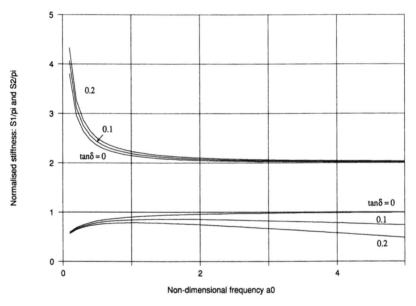

FIG. 5. Variation of stiffness coefficients S_1 and S_2 with frequency.

For undamped soil ($\delta = 0$), S_1 tends towards π at high frequencies, while S_2 tends towards 2π. At frequencies of typical interest for pile driving (a_0 in the range of 1–5), the value of S_1 may be taken in the range 2·5–3, depending on the amount of hysteretic damping considered appropriate for the soil. It will be shown later that the soil response prior to slip is dominated by inertial damping, with the spring stiffness contributing relatively little to the soil resistance. As such, a single value of $S_1 = 2.75$, as proposed by Simons and Randolph (1985), is not unreasonable. This value is nearly twice the corresponding static value (see eqn (34)).

Since S_1 and S_2 are relatively independent of frequency, the shear stress induced at the pile-soil interface depends linearly on the displacement, w, and velocity, v, but is independent of the local acceleration. As such, the soil response prior to slip may be represented by a simple spring in parallel with a dashpot, with a governing equation

$$T = K_s w + C_s v \qquad (42)$$

Note that the dashpot represents radiation (or inertial) damping and there is little effect of material damping in the soil mass. The spring and dashpot constants are

$$K_s = 2.75G \qquad (43)$$

$$C_s = \frac{2\pi r_0 G}{V_s} = 2\pi r_0 \sqrt{G\rho_s} \qquad (44)$$

It is necessary to consider carefully what happens when the pile slips past the soil. For a limiting (dynamic) skin friction, τ_d, the equation of motion of the soil slice is

$$C_s \frac{dw}{dt} + K_s w = 2\pi r_0 \tau_d \qquad (45)$$

If the skin friction is assumed to be independent of the relative velocity between pile and soil (that is, if there is no viscous damping), this equation may be integrated to give the subsequent motion (Simons and Randolph, 1985). However, it is generally more convenient to integrate the equation numerically within the normal time stepping algorithm.

It is a simple matter to allow for viscous damping, with the dynamic skin friction given by

$$\tau_d = \tau_s [1 + \alpha(\Delta v/v_0)^\beta] \qquad (46)$$

where $v_0 = 1$ m/s and Δv is the *relative* velocity between the pile and the soil. The quantity τ_s then represents a 'static' value of skin friction associated with shearing at low strain rates. It is more correct to use the relative velocity in eqn (46) rather than the absolute pile velocity, since the main viscous effects will be confined to the zone of high shear strain rate immediately adjacent to the pile. Typical values for the viscous parameters may be taken as $\beta = 0.2$ (following Gibson and Coyle (1968) and Litkouhi and Poskitt (1980)) and α in the range 0 for dry sand up to 1 or possibly higher for clay soils.

The model described above enables the effects of radiation damping to be quantified separately from those of viscous damping. Schematically, the model may be depicted as shown in Fig. 6. A plastic slider and viscous

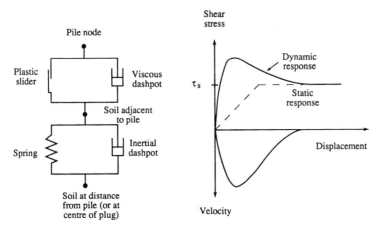

FIG. 6.　Revised soil model separating viscous and inertial damping (after Randolph and Simons, 1986).

dashpot, which together represent eqn (46), are in series with a spring and inertial dashpot which represent eqn (42). The intermediate node represents the soil immediately adjacent to the pile. It may be noted that this model effectively eliminates radiation damping from the pile as it slips past the soil (since the intermediate (soil) node moves no further).

It is necessary to keep track of both the pile displacement (and velocity) and also the displacement and velocity of the adjacent soil node. Where slip does not occur, the two velocities will be equal, and the displacements will thus differ by a constant amount. During slip, the velocities will differ and the relative displacement will change. The condition for rejoining of pile and soil is when

$$C_s v_p + K_s w_s < 2\pi r_0 \tau_s \tag{47}$$

where subscripts p and s denote pile and soil, respectively. Note that the static skin friction, τ_s, is appropriate in eqn (47), since if slip ceases, the relative velocity between pile and soil becomes zero.

3.3 Soil Model at the Pile Base

Along the pile shaft, the need to allow for slip between pile and soil, and yet still follow the motion of the adjacent soil during such slip, led to the introduction of an additional degree of freedom in the soil model (Fig. 6). The extra degree of freedom allows effects of radiation and viscous damping to be treated separately. At the base of the pile, the situation is rather different. The soil directly beneath the pile tip does not 'slip' past the pile in the way that occurs along the shaft. Also, during plastic penetration of the pile, there is no reason to suppose that energy is not still radiated into the far field. As such, the original model proposed by Smith (1960), and shown in Fig. 3 is appropriate. However, the spring and dashpot parameters need to be chosen with care.

The elastodynamic response of foundations has been studied extensively, and simplified 'one-dimensional' models have been proposed that give an adequate representation of the exact response. The model that is most widely adopted is that based on the work of Lysmer and Richart (1966), where the response of a circular footing of radius r_0 is given by

$$Q = K_b w + C_b v \tag{48}$$

where

$$K_b = \frac{4Gr_0}{1-v} \tag{49}$$

and

$$C_b = \frac{3 \cdot 4 r_0^2}{1-v} \frac{G}{V_s} = \frac{3 \cdot 4 r_0^2 \sqrt{G\rho_s}}{1-v} \tag{50}$$

The frequency independence of the spring and dashpot parameters has been demonstrated by Gazetas and Dobry (1984) through a simple cone model of the soil response.

In making use of the Lysmer and Richart analogue for pile driving, it should be borne in mind that the response at the base of a pile may be rather different from that of a shallow footing, particularly in respect of radiation damping. For a shallow footing, a high proportion of energy is radiated as Raleigh waves near the ground surface. For a pile, no Raleigh waves will be generated from the base. Further studies are needed to quantify the differences in dynamic response of shallow and deep footings.

In order to allow for plastic penetration of the pile tip, it is customary to limit the *static* component of Q to the bearing capacity of the pile tip. Thus the quantity $K_b w$ should be limited to

$$K_b w \leqslant Q_{max} = A_b q_b \qquad (51)$$

where A_b is the area of the pile base (the area of steel for an H section or open ended pipe pile) and q_b is the limiting end-bearing pressure. It should be noted that only the static component is limited to Q_{max}, since energy will continue to be radiated into the far field during plastic penetration. Thus there will be soil resistance from the dashpot representing the inertial resistance of the soil, in addition to the limiting static end-bearing capacity.

In principle, it would be possible to allow for viscous enhancement of the static end-bearing capacity, due to the high strain rates. However, in clay soils, where viscous effects may be significant, the radiation damping is also high and dominates the soil response at the pile tip. Allowing the radiation damping to continue, unaffected by local plasticity, is probably sufficient compensation for ignoring potential viscous effects. Further studies are needed in this area, perhaps by means of dynamic finite element analysis, in order to explore the relative magnitudes of viscous and radiation damping during plastic penetration of a rigid punch into soil.

There has been little work reported on the dynamic response of footings in a load range which causes plasticity in the soil. In an attempt to explore the effects of such plasticity, Randolph and Pennington (1988) have analysed the response of a spherical cavity under dynamic loading. They showed that the peak pressure occurs before significant plasticity, due to the high inertia of the soil. The maximum cavity pressure, p_{max}, expressed as a ratio of the shear modulus, G, of the soil, was largely independent of the soil shear strength, c_u. Treating cavity expansion as analogous to bearing capacity, the implication is that the peak load at the tip of a driven pile (the 'dynamic bearing capcity') is primarily governed by the inertia of the soil. The dynamic bearing capacity, Q_{max}/c_u will thus vary inversely with the rigidity index, G/c_u, of the soil.

The results presented by Randolph and Pennington (1988) also show that the peak cavity pressure in a rapidly expanded spherical cavity is a function of the rise time of the pressure pulse – with a shorter rise time giving a correspondingly higher peak pressure. The analogous result for pile driving is that, for a given hammer and pile combination (and thus rise time of the stress wave), the dynamic bearing capacity factor $N_d = Q_{max}/c_u$ will also tend to increase as the shear strength and stiffness of the soil decrease.

These observations are consistent with results from axisymmetric finite element results reported by Smith and Chow (1982), who show dynamic bearing capacity factors in medium strength soil that increase from about 10 at low values of rigidity index, to over 20 at high values. For very soft soil, dynamic bearing capacity factors as high as 40 were computed.

Overall, it appears that inertial effects dominate the immediate response at the base of the pile, and that the original model of Smith (1960) shown in Fig. 3 is adequate provided the dashpot parameter is chosen to model the inertial damping (eqn (50)).

3.4   Model for the Soil Plug Inside Pipe Piles

In spite of the widespread use of open ended steel pipe piles, particularly in the offshore industry, modelling of the soil plug response has received relatively little attention. Heerema and de Jong (1979) outlined an approach for modelling the soil plug, by treating it as a separate 'pile within a pile', with soil mass nodes connected by springs to adjacent soil nodes, and by standard 'Smith' elements to the pile nodes. This scheme has been extended by Randolph (1987) to allow for the shear stiffness of each horizontal disc of soil within the plug.

Close to the pile, the dynamic response of the soil plug is likely to be similar to that just outside the pile, since effects of curvature of the pile wall will be small. Within the central part of the soil plug, the shear wave transmitted at the pile wall must be transformed to a vertical stress wave propagating axially along the soil plug. This process may be represented by the model shown in Fig. 6, but where the far-field soil node now represents the soil at the central part of the plug. Figure 7 shows the various soil elements distributed along the pile shaft, and also at the base of the pile wall and the soil plug.

Randolph (1987) argues that the spring stiffness for the soil plug element should be approximately double that of the element outside the pile. The internal shear force per until length of pile may then be expressed as

$$T_i = 5 \cdot 5 G(w_{sp1} - w_{sp2}) + 2\pi r_i \frac{G}{V_s}(v_{sp1} - v_{sp2}) \tag{52}$$

where the subscripts sp1 and sp2 represent the soil nodes adjacent to the

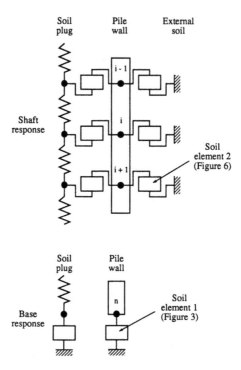

FIG. 7. Model of internal and external soil interaction with pile (after Randolph, 1987).

pile and at the centre of the soil plug respectively (see Fig. 6). The maximum internal shear force will be limited by the available skin friction on the inside of the pile.

The manner of modelling the soil plug described here is capable of capturing partial plugging of pipe piles, where the soil plug moves up within the pipe pile at a slower rate than the pile advances into the ground (Randolph, 1987).

3.5 Implications of Inertial Damping

The soil models depicted in Figs 3 and 5, together with spring and dashpot constants based on elastodynamic theory, have far reaching implications in pile driving analysis. Parameters for the models are given in terms of fundamental soil properties such as shear modulus and density, enabling the models to make due allowance for differences between dynamic and static loading. These differences are discussed below, in terms of both the ultimate soil resistance, and the pile displacement to mobilise that resistance.

Dynamic and Static Capacity

The limiting skin friction under dynamic conditions is given by eqn (46), in terms of the static skin friction, τ_s, and a viscous enhancement which depends on the relative velocity between pile and soil. The static skin friction may be assessed along conventional lines. In cohesive soil, prior to dissipation of any excess pore pressures generated during the driving process, typical values of static skin friction range from 20 to 60% of the undrained shear strength of the soil, with the lower value applicable to softer, normally consolidated soil. In non-cohesive soil, the skin friction is commonly estimated as some multiple of the in-situ vertical effective stress, with typical values ranging from 0·3 (loose soil) to 1 (dense soil) times the vertical effective stress. Where cone penetration results are available, reasonable estimates of the skin friction during continuous driving may be obtained directly from friction sleeve measurements.

It is generally assumed that the dynamic skin friction in relatively coarse grained material is similar to the static value, with the coefficient α in eqn (46) being taken as close to zero. In cohesive soil, the dynamic skin friction can be 1–3 times the static value. Adopting a β value of 0·2 in eqn (46), the value of α should be chosen accordingly, with typical values of about unity.

At the pile tip, it has been argued that enhancement of the static bearing capacity is primarily due to the inertia of the soil. For the proposed model, it is possible to estimate the dynamic bearing capacity directly for given values of the soil constants. The dashpot contribution can be written as

$$Q_d = A_b \frac{3·4\sqrt{G\rho_s}}{\pi(1-v)} v \tag{53}$$

where A_b is the tip area of the pile. This may be rewritten as

$$Q_d = 2·2 A_b I_r^{0·5} \sqrt{\rho_s c_u} \, v \tag{54}$$

where I_r is the rigidity index, G/c_u. Taking the static end-bearing pressure as $9c_u$, the ratio of dynamic to static bearing capacity may be written

$$\frac{Q_d}{Q_s} = 0·24 I_r^{0·5} \sqrt{\frac{\rho_s}{c_u}} v = 0·024 I_r^{0·5} \sqrt{\frac{\rho_s}{\rho_w}} \sqrt{\frac{p_a}{c_u}} \frac{v}{v_0} \tag{55}$$

where ρ_w is the density of water, p_a is atmospheric pressure and v_0 is a reference velocity of 1 m/s. Thus for a rigidity index of 200, $\rho_s = 2000\,kg/m^3$, $c_u = 100\,kPa$ and $v = 1\,m/s$, this ratio is equal to 0·48. The ratio increases in direct proportion to the velocity, proportionally to the square root of the rigidity index and soil density, and inversely proportionally to the square root of the undrained shear strength of the soil.

For non-cohesive soil, the static bearing capacity is generally expressed in terms of a bearing capacity factor, N_q, times the in-situ vertical effective stress. An analogous expression to eqn (55) can then be written:

$$\frac{Q_d}{Q_s} = \frac{0·22}{N_q} \sqrt{\frac{\rho_s}{\rho_w}} \sqrt{\frac{G}{\sigma_v'}} \sqrt{\frac{p_a}{\sigma_v'}} \frac{v}{v_0} \qquad (56)$$

Typical values for the ratio G/σ_v' (for the small strains associated with stress-wave propagation) are in the range of 300–1000. For a vertical effective stress of 200 kPa and $N_q = 40$, and other parameters as given above, the ratio of dynamic to static resistance at the pile tip would then lie in the range 0·08–0·14. Thus inertial effects may be expected to be significantly smaller for non-cohesive soil than for cohesive soil. This conforms with existing practice in the choice of Case damping value j_c, where the value for sandy soils is an order of magnitude smaller than for clay soils.

Dynamic and Static Stiffness

Elastodynamic theory provides guidance on the inertial contribution to soil resistance prior to slip, and in particular the pile displacement needed to mobilise the maximum soil resistance locally under dynamic conditions. This displacement is referred to as the 'quake'. It is common practice to deduce the static load-displacement response of a pile directly from the back-analysis of stress-wave measurements. This can only be achieved consistently if allowance is made for differences in the 'quake' under dynamic and static conditions.

For the pile shaft, the value of quake under static conditions is given by eqn (33) which, as has already been remarked, implies quake values of 0·5–1% of the pile diameter. For typical pile sizes in use onshore, with diameters commonly in the range 300–600 mm, the static quake would lie in the range 2–6 mm. For larger diameter piles such as are used offshore, the static quake would be correspondingly larger.

Under dynamic conditions, the relative contribution to the soil resistance from the (inertial) dashpot and the spring may be assessed from eqns (42)–(44). The ratio of dynamic resistance to static resistance may be written as

$$n = \frac{C_s v}{K_s w} = 2 \cdot 3 \frac{r_0 v}{V_s w} \tag{57}$$

The relationship between displacement and v may be written in terms of the rise time of the stress-wave. Thus, assuming a sinusoidal increase in velocity at a particular position down the pile, with a maximum velocity of v_m and rise time t_m, the ratio v/w may be represented approximately by the quantity $\pi/(2t_m)$. The ratio of dashpot to spring resistance is then

$$n \approx 3 \cdot 6 \frac{r_0}{V_s t_m} \tag{58}$$

Typical values for V_s may be taken in the range 50–200 m/s, while the rise time for a pile of diameter 0·3 m would typically be of the order of 1 ms (depending on the hammer and cushion properties). The ratio n would then lie in the range 3–10.

Although the above calculation involves a number of simplifying assumptions, it is clear that the inertial resistance of the soil dominates the initial response. This point has been made by Simons (1985), who comments that the spring component of resistance contributes typically only 20–40% of the total resistance during the passage of a stress-wave. The dynamic quake will be correspondingly lower than the static value.

At the pile tip, the displacement to mobilise the plastic slider may be calculated directly from eqn (49). Thus, for a limiting end-bearing pressure q_b, the displacement to cause plastic penetration is

$$w_p \approx \frac{q_b}{8Gr_0} = \frac{q_b}{4Gd} \tag{59}$$

where d is the pile diameter. From cone penetration testing, correlatons of shear modulus and cone resistance generally lie in the range 5–10, implying a displacement range of 2·5–5% of the pile diameter. This range is rather higher than the static (or dynamic) quake along the pile shaft. However, two further factors must be considered. Firstly, the maximum tip force will generally arise from the inertial resistance of the soil, at a smaller displacement than given above. Secondly, the actual tip displacement to cause plastic penetration will be reduced by residual forces locked in at the pile tip, which effectively maintain the tip force at or close to the full end-bearing resistance. This point is considered further later.

4 PILE DRIVABILITY

An important aspect of the design of a driven pile foundation is the assessment of what size and type of hammer is needed to drive the piles to the required penetration. This aspect of the design is referred to as a 'drivability' study. Such studies can take various forms, but the main objectives are to establish that the piles may be driven with a particular hammer without being subjected to excessive driving stresses, and to provide guidance on the penetration rate for a given (assumed) soil resistance. The latter result is often used for quality control during installation of the piles.

In order to conduct the drivability study, it is necessary to estimate the distribution down the pile of soil resistance and other parameters. It is also necessary to model the particular hammer under consideration.

4.1 Hammer Modelling

Various approaches may be used to model the impact between hammer and pile. These include:

(1) analytical solutions for simple configurations;
(2) numerical modelling of the ram, capblock, anvil and cushion system (or equivalent for a diesel hammer);
(3) the use of a 'signature' force-time response (generally supplied by the hammer manufacturer) for a given hammer and pile system; the force-time response represents the downward travelling wave only, and the force actually observed at the pile head would be modified by interaction with upward travelling waves due to soil resistance and reflection from the pile tip.

Analytical models of impact are necessarily confined to relatively simple hammer systems. However, they may be useful in conducting parametric studies, without the need for a full wave equation analysis. Figure 8 shows the main components of a typical hammer and pile system, with the ram and anvil treated as lumped masses of m_r and m_a, respectively. The cushion is represented by a spring of stiffness, k_c, while the pile is represented by a dashpot of coefficient, Z (the pile impedance).

For the limiting case of a very stiff (rigid) cushion and light anvil, the force at the pile head is given by the classical solution (Johnson, 1982)

$$F = Zv_i \exp(-Zt/m_r) \tag{60}$$

where v_i is the impact velocity. For a finite spring stiffness, the expression

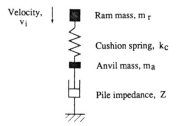

FIG. 8.   Idealised model of hammer components.

becomes

$$F = Zv_i \exp(-kt/2Z)\frac{\sinh(\mu t)}{Z\mu/k} \tag{61}$$

where $\mu = [(k^2/4Z^2)-(k/m_r)]^{0.5}$. In cases where the anvil mass is a significant proportion of the ram mass, an analytical solution may be achieved conveniently through the use of Laplace transforms. However, the usefulness of such solutions is limited by the boundary conditions of perfect contact between each of the components, whereas in reality a gap may occur between, say, the anvil and the pile, followed by re-striking.

Modern programs for pile driving analysis, which make use of the characteristic solutions of the wave equation, enable accurate simulation of the impact process to be achieved with the distributed mass of the ram correctly accounted for. As discussed by Middendorp and van Weele (1986), a relatively crude model of the hammer may suffice to give adequate results. Figure 9 shows an example of four increasingly sophisticated models of an MRBS 8000 steam hammer striking a pile of 1·83 m diameter and 48 mm wall thickness. The pile impedance is 10 880 kNs/m, and the impact velocity of the ram has been taken as 5·1 m/s. The mass of the hammer ram has been taken as 80 tonnes, the mass of the anvil as 38 tonnes, while the capblock has been modelled as 0·3 m high, with a Young's modulus of $E_c = 1$ GPa (resulting stiffness, $k_c = 10·5$ GN/m).

The first curve represents the ram hitting the pile directly, treating the ram as a lumped mass (eqn (60)). The rise to a peak force of 55·5 MN is immediate, followed by an exponential decay. The second and third curves then add the finite capblock stiffness and the anvil mass, respectively. The effect of the capblock is to give a finite rise time to the stress wave, at the same time reducing the maximum force by 20%. The addition of the anvil delays the peak force still further, but increases the

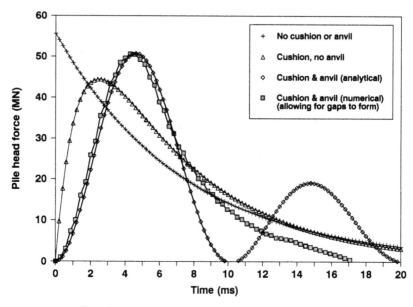

FIG. 9.   Numerical simulation of hammer response.

magnitude back to 50 MN, and introduces an oscillation with a period
of just over 10 ms. The fourth curve, obtained numerically, results from
allowing gaps to form between each of the hammer elements and the pile
(avoiding any tensile forces in the system). The curve follows the form of
the (analytical) third curve until past the initial peak. However, the
oscillation evident in that curve is removed by allowing gaps to form
between the elements.

   One of the main uses of hammer modelling is to explore the effects of
different parameters on the resulting stress wave. As an example, Fig. 10
shows the effect of variations in cushion modulus for the steam hammer
and pile considered above. The results were obtained numerically, using
the characteristic solution approach, with the ram and anvil being
modelled by eight and four elements, respectively, while the cushion was
represented by one to four elements, depending on its stiffness (and hence
the wave speed). Data for the hammer and typical modulus values for
the cushion have been taken from van Luipen and Jonker (1979), who
quote an initial modulus value of $E_c = 20$ GPa, reducing to between 1 and
5 GPa after a few hundred blows. The results in Fig. 10 show that the
rise time increases inversely with the square root of the cushion stiffness,

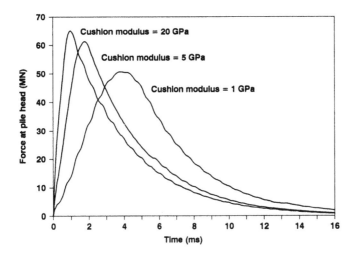

FIG. 10.   Parametric study of the effects of capblock stiffness.

from 1 ms for $E_c = 20$ GPa to just over 4 ms for $E_c = 1$ MPa.

Precise modelling of hammer impact can be very involved, particularly for diesel hammers. However, the main features may be simulated relatively simply by three or four components, with results that match field measurements well. As in many aspects of pile driving analysis, the ultimate success of the model depends heavily on the magnitude of key parameters, particularly in respect of the cushion or capblock. The example above demonstrates that the resulting form of the stress wave is very dependent on the cushion stiffness, and thus will vary in a real situation depending on the degree of wear of the cushion. It is therefore questionable whether it is appropriate to adopt too sophisticated a model of the hammer when conducting drivability studies.

4.2   Parametric Studies
The main outcome of a drivability study is a series of curves that give the penetration rate (or blow count) as a function of the assumed 'static' resistance of the pile. These curves may be used to choose an appropriate hammer, assess the time (and cost) of installation of each pile, and to provide a quality control on pile installation. The last aspect will generally be in the form of a required 'set' (or penetration per blow) specified to ensure that each pile has sufficient working capacity.

There are also a number of additional outcomes of a drivability study, which include assessment of maximum stress values (tensile and compressive) at any point in the pile, maximum acceleration levels (important if the pile is to carry any instrumentation), range of hammer stroke permitted (or required), and so forth.

Probably the most widely used program in commercial use is the WEAP program developed originally by Goble and Rausche (1976) for the Federal Highway Administration in North America. The program has been updated recently (Gobe and Rausche, 1986; Rausche *et al.*, 1988). A typical output from the program is shown in Fig. 11, with peak

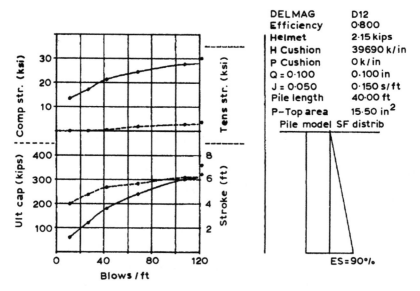

FIG. 11.   Typical output from WEAP (after Rausche *et al.*, 1988).

compressive and tensile stresses, the pile capacity, and hammer stroke, all plotted against penetration rate.

One of the motivations behind development of the WEAP program was to improve modelling of diesel hammers. The program uses a sophisticated approach for such hammers, which includes modelling of the combustion process. The program has an extensive library of hammer data, simplifying data input considerably. Whereas conventional drivability analyses assume very simple distributions of soil resistance with depth (generally either uniform or triangular shaft resistance with depth), the most recent version of WEAP allows irregular variation of shaft resistance and other parameters through the soil strata (Rausche *et al.* 1988).

The accuracy of a drivability study should be assessed by appropriate field measurements of hammer performance, pile penetration rates and, if possible, stress-wave data. Stress-wave data provide a check on the actual driving energy transmitted to the pile, and may also be analysed to provide a revised distribution of soil resistance, which may lead to changes in the foundation design. This aspect of stress wave analysis is considered further below.

5 INTERPRETATION OF FIELD MEASUREMENTS

Traditional pile driving formulae may be used to assess the pile capacity by means of balancing the energy transmitted to the pile from the hammer, and the elastic and plastic work performed on the pile. The uncertainties in applying such formulae centre around the overall efficiency of the driving system (that is, how much useful energy is transmitted), the elastic compression of the pile and other components of the system, and the effects of dynamic enhancement of the static pile capacity.

The use of field instrumentation to monitor dynamic force and velocity near the head of a driven pile can eliminate many of the uncertainties present in simple driving formulae. Analysis of stress-wave data may be considered in two steps:

(1) immediate analysis (in real time) in the field, which leads to blow by blow records of key data such as transmitted energy, maximum compressive and tensile stress levels, dynamic and (estimated) static pile capacity and so forth;

(2) subsequent analysis, either in the field or in an office environment, where the detailed stress-wave data are matched through numerical models, to arrive at estimates of the distribution and magnitude of soil resistance down the pile.

5.1 Real Time Analysis
The first stage of interpretation of stress-wave data is performed by what is commonly referred to as a 'Pile Driving Analyser'. Strain and acceleration data are processed, generally through electronic hardware, to obtain force and velocity data. From these data, various parameters may be derived. Thus, integration with time of the product of force and velocity up to the time at which the product becomes negative leads to a

figure for the maximum energy transmitted to the pile. This allows the overall operating efficiency of the hammer to be assessed in terms of its rated energy. If additional information is available on the ram velocity at impact, then the energy losses may be subdivided into mechanical losses in the hammer, and losses in the impact process due to inelasticity and bounce of the components.

In traditional pile driving formulae, one of the largest sources of error in estimating the overall pile capacity is uncertainty in the energy transmitted to the pile. Measurement of the actually transmitted energy allows use of simple pile driving formulae with increased confidence. Such formulae provide a means whereby information obtained on instrumented piles may be extrapolated in order to assess the quality of uninstrumented piles driven on the same site. Of course, for piles where stress-wave data are obtained, more sophisticated techniques may be used to assess the pile capacity.

The relationships developed in Section 2 may be used to obtain an estimate of the dynamic and static soil resistance from the stress-wave data. Equation (18) implies that, as the stress-wave travels down the pile, the magnitude of the force will decrease by half the total (dynamic plus static) shaft resistance, T_s. Thus, at the bottom of the pile, the downward travelling force is

$$F_d = F_0 - T_s/2 \qquad (62)$$

where F_0 is the original value at the pile head. Similarly, eqn (22) may be used to obtain the upward travelling force after reflection at the pile tip as

$$F_u = -Zv_u = -Z(v_d - Q_b/Z) = Q_b - F_d = Q_b + T_s/2 - F_0 \qquad (63)$$

On the way back up the pile, *provided the particle velocity at each position is still downwards, implying upward forces from the soil on the pile*, the upward travelling wave will be augmented by half the shaft resistance (again, see eqn (18)), to give a final return wave of

$$F_r = Q_b + T_s - F_0 \qquad (64)$$

where the subscript r refers to the return (upward travelling) wave at a time $2l/c$ later than the time at which the value of F_0 was obtained (l being the length of pile below the instrumentation point). The total dynamic pile capacity is then

$$R = Q_b + T_s = F_0 + F_r \qquad (65)$$

Equations (8) and (10) may be used to derive the upward and downward components of force from the net force and particle velocity at the instrumentation level, so that eqn (65) may be re-written

$$R = 0·5(F_0 + Zv_0) + 0·5(F_r - Zv_r) \qquad (66)$$

where subscripts 0 and r refer to times t_0 (generally close to the peak transmited force) and $t_r = t_0 + 2l/c$. This equation is the basis for estimating the total dynamic pile capacity directly from the stress-wave measurements. A search may be made for the value of t_0 which gives the largest value of capacity.

Since the dynamic capacity will be greater than the current static capacity, a simple method is needed to estimate the static capacity in the field, without the need for a full numerical analysis of the pile. In the Case approach, which has gained widespread acceptance, this estimate is made on the basis that all the dynamic enhancement of the capacity occurs at the pile tip, with a dynamic component of resistance that is proportional to the pile tip velocity, v_b. Thus the dynamic tip resistance is written as

$$(Q_b)_d = j_c Z v_b \qquad (67)$$

where j_c is the Case damping coefficient. These simplifying assumptions lead to an expression for the static pile capacity, R_s

$$R_s = 0·5(1 - j_c)(F_0 + Zv_0) + 0·5(1 + j_c)(F_r - Zv_r) \qquad (68)$$

The assumptions regarding the dynamic soil resistance are clearly an oversimplification, and the deduced static pile capacity can be very sensitive to the value adopted for the damping parameter, j_c (see case study later). However, the above expression can provide useful guidance on the static pile capacity where it is possible to calibrate the parameter j_c for a particular site. Where no static load tests are carried out, guidelines for j_c as given in Table 1 may be adopted (Rausche et al., 1985).

Equation (67) implies that, for a given tip velocity, v_b, and damping parameter, j_c, the dynamic tip resistance is proportional to the pile impedance. This does not seem particularly logical, and certainly conflicts with the form of eqn (48) which implies dynamic resistance that is independent of the pile impedance. In practice, the value of j_c adopted for any given set of stress-wave data tends to be determined by the operator of the pile driving analyser on an ad hoc basis, and the correlations suggested in Table 1 are of limited value.

5.2 Matching of Stress-Wave Data

Predictions of pile capacity directly from the stress-wave measurements using expressions such as eqns (66) and (68) are rarely used in isolation without calibration either through static load tests or by means of a full

M.F. RANDOLPH

TABLE 1
SUGGESTED VALUES FOR CASE DAMPING COEFFICIENT

Soil type in bearing strata	Suggested range of j_c	Correlation value of j_c
Sand	0·05–0·20	0·05
Silty sand/sandy silt	0·15–0·30	0·15
Silt	0·20–0·45	0·30
Silty clay/clayey silt	0·40–0·70	0·55
Clay	0·60–1·10	1·10

dynamic analysis of the pile and matching of the stress-wave data. This latter process is considerably more reliable as an estimate of pile capacity than the direct formulae given above.

The process of matching the measured stress-wave data is an iterative one, where the soil parameters for each element down the pile are varied until an acceptable fit is obtained between measurements and computed results. In order to avoid uncertainties in modelling the hammer, either the measured force signal or (more generally) the measured velocity signal is used as an upper boundary condition in the computer model. The fit is then obtained in terms of the other variable (generally measured and computed force). An example of the effect of varying different parameters is given by Goble et al. (1980), and reproduced as Fig. 12.

It is possible to automate the matching process, with the computer optimising the soil parameters in order to minimise some measure of the difference between measured and computed response (Dolwin and Poskitt, 1982). However, it has been found that computation time can become excessive, particularly for long piles, unless the search zone for each parameter is restricted by operator intervention. It is rather more straightforward to carry out the matching process manually. Experience soon enables assessment of where values of soil resistance, damping or stiffness need to be adjusted in order to achieve an improved fit. A satisfactory fit may generally be achieved after 5–10 iterations of adjusting the parameters and re-computing the response.

Limitations in the soil models used for pile driving analysis entail that the computer simulation will not match the real situation exactly. A consequence is that the final distribution of soil parameters should not be considered as unique, but rather as a best fit obtained by one particular operator. Generally, the total static resistance computed will show little variation provided a reasonable fit is obtained. However, the distribution of resistance down the pile, and the proportion of the

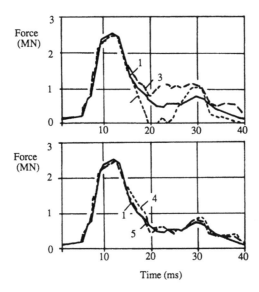

FIG. 12. Example of stress-wave matching using CAPWAP (after Goble *et al.*, 1980). 1, Measured force curve; 2, low damping; 3, high static resistance; 4, high static friction low end bearing; 5, final solution.

resistance at the pile base, may show considerable variation (Middendorp and van Weele, 1986).

An interesting investigation of operator dependence in the analysis of stress-wave measurements has been reported by Fellenius (1988). Eighteen operators were given four sets of stress-wave data to analyse, covering a range of pile types and soil conditions. One of the sets of data was from a re-drive of a pile that was subjected to a static load test the following day. All the operators were using the same computer program, CAP-WAP, which is one of the most widely utilised programs for such analyses, originating from the work of Goble and Rausche (1979). Some of the results from that study are reproduced here.

Figure 13 shows the four blows that were analysed, with the two-letter code for each pile. The deduced pile capacities and load distributions are shown in Figs 14 and 15, with the static load test result for pile AM also indicated. There is a good measure of agreement among the participants in the study, with the coefficient of variation being under 5–7% for piles JI, JA and AM (excluding the one very high prediction, which raises the coefficient to 13%), and 14% for pile LW. The static load test result for pile AM is well predicted by the mean of the dynamic analyses.

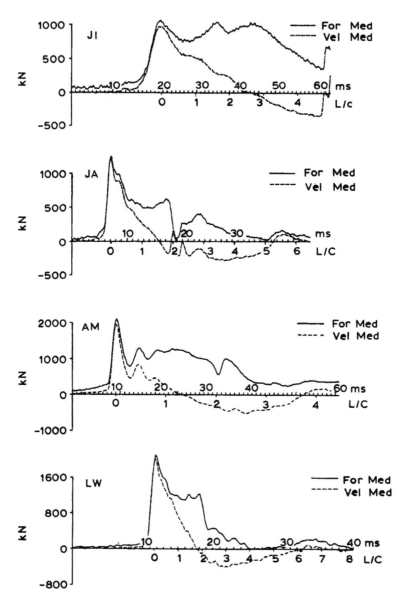

FIG. 13.   Stress-wave data used for investigaton of operator dependence (after
Fellenius, 1988).

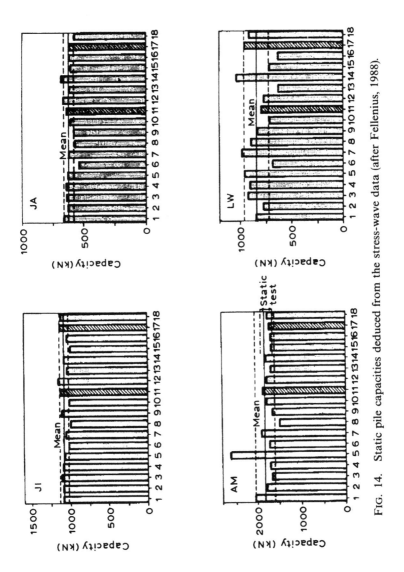

FIG. 14.   Static pile capacities deduced from the stress-wave data (after Fellenius, 1988).

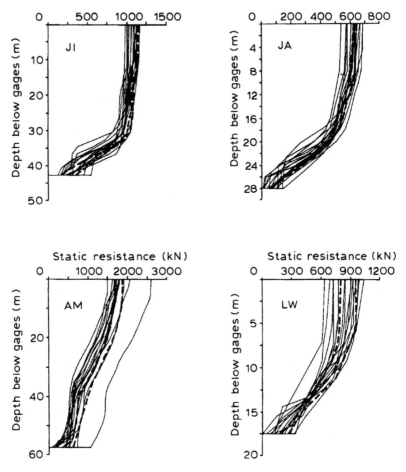

FIG. 15.   Load distributions deduced from the stress-wave data (after Fellenius, 1988).

Fellenius' study also considered predictions of penetration rate from the stress-wave data. For piles JI, JA and LW, the mean of the predictions was generally on the high side (by 0–22%), with coefficients of variation between 13 and 22%. However, there was a surprising – and somewhat alarming – variation among the predictions for pile AM, with a range of 278—5577 blows/m compared with the observed value of 330 blows/m. There was no obvious correlation between predicted blow rate and static capacity, with three predicted blow rates that were in excess of three times the observed one corresponding to good estimates of the static capacity, while the very high predicted capacity (see Fig. 14) corresponded to a reasonable blow rate.

Variations in predicted static capacity often result from different assessment of the amount of damping present. This was certainly true in the above study, with Case damping parameters assumed along the shaft of pile AM ranging from 0·10 to 1·27, and toe damping parameters for pile LM ranging from 0·06 to 0·80 (these were the two largest ranges). Separation of damping into viscous and inertial components, and reducing the reliance on empirical parameters such as J and j_c should allow such scatter to be reduced, improving the accuracy of capacity predictions.

In addition to the CAPWAP analyses conducted on the stress-waves, it is also interesting to consider the use of the Case formula for estimating the static capacity (eqn (68)). Taking t_0 as the time at peak force and velocity (a slight over-simplication), the values of Case damping parameter, j_c, needed to achieve the average static capacities predicted using CAPWAP, are given in Table 2. For comparison, average values adopted in the CAPWAP analyses are also given. Although the values for piles JA and AM look reasonable by comparison with those from the CAPWAP analyses, the values for piles JI and LW seem rather high, particularly in view of the soil conditions and the guidelines given in Table 1.

TABLE 2

VALUES OF CASE DAMPING COEFFICIENT

Pile code	Predominant soil type	Values of Case damping parameter, j_c		
		Eqn (68)	CAPWAP values Shaft	Base
JI	Silty clay and clayey silt	0·7	0·33	0·28
JA	Sand, some silt layers	0·4	0·68	0·18
AM	Silty clay (shaft) silt/sand (base)	0·4	0·62	0·34
LW	Weathered sandstone	0·9	0·97	0·37

5.3 Sources of Error in Static Capacity Deduced from Stress–Wave Data

There are a number of factors which can contribute to errors and uncertainty in the static pile capacity deduced from stress-wave measurements. One of the major pitfalls to be avoided is trying to estimate the capacity from hammer blows of insufficient energy to fail the pile. A classic case of this type has been described by Nguyen *et al.* (1988) (see also Nguyen, 1987). A closed ended steel pipe pile of 812 mm diameter was driven 32 m to bear in dense sand and gravel. The capacity calculated from conventional soil mechanics was 7150 kN, made up of 2320 kN shaft resistance and 4830 kN base resistance.

The results of stress-wave measurements and computer simulation are shown in Fig. 16, for a re-drive blow some 50 days after initial driving.

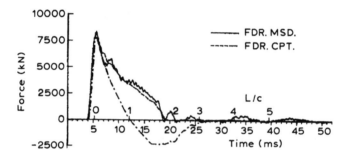

FIG. 16. Computer simulation of stress-wave data (after Nguyen *et al.*, 1988). ———, Measured force; – – –, computed force; – –, computed velocity.

The blow was from a 6 tonne hammer falling through 3 m, and gave rise to permanent displacement of the pile of 3·5 mm. The static capacity deduced from the dynamic analysis was 3520 kN (3040 kN shaft, and 480 kN base). This compares with a capacity of 7300 kN obtained from a static load test. The deduced base resistance is clearly too low in view of the soil conditions. This example emphasises the need to use engineering judgement when interpreting the results of dynamic analyses.

It is well known that the capacity of a driven pile increases with time following installation. This phenomenon, referred to as 'set-up', is generally attributed to dissipation of excess pore pressures generated during installation. Case studies have shown that, particularly in soft cohesive soil, the pile capacity may increase by a factor of 4–5 following installation, over a time period of several weeks or months, depending on the consolidation characteristics of the soil and the pile diameter (Randolph *et al.*, 1979). Where the pile capacity is to be estimated from dynamic measurements, it is necessary to allow for such set-up. This may be achieved by 're-striking' the pile after an appropriate time delay. It is important to ensure that the full pile capacity is mobilised in such a re-strike within the first few blows, so as not to reduce the long term performance of the pile.

Use of the Case formula to estimate the static capacity of a pile is not recommended without a full numerical matching of the stress-wave data as corroboration. Even where this has been done, and a reasonable estimate of damping parameter, j_c, is available, care should be taken to ensure that the conditions assumed in deriving the relationship are met in practice. Two particular conditions are (a) that there are no major changes in cross-section (or impedance) of the pile along its length, and (b) that the pile velocity remains positive (downwards) over the major part of the return time of the impact wave.

Effects of changes in pile cross-section have been discussed in Section 2.1. The relationships may be used to show the effect of a change in impedance (from Z_1 to Z_2) occurring at some stage along the pile. Assume that the change in section occurs above the zone where most of the soil resistance acts. Equations (16) and (17) may be used to show that, for an impact force F_0 and soil resistance R, the return wave in the upper section is

$$F_u = \frac{2Z_1}{Z_1+Z_2} R - \frac{4Z_1 Z_2}{(Z_1+Z_2)^2} F_0 \qquad (69)$$

whence

$$R = 0 \cdot 5 \frac{2Z_2}{Z_1+Z_2} (F_0 + Zv_0) + 0 \cdot 5 \frac{Z_1+Z_2}{2Z_1} (F_r - Zv_r) \qquad (70)$$

Comparing this expression with eqn (66), it is clear that changes in cross-section may lead to significant errors in the estimated soil resistance.

The derivation of the Case formulae (eqns (66) and (68)) rests on the assumption that the soil resistance continues to act upwards (opposing downward movement of the pile) during the whole period in which the impact wave is returning up the pile. In many cases, this assumption is not fulfilled. For example, Fig. 13 shows that the velocity traces for two of the piles become negative well before a time of $2l/c$. Such rebound at the pile head does not necessarily entail rebound further down the pile at the time when the impact wave was returning. This depends on the location of the instrumentation relative to the main soil resistance. However, any program that calculates the pile resistance using the Case relationship should include a check that the particle velocity remains positive at each point down the pile during passage of the impact wave.

The two main causes of differences between static and dynamic performance of the pile are (a) viscous damping and (b) inertial damping. The dynamic shaft capacity of a pile may exceed the static capacity by a factor of 2 or more, due to viscous effects. Similarly, the dynamic base capacity may be 2–4 times the static capacity due to inertial effects alone. In both cases, differences between dynamic and static capacity are greater in soft cohesive soil, than in stiffer or coarser material.

Allowance for damping is made by appropriate choice of damping parameters in the dynamic analysis. However, the deduced static capacity is relatively sensitive to the choice of damping constant, and it is clear that further research is needed in order to provide better guidance on damping parameters for different soil types. At this stage, it is strongly recommended that at least one static load test be performed on a given site in order to calibrate the dynamic analyses.

Base Capacity of H-Piles and Open Ended Pipe Piles
One area which has received insufficient attention is the different response of H-section piles and open ended pipe piles under dynamic and static conditions. It may be shown that both types of pile tend to drive in an 'unplugged' condition, with soil moving up the inside of the pipe, or filling the space between the flanges of the H-pile. However, during a static load test the reverse is true. The frictional resistance of the soil plug is such that both types of pile will tend to fail as a solid body. Thus, during a dynamic test, these piles will show relatively high shaft friction but low end-bearing. By contrast, during a static test, the shaft capacity will be just that on the outside of the pile, while the end-bearing resistance will act over the gross area of the pile. It is essential that estimates of the static capacity of such piles take account of differences in the failure modes during dynamic and static penetration.

Residual Stresses
One final consideration, which has implications in pile drivability studies as well as in the analysis of stress-wave data, is the influence of residual stresses. It is a straightforward matter to allow for residual stresses in pile driving analysis, allowing the dynamic waves to dissipate at the end of each hammer blow by means of a static analysis (e.g. Goble and Hery, 1984). Modern models of the soil response, based on elastodynamic theory of the continuum, are particularly appropriate for assessing residual stress conditions.

Residual stresses acting down the length of the pile can have a significant effect on the calculated pile response under dynamic conditions (Simons, 1985). This may be illustrated by Fig. 17, which shows an idealised response at the base of a pile. For a pile 'wished into place', the

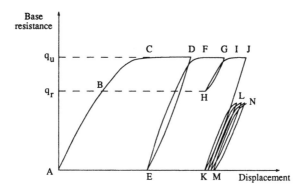

FIG. 17.   Schematic diagram of base response during pile driving.

base response might be ABCD, with a limiting end-bearing pressure of q_b reached at a displacement of 5–10% of the pile diameter. During driving, assuming no residual stress builds up at the base of the pile, each new blow will follow a path such as EFG. The elastic displacement range (from E to F) may be estimated from the elastic stiffness of a rigid punch.

$$\frac{\Delta q}{\Delta w} = \frac{4}{\pi} \frac{G}{1-\nu} \frac{1}{r_0} \tag{71}$$

Thus, for a ratio G/q_b of about 10, the displacement to re-mobilise full end-bearing pressure will be about 4% of the pile radius (2% of the diameter). For a pile of 1 m in diameter, this would be 20 mm. If the pile tip penetrates plastically by a further 10 mm, only half the energy arriving at the base is useful in penetrating the pile, the other half is lost in elastic energy.

In practice, residual stresses will generally build up along the pile, with some locked in end-bearing stress, q_r (see Fig. 17), balanced by negative shear stresses acting along the pile shaft. The energy required to re-mobilise the end-bearing capacity will be reduced, although slightly greater energy will be required to re-mobilise the full (positive) skin friction along the shaft.

Two observations may be made. Firstly, it is clear that residual stresses may change the distribution of where energy is absorbed by the soil during driving. This will affect the driving performance and, for an instrumented pile, will alter the form of the stress-waves reflected from interaction with the soil. Any program that is used to estimate the pile capacity, by matching computed and measured stress-waves, must allow for the effects of residual stresses on the computed response.

The second observation concerns the driving performance of a pile. Since the local displacement to mobilise full skin friction along the pile shaft is significantly less than that to mobilise the end-bearing resistance at the pile base, the existence of residual load at the pile base reduces the overall amount of elastic energy needed before the pile penetrates plastically. Where piles are driven through soft soil onto a hard stratum, the absence of significant residual stresses can lead to large 'quakes' at the pile tip, and difficulty in driving (essentially, the pile 'bounces' following paths such as KLMN in Fig. 17). Quakes that are larger than normal are to be expected in any situation where the pile capacity is concentrated at the tip of the pile.

6 CONCLUSIONS

This chapter has outlined the theoretical basis for analysis of the dynamic response of piles during driving. The pile has been treated as an elastic rod, with one-dimensional propagation of stress-waves up and down the rod. The characteristic solutions of the wave equation provide an efficient and numerically stable basis for analysis of the dynamic interaction between pile and soil. Elastodynamic theory has been used to develop simple spring and dashpot models for the soil response, which take due account of inertial damping of the stress-waves.

Developments in instrumentation techniques over the last decade or so have enabled high speed acquisition of force and acceleration data at the pile head to become routine. Such 'stress-wave' data provide a major advance over traditional pile driving formulae for estimating the soil resistance during driving. At the simplest level, the data enable the energy transmitted to the pile to be determined accurately, providing a direct measure of the hammer efficiency. With this information, traditional pile driving formulae may be used with increased confidence to assess the capacity of uninstrumented piles driven under the same conditions.

The pile capacity may be estimated directly from the stress-wave data, using expressions such as the Case formulae, or may be assessed more accurately by means of a numerical solution of the wave equation. The latter approach requires iterative adjustment of soil parameters over the depth of penetration of the pile, until a satisfactory match is obtained between the measured results and those computed by the program.

Shortcomings in the soil models limit the quality of the match obtained, and the estimated distribution of soil resistance down the pile is by no means unique.

There are many areas where further research and development is needed in order to improve the accuracy of pile driving analysis, both in terms of drivability studies and in assessing the working performance of piles from dynamic measurements. Relatively little guidance is available on the choice of unit skin friction along the shaft and end-bearing resistance at the pile tip, during initial driving of a pile. Still less is known about internal friction within open ended pipe piles or along the web and inner flanges of H-section piles. Empirical factors are currently used to quantify damping effects, lumping together inertial and viscous effects. These factors cannot be extrapolated reliably from one site to another, or even between different hammer blows on the same site. Separation of damping effects into inertial damping, that may be quantified in terms of the stiffness and density of the surrounding soil, and viscous damping that will depend only on soil type, is essential if the science of pile driving analysis is to progress.

Relationships presented in this chapter have shown that inertial effects lead to smaller values of quake and higher end-bearing resistance under dynamic conditions than for static loading. For H-section piles or open ended pipe piles, the inertia of the soil will generally lead to failure along the steel-soil interface under dynamic penetration, in contrast to the static failure mode where a plug of soil will form within the pile. Such differences in failure mode are important when assessing the static capacity of a pile from dynamic measurements. Differences in values of quake under static and dynamic conditions must be allowed for when estimating the static response of piles, particularly in respect of the pile head stiffness. Soil models based on elastodynamic theory automatically allow for such differences, and should lead to more accurate assessment of the static pile response.

REFERENCES

BARANOV, V.A. (1967). On the calculation of an embedded foundation (in Russian), Voprosy Dinamiki i Prochnosti, No. 14, Polytechnical Institute of Riga, Latvia, pp. 195–209.

BATHE, K.J. and WILSON, E.L. (1976). *Numerical Methods in Finite Element Analysis*, Prentice Hall, Englewood Cliffs, NJ.

CHOW, Y.K., WONG, K.Y., KARUNATNE, G.P. and LEE, S.L. (1988). Prediction of pile capacity from stress-wave measurements: Some numerical aspects, *Int. J. Numer. Anal. Methods Geomech.*, **12**(5), 505–12.

COYLE, H.M. and REESE, L.C. (1966). Load transfer for axially loaded piles in clay, *J. Soil Mech. Found. Eng. Div. ASCE*, **92**(SM2), 1–26.

COYLE, H.M., LOWERY, L.L. and HIRSCH, T.J. (1977). Wave equation analysis of piling behaviour. In *Numerical Methods in Geotechnical Engineering*, McGraw-Hill, New York, 272–96.

DOLWIN, J.D. and POSKITT, T. (1982). An optimisation method for pile driving analysis, *Proc. 2nd Int. Conf. on Numer. Methods in Offshore Piling*, Austin, pp. 91–106.

FELLENIUS, B.H. (1988). Variation of CAPWAP results as a function of the operator, *Proc. 3rd Int. Conf. on Application of Stress-wave Theory to Piles*, Ottawa, pp. 814–25.

FISCHER, H.C. (1984). Stress-wave theory for pile driving applications, *Lecture at 2nd Int. Conf. on Application of Stress-Wave Theory on Piles*, Stockholm.

FOX, E.N. (1932), Stress phenomena occurring in pile driving, *Engineering*, **134** (September), 263–5.

GAZETAS, G. and DOBRY, R. (1984). Simple radiation damping model for piles and footings, *J. Eng. Mech. Div. ASCE*, **110**, 937–56.

GIBSON, G.C. and COYLE, H.M. (1968). Soil damping constants related to common soil properties in sands and clays, Report No. 125-1, Texas Transport Institute, Texas A and M University, Houston.

GLANVILLE, W.H., GRIME, G., FOX, E.N. and DAVIES, W.W. (1938). An investigation of the stresses in reinforced concrete piles during driving, British Building Research Board, Technical Paper No. 20, London.

GOBLE, G.G. and HERY, P. (1984). Influence of residual force on pile driveability, *Proc. 2nd Int. Conf. on Stress-Wave Theory on Piles*, Stockholm, pp. 154–61.

GOBLE, G.G. and RAUSCHE, F. (1976). Wave equation analysis of pile driving–WEAP program, US Department of Transportation, Federal Highway Administration, Implementation Division, Office of Research and Development, Washington DC 20590.

GOBLE, G.G. and RAUSCHE, F. (1979). Pile drivability predictions by CAPWAP, *Proc. Conf. on Numer. Methods in Offshore Piling*, ICE, London, pp. 29–36.

GOBLE, G.G. and RAUSCHE, F. (1986). WEAP86 program documentation in 4 Vols, Federal Highway Administration, Office of Implementation, Washington DC 20590.

GOBLE, G.G., RAUSCHE, F. and LIKINS, G.E. (1980). The analysis of pile driving – A state-of-the-art, *Proc. Int. Conf. on Stress-Wave Theory on Piles*, Stockholm, pp. 131–61.

HEEREMA, E.P. and DE JONG, A. (1979). An advanced wave equation computer program which simulates dynamic pile plugging through a coupled mass-spring system, *Proc. Int. Conf. on Numer. Methods in Offshore Piling*, ICE, London, pp. 37–42.

JOHNSON, W. (1972). *Impact Strength of Materials*, Arnold, London.

KRAFT, L.M., RAY, R.P. and KAGAWA, T. (1981). Theoretical t–z curves, *J. Geotech. Eng. Div. ASCE*, **107**(GT11), 1543–61.

LITKOUHI, S. and POSKITT, T.J. (1980). Damping constant for pile driveability calculations, *Geotechnique*, **30**(1), 77–86.

VAN LUIPEN, P. and JONKER, G. (1979). Post-analysis of full-scale pile driving tests, *Proc. Int. Conf. on Numer. Methods in Offshore Piling*, ICE, London, pp. 43–52.

LYSMER, J. and RICHART, F.E. (1966). Dynamic response of footing to vertical loading, *J. Soil Mech. Found. Eng. Div. ASCE*, **98**, 85–105.

MIDDENDORP, P. and VAN WEELE, A.F. (1986). Application of characteristic stress wave method in offshore practice, *Proc. 3rd Int. Conf. on Numer. Methods in Offshore Piling*, Nantes, Supplement, pp. 6–18.

NGUYEN, T.T. (1987). Dynamic and static behaviour of driven piles, PhD Thesis, Chalmers University of Technology, Report No. 33, Swedish Geotechnical Institute.

NGUYEN, T.T., BERGGREN, B. and HANSBO, S. (1988). A new soil model for pile driving and drivability analysis, *Proc. 3rd Int. Conf. on Application of Stress-Wave Theory to Piles*, Ottawa, pp. 353–367.

NOVAK, M., NOGAMI, T. and ABOUL-ELLA, F. (1978). Dynamic soil reactions for plane strain case, *J. Eng. Mech. Div. ASCE*, **104**(EM4), 953–9.

RANDOLPH, M.F. (1977). A theoretical study of the performance of piles, PhD Thesis, University of Cambridge.

RANDOLPH, M.F. (1986). RATZ – Load transfer analysis of axially loaded piles, Report No. Geo:86033, Department of Civil Engineering, The University of Western Australia.

RANDOLPH, M.F. (1987). Modelling of the soil plug response during pile driving, *Proc. 9th S.E. Asian Geotechnical Conf.*, Bangkok, 6·1–6·14.

RANDOLPH, M.F. and SIMONS, H.A. (1986). An improved soil model for one-dimensional pile driving analysis, *Proc. 3rd Int. Conf. on Numer. Methods in Offshore Piling*, Nantes, pp. 1–17.

RANDOLPH, M.F. and PENNINGTON, D.S. (1988). A numerical study of dynamic cavity expansions, *Proc. 6th Int. Conf. on Numer. Methods in Geomechanics*, Innsbruck, Vol. **3**, pp. 1715–21.

RANDOLPH, M.F. and WROTH, C.P. (1978). Analysis of deformation of vertically loaded piles, *J. Geotech. Eng. Div. ASCE*, **104**(GT12), 1–17.

RANDOLPH, M.F., CARTER, J.P. and WROTH, C.P. (1979). Driven piles in clay – the effects of installation and subsequent consolidation, *Geotechnique*, **29**(4), 361–93.

RAUSCHE, F. (1984).Stress-wave measuring in practice, *Lecture at 2nd Int. Conf. on Application of Stress-Wave Theory on Piles*, Stockholm.

RAUSCHE, F. GOBLE, G.G. and LIKINS, G.E. (1985). Dynamic determination of pile capacity, *J. Geotech. Eng. Div. ASCE*, **111**, 367–83.

RAUSCHE, F. GOBLE, G.G. and LIKINS, G.E. (1988). Recent WEAP developments, *Proc. 3rd Int. Conf. on Application of Stress-Wave Theory to Piles*, Ottawa, pp. 164–73.

SIMONS, H.A. (1985). A theoretical study of pile driving, PhD Thesis, University of Cambridge.

SIMONS, H.A. and RANDOLPH, M.F. (1985). A new approach to one-dimensional pile driving analysis, *Proc. 5th Int. Conft. on Numer. Methods in Geomechanics*, Nagoya, Vol. **3**, pp. 1457–64.

SMITH, E.A.L. (1960). Pile driving analysis by the wave equation, *J. Soil Mech. Found. Eng. Div. ASCE*, **86**, 35–61.

SMITH, I.M. and CHOW, Y.K. (1982). Three-dimensional analysis of pile driveability, *Proc. 2nd Int. Conf. on Numer. Methods in Offshore Piling*, Austin, pp. 1–19.

SMITH, I.M., TO, P. and WILLSON (1986). Plugging of pipe piles, *Proc. 3rd Int. Conf. on Numer. Methods in Offshore Piling*, Nantes, pp. 53–73.

TIMOSHENKO, S. and GOODIER, J. (1970). *Theory of Elasticity*, 3rd edn., McGraw-Hill, New York.

Chapter 7

FINITE LAYER METHODS IN GEOTECHNICAL ANALYSIS

J.C. Small and J.R. Booker

The University of Sydney, Sydney,
Australia

ABSTRACT

For certain classes of problems, the use of the finite layer method can dramatically reduce the computational and data preparation time as compared to that required for obtaining a solution using conventional numerical techniques such as finite element or finite difference methods. This saving is especially evident for problems which are three dimensional in nature, as such problems can be reduced so that they involve only one spatial dimension.

The method relies upon being able to represent the field quantities, such as the displacements, stresses etc., by an orthogonal series or on being able to transform them by the use of integral transforms. The only restriction to doing this is that the material properties do not vary in one or two spatial directions. One of the simplest forms of orthogonal series is the Fourier series, and this is commonly used in the solution of problems using the finite layer method.

In this chapter, the basic theory of the finite layer method is presented for both series and integral transforms. The application of the method to many different types of problems in geomechanics is then demonstrated with examples of problems involving stress analysis, settlement of foundations, and soil-structure interaction, as well as time dependent problems involving settlement, viscoelasticity and thermoelasticity.

There are many other problems to which the finite layer method may be applied beside those which have been presented here, and it is hoped that this chapter may serve as an introduction to the method which engineers and researchers working in the field of geomechanics may wish to adapt or use for solving new or complex problems in their own field of interest.

1 INTRODUCTION

With the improvements in computer technology which have occurred in recent years, numerical analysis of engineering problems has become commonplace and has found applications in many fields. The finite element method has proved to be a very powerful tool, allowing analysis of three-dimensional problems with complex geometries, material properties and boundary conditions.

However, there are certain problems for which a full finite element analysis is not necessary and use of the method is inefficient and costly. Similarly the large numbers of equations which result from finite element analyses may also mean that fast in-core solution may not be possible on the current generation of microcomputers and hence it is desirable to pursue alternative methods of analysis for certain types of problems which have some simplifying feature.

Problems which fall into the above category are those for which the geometry and material properties do not vary in one or two spatial directions. This situation often occurs in the field of geomechanics since sedimentary soil and rock tend to be horizontally layered because of the process of deposition. In such cases it is possible to use the finite layer approach (Cheung, 1976). This method dramatically reduces the size of the sets of equations which have to be solved, as well as greatly reducing the amount of data preparation needed. As a consequence, even problems which involve three spatial dimensions and the time dimension can be analysed on small microcomputers with little cost.

In the following sections, the ideas of finite layer analysis are presented and illustrated by examples drawn from the many fields of application in geomechanics.

2 OUTLINE OF METHOD

In order to convey the general concepts of the method, suppose a problem which is common in geomechanics, that of a strip loading, is taken as an example. Such a problem is shown schematically in Fig. 1 where a spatially periodic loading (of period L) is applied to the surface of a horizontally layered profile of an elastic soil.

It is well known that for such a periodic loading (or loading function), a Fourier series representation may be used. For instance, if the Cartesian coordinate in the horizontal direction is x and the loading function is $p(x)$

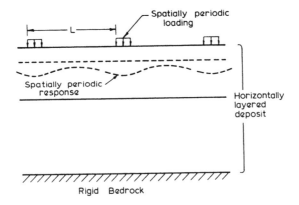

FIG. 1.   Spatially periodic loading on a horizontally layered deposit.

then it is found that

$$p(x) = \sum_{n=0}^{\infty} P^{(n)} \cos \alpha_n x \qquad (1)$$

where

$$P^{(n)} = \frac{2}{L} \int_0^L p(x) \cos \alpha_n x \, dx \quad (n > 0)$$

$$P^{(n)} = \frac{1}{L} \int_0^L p(x) \, dx \quad (n = 0)$$

and

$$\alpha_n = 2n\pi/L$$

The loading has therefore been represented by the sum of periodic functions (in this case cosine functions because the loading function $p(x)$ was chosen to be an even function of x).

It may also be observed that, for such a spatially periodic loading, the displacements in the soil below are also periodic. That is to say that the deflections beneath the centre of each loaded area are equal, similarly the deflections midway between the loads are equal and so on. If the displacements are periodic, then so too are the strains and stresses in the soil and it is therefore possible to write the displacements as a series of periodic functions, for example

$$u_x = \sum_{n=0}^{\infty} U^{(n)} \sin \alpha_n x$$

$$u_z = \sum_{n=0}^{\infty} W^{(n)} \cos \alpha_n x \qquad (2)$$

It will be observed that whereas a series of cosine functions is used to represent the displacement in the z direction (u_z), a sine series is used for the lateral displacements in the lateral or x direction (u_x). This follows from the observation that the lateral displacements are anti-symmetric about the axis of symmetry of each loaded area. The coefficients in the above series $U^{(n)}$, $W^{(n)}$ are of course different at different depths z since both the displacement components vary throughout the depth of the layer. Hence it is more correct to write

$$u_x(x, z) = \sum_{n=0}^{\infty} U^{(n)}(z) \sin \alpha_n x$$

$$u_z(x, z) = \sum_{n=0}^{\infty} W^{(n)}(z) \cos \alpha_n x \qquad (3)$$

However the shorthand notation of eqn (2) will be used in situations in which no confusion arises. These coefficients are not initially known and need to be determined before a solution can be found.

Suppose that we now take one term of the cosine series representing the loading function $p(x)$, and obtain the solution to the problem associated with this single sinusoidal load applied to the surface of the soil layer. It is not difficult to establish that if

$$p(x) = P_n \cos \alpha_n x$$

then

$$u_x = U_n(z) \sin \alpha_n x$$

$$u_z = W_n(z) \cos \alpha_n x \qquad (4)$$

This is shown in Fig. 2. Hence if this component problem can be solved the principle of superposition can be used to show that the solutions for each component can be added to synthesize the solutions for any loading that can be expressed in terms of the periodic series.

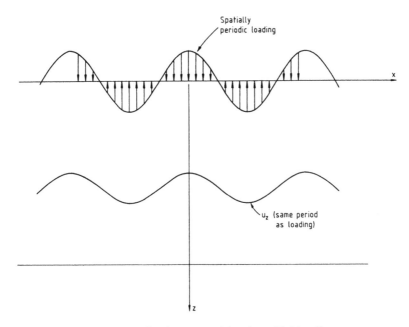

FIG. 2.   Deflections caused by sinusoidal loading.

3   DEVELOPMENT OF LAYER STIFFNESS MATRICES

Let us now turn our attention to the elastic horizontally layered material shown in Fig. 1 which is subjected to a symmetric spatially periodic loading with period L. For the sake of simplicity it will be assumed that conditions of plane strain prevail so that there is no displacement in the y direction and no variation of the field quantities with y. It was shown in the previous section that it is only necessary to consider stress and displacement fields having the form

$$u_x = U(\alpha, z) \sin \alpha x \tag{5a}$$

$$u_z = W(\alpha, z) \cos \alpha x \tag{5b}$$

$$\sigma_{xz} = T(\alpha, z) \sin \alpha x \tag{5c}$$

$$\sigma_{zz} = N(\alpha, z) \cos \alpha x \tag{5d}$$

$$\sigma_{xx} = H(\alpha, z) \cos \alpha x \tag{5e}$$

$$\sigma_{yy} = M(\alpha, z) \cos \alpha x \tag{5f}$$

since the solution for more complex load cases can be obtained by superposition of components given by eqns (5) and where it has been assumed that α stands for any particular value of $\alpha_n = 2n\pi/L$ and $U^{(n)} = U(\alpha_n, z)$, etc.

Let us now consider the determination of the field quantities in a particular layer l of the material. If eqns (5) are substituted into the equilibrium equations it is found that

$$-\alpha H + \frac{\partial T}{\partial z} = 0$$

$$\tag{6}$$

$$+\alpha T + \frac{\partial N}{\partial z} = 0$$

Similarly if eqns (5) are substituted into Hooke's Law it is found that

$$\alpha U = \frac{1}{E_l}[H - v_l(M + N)]$$

$$\frac{\partial W}{\partial z} = \frac{1}{E_l}[N - v_l(H + M)]$$

$$\tag{7}$$

$$\frac{\partial U}{\partial z} - \alpha W = \frac{2(1 + v_l)T}{E_l}$$

$$M = v_l(H + N)$$

where E_l, v_l denote the values of Young's Modulus and Poisson's ratio for layer l.

Equations (6), (7) allow the non-zero stress and displacement to be expressed in terms of N and thus

$$H = -\frac{\partial^2 N}{\partial Z^2}$$

$$T = -\frac{\partial N}{\partial Z}$$

$$M = v_l\left(N - \frac{\partial^2 N}{\partial Z^2}\right) \tag{8}$$

$$\alpha E^* U = -\frac{\partial^2 N}{\partial Z^2} - v^* N$$

$$\alpha E^* W = -\frac{\partial^3 N}{\partial Z^3} + (2 + v^*)N$$

where

$$E^* = \frac{E_l}{1 - v_l^2}$$

$$v^* = \frac{v_l}{1 - v_l}$$

$$Z = \alpha z$$

and

$$\frac{\partial^4 N}{\partial Z^4} - 2\frac{\partial^2 N}{\partial Z^2} + N = 0 \tag{9a}$$

If eqn (9) is now solved, we obtain

$$N = X_1 C + X_2 Z + X_3 S + X_4 ZC \tag{9b}$$

The complete solution is given in Table 1.

TABLE 1

	X_1	X_2	X_3	X_4
N	C	ZS	S	ZC
T	$-S$	$-(ZC+S)$	$-C$	$-(ZS+C)$
H	$-C$	$-(ZS+2C)$	$-S$	$-(ZC+2S)$
$\alpha E^* U$	$-(1+v^*)C$	$-(1+v^*)ZS$ $-2C$	$-(1+v^*)S$	$-(1+v^*)ZC$ $-2S$
$\alpha E^* W$	$+(1+v^*)S$	$+(1+v^*)ZC$ $-(1-v^*)S$	$(1+v^*)C$	$+(1+v^*)ZS$ $-(1-v^*)C$

where $C = \cosh Z$ and $S = \sinh Z$.

The fundamental step in the finite layer technique is to determine the four constants $X_1, \ldots, X_4$ appearing in Table 1 in terms of boundary quantities. To be more precise, suppose that layer l is bounded by the node planes $(z=z_l)$ and $(z=z_m)$ where $m=l+1$ and that the subscripts l, m indicate the value of a particular quantity on the indicated node plane (see Fig. 3). Then the solution given in Table 1 can be used to determine $X_1,...,X_4$ in terms of U_l, W_l, U_m, W_m. Once $X_1,...,X_4$ are known they can be used to evaluate T_l, N_l, T_m, N_m and so establish a relationship of

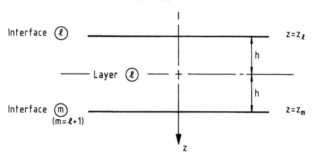

FIG. 3.   Single material layer.

the form

$$
\begin{bmatrix} -T_l \\ -N_l \\ +T_m \\ +N_m \end{bmatrix} = \begin{bmatrix} k_{11} & k_{12} & k_{13} & k_{14} \\ k_{21} & k_{22} & k_{23} & k_{24} \\ k_{31} & k_{32} & k_{33} & k_{34} \\ k_{41} & k_{42} & k_{43} & k_{44} \end{bmatrix} \begin{bmatrix} U_l \\ W_l \\ U_m \\ W_m \end{bmatrix}
\tag{10}
$$

The matrix occurring in eqn (10) is called the layer stiffness matrix of the particular layer (layer l in this case). Explicit details of its derivation are given in Appendix 1. Layer stiffness matrices can be used to construct solutions for layered deposits in exactly the same way as element stiffness matrices are used in conventional applications of the finite element method.

To illustrate this, consider the specific case of a deposit consisting of three distinct elastic layers overlaying rigid bedrock and subject to the surface loading shown in Fig. 4. It has been shown previously that it is only necessary to consider loads of the type

$$\sigma_{xz} = Q \sin \alpha x$$

$$\sigma_{zz} = P \cos \alpha x$$

and that then the displacements will have the form

$$u_x = U \sin \alpha x$$

$$u_z = W \cos \alpha x$$

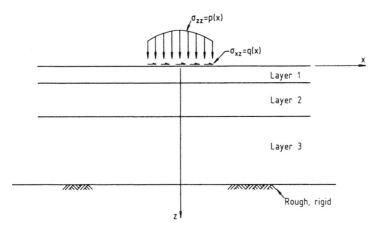

FIG. 4.   Normal and shear loading applied to a matrial having three layers.

The layer matrices of each of the layers can be computed in a straight-forward fashion, to obtain three relationships of the type

$$
\begin{bmatrix} -T_1 \\ -N_1 \\ +T_2 \\ +N_2 \end{bmatrix} = \begin{bmatrix} a_{11} & a_{12} & a_{13} & a_{14} \\ a_{21} & a_{22} & a_{23} & a_{24} \\ a_{31} & a_{32} & a_{33} & a_{34} \\ a_{41} & a_{42} & a_{43} & a_{44} \end{bmatrix} \begin{bmatrix} U_1 \\ W_1 \\ U_2 \\ W_2 \end{bmatrix} \tag{11a}
$$

$$
\begin{bmatrix} -T_2 \\ -N_2 \\ +T_3 \\ +N_3 \end{bmatrix} = \begin{bmatrix} b_{11} & b_{12} & b_{13} & b_{14} \\ b_{21} & b_{22} & b_{23} & b_{24} \\ b_{31} & b_{32} & b_{33} & b_{34} \\ b_{41} & b_{42} & b_{43} & b_{44} \end{bmatrix} \begin{bmatrix} U_2 \\ W_2 \\ U_3 \\ W_3 \end{bmatrix} \tag{11b}
$$

$$
\begin{bmatrix} -T_3 \\ -N_3 \\ +T_4 \\ +N_4 \end{bmatrix} = \begin{bmatrix} c_{11} & c_{12} & c_{13} & c_{14} \\ c_{21} & c_{22} & c_{23} & c_{24} \\ c_{31} & c_{32} & c_{33} & c_{34} \\ c_{41} & c_{42} & c_{43} & c_{44} \end{bmatrix} \begin{bmatrix} U_3 \\ W_3 \\ U_4 \\ W_4 \end{bmatrix} \tag{11c}
$$

These equations may be combined and thus recalling the boundary conditions

$$
T_1 = Q, \quad N_1 = P, \quad U_4 = 0, \quad W_4 = 0 \tag{12}
$$

it is found that the unknown displacements satisfy an equation

$$
\mathbf{KA} = \mathbf{F} \tag{13}
$$

where

$$A = [U_1 \quad W_1 \quad U_2 \quad W_2 \quad U_3 \quad W_3]^T$$

is the vector of unknown displacement amplitudes,

$$F = [-Q \quad -P \quad 0 \quad 0 \quad 0 \quad 0]^T$$

is the vector of applied tractions and

$$K = \begin{bmatrix} a_{11} & a_{12} & a_{13} & a_{14} & & 0 \\ a_{21} & a_{22} & a_{23} & a_{24} & & 0 \\ a_{31} & a_{32} & a_{33}+b_{11} & a_{34}+b_{12} & b_{23} & b_{14} \\ a_{41} & a_{42} & a_{34}+b_{21} & a_{43}+b_{22} & b_{14} & b_{24} \\ 0 & 0 & b_{31} & b_{32} & b_{33}+c_{11} & b_{34}+c_{12} \\ 0 & 0 & b_{41} & b_{42} & b_{43}+c_{21} & b_{44}+c_{22} \end{bmatrix}$$

The matrix K is called the total layer stiffness matrix and can be assembled using the same assembly procedure as the finite element method.

In order to solve for a case in which the load can be adequately approximated by

$$\sigma_{xz} = \sum_{n=0}^{N} Q^{(n)} \sin \alpha_n x$$

$$\sigma_{zz} = \sum_{n=0}^{N} P^{(n)} \cos \alpha_n x \tag{14}$$

it is only necessary to set

$$\alpha = \alpha_n, \quad P = P^{(n)}, \quad Q = Q^{(n)}$$

in eqn (13) to obtain the component solution and then to superimpose these solutions for $n=0, \ldots, N$ to obtain the complete solution.

This procedure may be written formally by saying that when $\alpha = \alpha_n$, eqn (13) becomes

$$K^{(n)} A^{(n)} = F^{(n)}, \quad n = 0, \ldots, N \tag{15}$$

and thus if the deflections on a particular node plane $z = z_p$ are denoted by $u_x = u_{xp}$, $u_z = u_{zp}$ it is found that

$$u_x = \sum_{n=0}^{N} U_p^{(n)} \sin(\alpha_n x)$$

$$(16)$$

$$u_z = \sum_{n=0}^{N} W_p^{(n)} \cos(\alpha_n x)$$

The derivation given in this section is quite different to the original derivation developed by Cheung (1976), in that it employs solutions which satisfy the governing equations exactly in contrast to Cheung (1976) who employs a virtual work approximation. Cheung's (1976) approach is more widely applicable since it can be used in situations where an exact solution cannot be found; however, the approach developed in this section has the advantage of requiring a much smaller number of layers to obtain an adequate approximation.

4 INTEGRAL TRANSFORMS

The development discussed in the previous section is only valid when the applied loads are spatially periodic. One way of obtaining an analysis for a single load is to take a very large value of L so that in regions close to one of the loaded areas, the contribution due to the other loaded areas is very small. This is equivalent to employing a Fourier transform (see for example Sneddon (1951), Small and Booker (1984)).

Again attention will be restricted to plane strain. The Fourier transforms of the displacements and stresses are defined by the relations

$$U_k = \int_{-\infty}^{\infty} e^{-i\alpha x} u_k \, dx$$

$$(17)$$

$$S_{jk} = \int_{-\infty}^{\infty} e^{-i\alpha x} \sigma_{jk} \, dx$$

where j, k denote any value of the index set $\{x, y, z\}$.

The inverse transform, is found using Fourier's integral theorem and thus

$$u_k = \frac{1}{2\pi} \int_{-\infty}^{\infty} e^{+i\alpha x} U_k \, d\alpha$$

$$\sigma_{jk} = \frac{1}{2\pi} \int_{-\infty}^{\infty} e^{+i\alpha x} S_{jk} \, d\alpha \tag{18}$$

Equations (18) provide a means of representing the field quantities and if these are introduced into the equilibrium equations and Hooke's law, it is found, in a paticular layer l, that

$$i\alpha S_{xx} + \frac{\partial S_{xz}}{\partial z} = 0$$

$$i\alpha S_{xz} + \frac{\partial S_{zz}}{\partial z} = 0 \tag{19}$$

and

$$i\alpha U_x = \frac{1}{E_l} [S_{xx} - v_l(S_{yy} + S_{zz})]$$

$$\frac{\partial U_z}{\partial z} = \frac{1}{E_l} [S_{zz} - v_l(S_{xx} + S_{yy})]$$

$$\frac{\partial U_x}{\partial z} + i\alpha U_z = \frac{2(1 + v_l)}{E_l} S_{xz} \tag{20}$$

$$S_{yy} = v_l(S_{xx} + S_{zz})$$

If we now introduce

$$T = iS_{xz}, \quad H = S_{xx}, \quad N = S_{zz}$$

$$M = S_{yy}, \quad U = iU_x, \quad W = U_z$$

it is found that eqns (19), (20) are identical to eqns (6), (7) and thus it is possible to set up layer stiffness matrices in precisely the same way as described in the previous section and to assemble individual layer stiffness matrices to obtain a set of equations

$$\mathbf{K}(\alpha)\mathbf{A} = \mathbf{F} \tag{21}$$

where the stiffness matrix is precisely the one developed in the previous section (eqn (13)), the vector **A** is the vector of the transformed node plane deflections defined in the previous section and the vector **F** is the vector of transformed applied tractions.

Suppose for example we consider the three layer system shown in Fig. 4, subjected to surface loading

$$\sigma_{zz} = p(x) \quad \text{when } z = 0$$

$$\sigma_{xz} = q(x) \quad \text{when } z = 0$$

(22)

then

$$\mathbf{F} = [-Q \quad -P \quad 0 \quad 0 \quad 0 \quad 0]^{\mathsf{T}}$$

where

$$Q = i \int_{-\infty}^{\infty} e^{-i\alpha x} q(x) \, dx$$

$$P = \int_{-\infty}^{\infty} e^{-i\alpha x} p(x) \, dx$$

Suppose for example, the normal load had a triangular distribution

$$\sigma_{zz} = \begin{cases} p_{max}(a - |x|)/a, & 0 < |x| < a \\ 0, & \text{elsewhere} \end{cases}$$

and the shear distribution was linear

$$\tau_{xz} = \begin{cases} \tau_{max} x/a, & 0 < |x| < a \\ 0, & \text{elsewhere} \end{cases}$$

Then

$$Q = i \int_{-\infty}^{\infty} e^{-i\alpha x} q(x) \, dx$$

$$= \frac{i\tau_{max}}{a} \int_{-a}^{a} e^{-i\alpha x} x \, dx$$

$$= 2\tau_{max} \left[\frac{\cos \alpha a}{\alpha} - \frac{\sin \alpha a}{a\alpha^2} \right]$$

J.C. SMALL and J.R. BOOKER

$$P = \int_{-\infty}^{\infty} e^{-i\alpha x} p(x) \, dx$$

$$= \frac{2p_{max}}{a} \int_{0}^{a} \cos \alpha x [a - x] \, dx$$

$$= \frac{2p_{max}(1 - \cos \alpha a)}{a\alpha^2}$$

It will be observed, referring to eqn (21), that both the stiffness matrix **K** and the load vector **F** depend upon the integration parameter α, thus eqn (21) determines the vector of node plane displacements as a function of α and so the actual displacements can be evaluated by inverting the transforms using eqn (18). For example suppose we wished to evaluate the displacements on a paticular node plane $z = z_p$.

Then

$$u_x = \frac{1}{2\pi i} \int_{-\infty}^{\infty} e^{i\alpha x} U_p(\alpha) \, d\alpha$$

$$u_z = \frac{1}{2\pi} \int_{-\infty}^{\infty} e^{i\alpha x} W_p(\alpha) \, d\alpha$$

(23)

These integrals cannot be evaluated analytically except in very simple cases and so it is necessary to employ a numerical integration scheme. When this is done an integral of the form

$$I = \int_{a}^{b} f(x) \, dx$$

is approximated by a sum

$$I \simeq \sum_{j=1}^{M} w_j f(\alpha_j)$$

where α_j are sample points and w_j are the associated weights.

It thus follows that

$$u_x \simeq \frac{1}{2\pi i} \sum_{j=1}^{M} w_j e^{i\alpha_j x} U_p(\alpha_j)$$

$$u_z \simeq \frac{1}{2\pi} \sum_{j=1}^{M} w_j e^{i\alpha_j x} W_p(\alpha_j)$$

(24)

and so the solution process involves the solution of eqn (21) for each of the values $\alpha = \alpha_j$ and then a summation of the results and thus closely follows the process for periodically spaced loads.

5 THREE-DIMENSIONAL ANALYSIS

The discussion so far has been restricted to the case of plane strain, however it is possible to extend the method to a full three-dimensional analysis of horizontally layered systems.

In order to do this the repeated Fourier transforms

$$U_k = \int_{-\infty}^{\infty} \int_{-\infty}^{\infty} e^{-i(\alpha x + \beta y)} u_k \, dx \, dy$$

$$S_{jk} = \int_{-\infty}^{\infty} \int_{-\infty}^{\infty} e^{-i(\alpha x + \beta y)} \sigma_{jk} \, dx \, dy$$

(25)

with the corresponding inverse transforms

$$u_k = \frac{1}{4\pi^2} \int_{-\infty}^{\infty} \int_{-\infty}^{\infty} e^{+i(\alpha x + \beta y)} U_k \, d\alpha \, d\beta$$

$$\sigma_{jk} = \frac{1}{4\pi^2} \int_{-\infty}^{\infty} \int_{-\infty}^{\infty} e^{+i(\alpha x + \beta y)} S_{jk} \, d\alpha \, d\beta$$

(26)

are introduced.

The equations of equilibrium become

$$i\alpha S_{xx} + i\beta S_{xy} + \frac{\partial S_{xz}}{\partial z} = 0$$

$$i\alpha S_{xy} + i\beta S_{yy} + \frac{\partial S_{yz}}{\partial z} = 0$$

(27)

$$i\alpha S_{xz} + i\beta S_{yz} + \frac{\partial S_{zz}}{\partial z} = 0$$

and Hooke's Law trasforms to

$$i\alpha U_x = \frac{1}{E}[S_{xx} - v(S_{yy} + S_{zz})]$$

$$i\beta U_y = \frac{1}{E}[S_{yy} - v(S_{zz} + S_{xx})] \tag{28}$$

$$\frac{\partial U_z}{\partial z} = \frac{1}{E}[S_{zz} - v(S_{xx} + S_{yy})]$$

$$i\alpha U_z + \frac{\partial U_x}{\partial z} = \frac{2(1+v)}{E}S_{xz}$$

$$i\beta U_z + \frac{\partial U_y}{\partial z} = \frac{2(1+v)}{E}S_{yz} \tag{29}$$

$$i\alpha U_y + i\beta U_x = \frac{2(1+v)}{E}S_{xy}$$

These equations can be simplified considerably by introducing the following change of variable defined by the relations:

$$\begin{bmatrix} U_\xi \\ U_\eta \\ U_z \end{bmatrix} = \begin{bmatrix} C & S & 0 \\ -S & C & 0 \\ 0 & 0 & 1 \end{bmatrix} \begin{bmatrix} U_x \\ U_y \\ U_z \end{bmatrix} \tag{30}$$

and

$$\begin{bmatrix} S_{\xi\xi} & S_{\xi\eta} & S_{\xi z} \\ S_{\eta\xi} & S_{\eta\eta} & S_{\eta z} \\ S_{z\xi} & S_{z\eta} & S_{zz} \end{bmatrix} = \begin{bmatrix} C & S & 0 \\ -S & C & 0 \\ 0 & 0 & 1 \end{bmatrix} \begin{bmatrix} S_{xx} & S_{xy} & S_{xz} \\ S_{yx} & S_{yy} & S_{yz} \\ S_{zx} & S_{zy} & S_{zz} \end{bmatrix} \begin{bmatrix} C & -S & 0 \\ S & C & 0 \\ 0 & 0 & 1 \end{bmatrix} \tag{31}$$

where

$$C = \cos \varepsilon = \frac{\alpha}{\rho}$$

$$S = \sin \varepsilon = \frac{\beta}{\rho}$$

$$\rho = (\alpha^2 + \beta^2)^{1/2}$$

The equations then reduce to two groups, the first being

$$-\rho H + \frac{\partial T}{\partial z} = 0$$

$$+\rho T + \frac{\partial N}{\partial z} = 0 \tag{32}$$

$$\rho U = \frac{1}{E}[H - v(N + M)]$$

(33)

$$\frac{\partial W}{\partial z} = \frac{1}{E}[N - v(H + M)]$$

$$M = v(N + H)$$

where it has been convenient to introduce

$$T = iS_{\xi z}$$
$$H = S_{\xi \xi}$$
$$N = S_{zz}$$
$$M = S_{\eta \eta}$$
$$U = iU_{\xi}$$
$$W = U_{z}$$

(34)

These equations are completely analogous to those developed for the case of plane strain (eqns (6), (7) and eqns (19), (20)) with α replaced by ρ. Thus the method of analysis developed in the previous sections can be applied immediately and so the node plane deflections satisfy the equation

$$\mathbf{K}(\rho)\mathbf{A} = \mathbf{F}$$

(35)

where $\mathbf{K}(\rho)$ is precisely the layer matrix defined in eqn (21) with α replaced by ρ, $\mathbf{A}$ is the vector of transformed node plane displacements and $\mathbf{F}(\alpha, \beta)$ is the vector of transformed applied tractions. For the three layer case shown in Fig. 4, the vector $\mathbf{F}$ is given by

$$\mathbf{F} = [-Q \quad -P \quad 0 \quad 0 \quad 0 \quad 0]^{\mathsf{T}}$$

where

$$Q = \frac{1}{4\pi^2 i} \int_{-\infty}^{\infty} \int_{-\infty}^{\infty} (\cos \varepsilon \sigma_{xz} + \sin \varepsilon \sigma_{yz}) e^{-i(\alpha x + \beta y)} \, dx \, dy$$

$$P = \frac{1}{4\pi^2} \int_{-\infty}^{\infty} \int_{-\infty}^{\infty} \sigma_{zz} e^{-i(\alpha x + \beta y)} \, dx \, dy$$

these integrals being evaluated at the surface.

If for example the three layer system was subjected to the following surface loading,

$$\sigma_{zz} = \begin{cases} p_0, & -a<x<a, \ -b<y<a \\ 0, & \text{elsewhere on } z=0 \end{cases}$$

$$\sigma_{xz} = \begin{cases} q_0, & -a<x<a, \ -b<y<b \\ 0, & \text{elsewhere on } z=0 \end{cases} \tag{36}$$

then

$$Q = \frac{1}{4\pi^2 i} \int_{-a}^{a} \int_{-b}^{b} \frac{\alpha}{\rho} q_0 \, e^{-i(\alpha x + \beta y)} \, dx \, dy$$

$$= \frac{q_0}{\pi^2 i} \frac{\sin \alpha a \ \sin \beta b}{\rho \beta}$$

$$P = \frac{1}{4\pi^2} \int_{-a}^{a} \int_{-b}^{b} p_0 \, e^{-i(\alpha x + \beta y)} \, dx \, dy$$

$$= \frac{p_0}{\pi^2} \frac{\sin \alpha a \ \sin \beta b}{\alpha \beta}$$

The second set of equations is

$$\frac{\partial S}{\partial z} + i\rho S_{\xi\eta} = 0$$

$$\frac{\partial V}{\partial z} = \frac{2(1+v)}{E} S \tag{37}$$

$$i\rho V = \frac{2(1+v)}{E} S_{\xi\eta}$$

where it has been convenient to introduce the notation

$$S = S_{\eta z}, \quad V = U_\eta$$

Equations (37) can be used to construct layer matrices in similar fashion to the method detailed in Appendix 1 and so for layer l it is found that

$$\begin{bmatrix} -S_l \\ \\ +S_m \end{bmatrix} = \frac{\rho E_l}{2(1+v_l)} \begin{bmatrix} \dfrac{\cosh 2\rho h}{\sinh 2\rho h} & -\dfrac{1}{\sinh 2\rho h} \\ \\ -\dfrac{1}{\sinh 2\rho h} & \dfrac{\cosh 2\rho h}{\sinh 2\rho h} \end{bmatrix} \begin{bmatrix} V_l \\ \\ V_m \end{bmatrix} \tag{38}$$

where $2h$ is the depth of the layer.

These layer matrices can be assembled in precisely the same way as discussed previously. Thus if considering the three layer problem shown in Fig. 4, it is found that the first layer will have a stiffness relation of the form

$$\begin{bmatrix} -S_1 \\ S_2 \end{bmatrix} = \begin{bmatrix} a_{11}^* & a_{12}^* \\ a_{21}^* & a_{22}^* \end{bmatrix} \begin{bmatrix} V_1 \\ V_2 \end{bmatrix}$$

the second layer has a stiffness relation having the form

$$\begin{bmatrix} -S_2 \\ S_3 \end{bmatrix} = \begin{bmatrix} b_{11}^* & b_{12}^* \\ b_{21}^* & b_{22}^* \end{bmatrix} \begin{bmatrix} V_2 \\ V_3 \end{bmatrix}$$

and the third layer has a stiffness relation of the form

$$\begin{bmatrix} -S_3 \\ S_4 \end{bmatrix} = \begin{bmatrix} c_{11}^* & c_{12}^* \\ c_{21}^* & c_{22}^* \end{bmatrix} \begin{bmatrix} V_3 \\ V_4 \end{bmatrix}$$

Now recalling that the base is assumed rigid so that $V_4 = 0$ it follows

$$\begin{bmatrix} -S_1 \\ 0 \\ 0 \end{bmatrix} = \begin{bmatrix} a_{11}^* & a_{12}^* & 0 \\ a_{21}^* & a_{22}^* + b_{11}^* & b_{12}^* \\ 0 & b_{21}^* & b_{22}^* + c_{11}^* \end{bmatrix} \begin{bmatrix} V_1 \\ V_2 \\ V_3 \end{bmatrix}$$

The quantity S_1 can be calculated from a knowledge of the surface tractions,

$$S_1 = \frac{1}{4\pi^2} \int_{-\infty}^{\infty} \int_{-\infty}^{\infty} (-\sin \varepsilon \sigma_{xz} + \cos \varepsilon \sigma_{yz}) e^{-i(\alpha x + \beta y)} \, d\alpha \, d\beta$$

and so for the loading defined by eqns (36),

$$S_1 = -\frac{1}{4\pi^2} \int_{-a}^{a} \int_{-b}^{b} \frac{\beta}{\rho} q_0 e^{-i(\alpha x + \beta y)} \, dx \, dy$$

$$= -\frac{q_0}{\pi^2} \frac{\sin \alpha a \sin \beta b}{\alpha \beta}$$

By solving the two sets of layer equations it is possible to determine the quantities U_ξ, U_η, U_z, $S_{\xi\xi}$, ..., $S_{\eta\xi}$. These can be used to determine the more directly useful variables U_x, U_y, U_z, S_{xx}, ..., S_{yz} by inverting eqns

(30), (31) so that

$$
\begin{bmatrix} U_x \\ U_y \\ U_z \end{bmatrix} = \begin{bmatrix} C & -S & 0 \\ S & C & 0 \\ 0 & 0 & 1 \end{bmatrix} \begin{bmatrix} U_\xi \\ U_\eta \\ U_z \end{bmatrix} \tag{39}
$$

$$
\begin{bmatrix} S_{xx} & S_{xy} & S_{xz} \\ S_{yx} & S_{yy} & S_{yz} \\ S_{zx} & S_{zy} & S_{zz} \end{bmatrix} = \begin{bmatrix} C & -S & 0 \\ S & C & 0 \\ 0 & 0 & 1 \end{bmatrix} \begin{bmatrix} S_{\xi\xi} & S_{\xi\eta} & S_{\xi z} \\ S_{\eta\xi} & S_{\eta\eta} & S_{\eta z} \\ S_{z\xi} & S_{z\eta} & S_{zz} \end{bmatrix} \begin{bmatrix} C & S & 0 \\ -S & C & 0 \\ 0 & 0 & 1 \end{bmatrix} \tag{40}
$$

Once these quantities are determined, the Cartesian displacements and stresses can be determined by numerical integration as shown in eqns (26).

Numerical inversion may be carried out by use of a double summation, e.g.

$$
u_z \simeq \frac{1}{4\pi^2} \sum_{l=1}^{M} \sum_{m=1}^{N} w_l w_m \, e^{[i(\alpha_l x + \beta_m y)]} U_z \tag{41}
$$

where Gaussian quadrature has been used and w_l, w_m are the weights and α_l, β_m are the sample points.

This method is fairly time consuming, however computational efficiency can be improved by making the substitutions

$$
\alpha = \rho \cos \varepsilon, \quad \beta = \rho \sin \varepsilon, \quad d\alpha \, d\beta = \rho \, d\rho \, d\varepsilon
$$

in eqns (26).

For u_z we would obtain

$$
u_z = \frac{1}{4\pi^2} \int_0^{2\pi} \int_{-\infty}^{+\infty} U_z \, e^{[i\rho(x \cos \varepsilon + y \sin \varepsilon)]} \rho \, d\rho \, d\varepsilon \tag{42}
$$

The numerical integration may now be carried out with respect to ρ, ε, e.g.

$$
u_z \simeq \frac{1}{4\pi^2} \sum_{l=1}^{M} \sum_{m=1}^{N} w_l w_m \, e^{[i\rho_l(x \cos \varepsilon_m + y \sin \varepsilon_m)]} \rho_l U_z \tag{43}
$$

where ε_m, ρ_l are sample points and w_l, w_m are the associated weights. This is generally more efficient as the integration with respect to ε needs only to be carried out from 0 to 2π, and can be done with fewer Gauss points (values of ε_m).

5.1 Axial Symmetry

The expressions developed in the previous section can be simplified greatly in the case of axial symmetry. Suppose for example a horizontally layered system is subject to an axially symmetric distribution of surface traction

$$\sigma_{zz} = p(r), \quad \sigma_{rz} = q(r), \quad \sigma_{\theta z} = 0 \tag{44}$$

where r, θ, z denote cylindrical polar coordinates. Clearly the solution must be axially symmetric so that

$$\begin{aligned} u_x &= u_r(r, z)\cos\theta \\ u_y &= u_r(r, z)\sin\theta \\ u_z &= u_z(r, z) \end{aligned} \tag{45}$$

It thus follows, introducing polar coordinates, that the transformed stresses are given by

$$U = \int_0^\infty \int_0^{2\pi} e^{-i\rho r\cos(\theta-\varepsilon)} u_r(r, z)\cos(\theta-\varepsilon) r \, dr \, d\theta$$

$$V = \int_0^\infty \int_0^{2\pi} e^{-i\rho r\cos(\theta-\varepsilon)} u_r(r, z)\sin(\theta-\varepsilon) r \, dr \, d\theta \tag{46}$$

$$W = \int_0^\infty \int_0^{2\pi} e^{-i\rho r\cos(\theta-\varepsilon)} u_z(r, z) r \, dr \, d\theta$$

where upon, using Poisson's integral, it is found that

$$U = 2\pi \int_0^\infty r J_1(\rho r) u_r(r, z) \, dr$$

$$V = 0 \tag{47}$$

$$W = 2\pi \int_0^\infty r J_0(\rho r) u_z(r, z) \, dr$$

so that the repeated Fourier transforms are replaced by Hankel Transforms.

The transforms of the tractions acting on any plane $z = \text{constant}$ can

J.C. SMALL and J.R. BOOKER

be similarly evaluated and thus it is found that

$$T = 2\pi \int_0^\infty rJ_1(\rho r)\sigma_{rz}(r, z)\, dr$$

$$S = 0 \tag{48}$$

$$N = 2\pi \int_0^\infty rJ_0(\rho r)\sigma_{zz}(r, z)\, dr$$

In particular the transformed stresses corresponding to the applied tractions are

$$T = Q = 2\pi \int_0^\infty rJ_1(\rho r)q(r)\, dr$$

$$S = 0 \tag{49}$$

$$N = P = 2\pi \int_0^\infty rJ_0(\rho r)p(r)\, dr$$

The procedure for analysis is precisely the same as the full three-dimensional case and leads to the set of equations

$$\mathbf{K}(\rho)\mathbf{A}(\rho) = \mathbf{F}(\rho) \tag{50}$$

It is important to note that the solution of this equation does not depend upon the values of both the transform parameters α, β but only on $\rho = (\alpha^2 + \beta^2)^{1/2}$.

Once the transformed displacements have been determined from the above equation it is not difficult to show that the repeated Fourier transforms expressing the displacement field can be reduced to Hankel transforms and thus on node plane p,

$$u_r(r, z_p) = \frac{1}{2\pi} \int_0^\infty \rho J_1(\rho r)U_p(\rho)\, d\rho$$

$$u_z(r, z_p) = \frac{1}{2\pi} \int_0^\infty \rho J_0(\rho r)W_p(\rho)\, d\rho \tag{51}$$

It is seldom possible to evaluate these integrals (Hankel Transforms) analytically and so a numerical approach is adopted and it is found that

$$u_r = \sum_{j=1}^N w_j\rho_j J_1(\rho_j r)U_p(\rho_j)$$

$$u_z = \sum_{j=1}^{N} w_j \rho_j J_0(\rho_j r) W_p(\rho_j)$$

where ρ_j are the sample points of a numerical integration scheme and w_j are the associated weights

6 SETTLEMENT ANALYSIS FOR AN ELASTIC SOIL

As outlined in the previous sections, finite layer solutions may make use of orthogonal series such as Fourier or Fourier Bessel series or involve integral transforms such as Fourier or Hankel transforms.

In this section examples are given of the great range of problems involving stress analysis in elastic materials that may be solved by use of the finite layer method. The examples are divided into those which make use of integral transforms and those that make use of series.

6.1 Series Solutions

Single Fourier Series have been used by Cheung *et al.* (1976) to obtain solutions to the problem of loadings applied to layered pavements. Use is made of a 'finite prism' method, in which a two-dimensional finite element mesh is used to approximate the Fourier coefficients of the field variables in a plane and a Fourier series is used to represent the field quantities in the third dimension.

Cheung and Fan (1979) extended this approach, by employing a double Fourier series to approximate field variables. They demonstrated the effectiveness of their method by analysing the behaviour of horizontally layered pavements subjected to a rectangular surface loading. Various one-dimensional element types were tried as a means of approximating the Fourier coefficients which vary (in this case) with depth and the conclusion was reached that the 'lowest order element', presumably one using linear interpolation functions, was sufficient to provide results of acceptable accuracy.

Since the use of series implies that an artificial boundary, i.e. one on which the shear stress and the lateral displacements are zero, exists between adjacent loaded areas, the period of the loading must be made large enough so that an isolated loading may be accurately modelled if this is desired. Cheung and Fan (1979) examined the effect of increasing the period of the loading and showed that as the period became larger (i.e. the spacing between the loadings became larger) more terms were required in the series to obtain an adequate approximation.

They also noted that convergence of series for the stresses and displacements at locations near the surface (i.e. closest to the applied loading) were much slower than for points at some depth below the surface. This has also been reported by Maier and Novati (1988).

Figure 5 shows results obtained by Cheung and Fan (1979) for a square

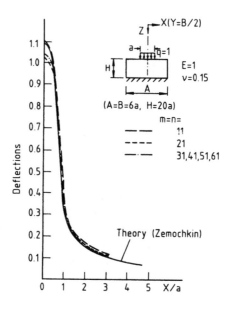

FIG. 5. Surface deflection due to square loading (after Cheung and Fan, 1979).

loading applied to the surface of a uniform soil layer in which the spatial period of loading was chosen so that the distance between the artificial side boundaries A (see inset to Fig. 5) was six times the footing width. An increasing number of terms was used in the double Fourier series (these are designated by m, n on the figure) and it can be seen that approximately 11 terms in each summation will provide reasonably accurate results.

6.2 Integral Transforms
Integral transform techniques have been applied to a number of problems in the field of Geomechanics where it was possible to treat the soil profile as being horizontally layered in one or two coordinate directions. Solutions for an elastic soil have been obtained for problems involving

soil profiles subjected to two- and three-dimensional loadings, anisotropic material properties, properties which vary with depth, and problems involving soil-structure interaction. Examples of these applications are given in the following sections.

6.2.1 Strip, Circular or Rectangular Loading Applied to Horizontally Layered Soil

In the case where the soil may be considered to be horizontally layered, integral transforms may be used to dramatically simplify the problem. In the case of a strip loading, a single Fourier transform may be used (see Section 4), for a circular loading a Hankel transform is applied (see Section 5.1) and for a general shaped loading, a double Fourier transform is required (see Section 5).

Small and Booker (1984, 1986a) have presented solutions for the case where uniformly distributed loads are applied to the surface of layered soil profiles. The method employs a flexibility formulation rather than a stiffness formulation and this has the advantage that solutions may be found for incompressible materials (or those for which Poisson's ratio is equal to 1/2). This is very useful in geotechnical applications as clays deform at constant volume under undrained loading conditions.

Furthermore, the transforms of the field quantities, which are a function of depth are determined explicitly within each material layer and do not have to be approximated by use of interpolation functions thus leading to the necessity of using additional nodes within each physical material layer.

Results from such an analysis are shown in Fig. 6. A circular, strip or a rectangular loading has been applied to the surface of a soil layer which is made up of two sublayers; sublayer (A) is of depth H_A while the lower layer (B) is of depth H_B. The loadings were chosen so that they had the same minimum dimension, i.e. the loading q was applied over the region $|x| < a$ (strip), $0 < r \leqslant a$ (circle), $|x| < a$, $|y| < b$ (rectangle). The material was assumed to be anisotropic. For layer A, $E_h/E_v = 1.5$, $G/E_v = 0.45$, $v_h = 0.25$, $v_{hv} = 0.3$, $v_{vh} = 0.2$ and for layer B, $E_h/E_v = 33$, $G/E_v = 0.5$, $v_h = 0.1$, $v_{hv} = 0.9$, $v_{vh} = 0.3$. (The subscripts h, v indicate horizontal and vertical directions, respectively.)

The plots shown in Fig. 6 are for the vertical σ_{zz} (Fig. 6(a)) and horizontal σ_{rr} or σ_{xx} (Fig. 6(b)) stresses along the axis of the loading ($x = y = r = 0$). There is a large difference in the vertical stress computed for each of the loading types, however there is less difference in the horizontal stresses for this particular soil profile.

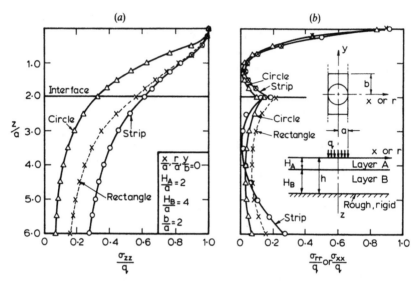

FIG. 6.   Stresses in a soil consisting of two anisotropic layers. Strip, circular and rectangular loading patterns.

6.2.2 Strip or Circular Loading Applied to a Soil with a Modulus which Increases Linearly with Depth

If the conventional finite layer method described in the previous section is used for problems in which the soil profile increases linearly with depth, the soil profile must be broken up into a number of sublayers and a step wise approximation made to the linear variation of modulus. This is clearly inefficient but may be overcome if the modulus within each layer is assumed to vary exponentially, e.g.

$$E(z) = E_{0k} \, exp[2\mu_k(z - z_k)] \tag{52}$$

E_{0k} is the modulus at $z = z_k$ and μ_k describes the variation of modulus with depth. The choice of an exponential function greatly simplifies the solution of the finite layer equations, and by careful choice of the parameters E_{0k} and μ_k the variation of modulus with depth z (i.e. $E(z)$) may be approximated closely (see Fig. 7).

The use of such a scheme has been demonstrated by Rowe and Booker (1981a,b), who presented parametric results for soils which have a modulus increasing with depth and/or a surface crust of stiffer material.

Shown in Fig. 8 is the result of an analysis of a strip footing of breadth B on a soil with a crust. The variation of the modulus of the soil is shown

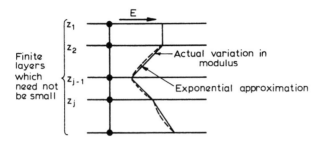

FIG. 7.   Typical non-homogeneous soil profile with exponential approximation
(after Rowe and Booker, 1981a).

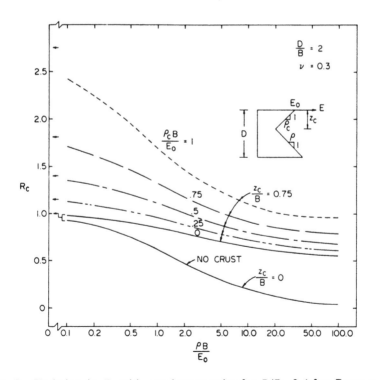

FIG. 8.   Variation in R_c with non-homogeneity for $D/B = 2$ (after Rowe and
Booker, 1981a).

in the inset on the figure. The deflection at the centre of the loading may be calculated by use of the correction factor R_c which is defined as the ratio of the settlement (at the centre) of a strip loading on a non-homogeneous soil to that for a homogeneous soil with the same surface modulus E_0 and Poisson's ratio v. In this figure, B is the full width of the strip loading, z_c is the crust thickness and D is the layer depth.

It may be seen that the factor R_c can become quite large for large values of ρ_c (the parameter describing the rate of decrease of the crust modulus). Thus the actual deflections would be far greater than those obtained for a strip footing where it was assumed that the Young's modulus was the same as or greater than the surface modulus.

6.2.3   Solutions for Profiles or Geometries which are Constant in One Coordinate Direction Only

If the soil profile is such that it may be approximated as horizontally layered in one coordinate direction or has a geometry which is constant in that direction (see Fig. 9), then a single integral transform may be used to reduce the three-dimensional problem to one which essentially involves solving a problem involving only two spatial dimensions.

The solution method is the integral transform equivalent of the finite prism method of Cheung et al. (1976) and has been demonstrated by Small and Wong (1988). The solution process involves applying a Fourier transform to the field quantities, effectively eliminating one spatial dimension (in this case y) and then solving for the transformed quantities in the plane perpendicular to this direction. Since an analytic solution for the transformed quantities is difficult to find, finite element techniques are used to approximate nodal values in this plane, and so the solution involves determining nodal values of the transformed field quantities.

Numerical integration is then used to invert the transformed field quantities in order to obtain the actual displacements, stresses, etc. at any position. The numerical integration process has been explained in Section 4.

As an example of where such an analysis can be most effective, the problem of a building constructed over an existing tunnel is examined. The problem is shown schematically in the inset to Fig. 10(a); a uniform vertical loading q is applied over a square region $|x| < a$, $|y| < a$ of the surface ($z = 0$). The tunnel lies in a uniform elastic soil or rock with a Young's modulus E and Poisson's ratio $v = 0.3$ and at a depth $2a$ beneath the surface.

For this particular problem discretization using triangular elements

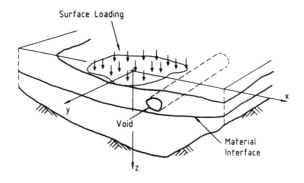

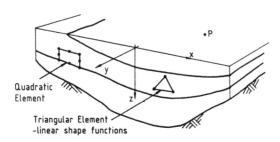

FIG. 9.   Problems involving constant soil cross-section in one axial direction.

has been carried out in the $x - z$ plane and the mesh (shown deformed on an exaggerated scale) is presented in Fig. 10(b). Using this mesh, the computed vertical displacements u_z along the crown of the tunnel are as shown in Fig. 10(a), where it may be seen that fairly substantial deflections are occurring in the tunnel roof for at least a distance of one loading width past the edge of the loading.

6.2.4   Problems Involving Soil-Structure Interaction
Soil-structure interaction problems may be dealt with in a straightforward manner (Cheung *et al.*, 1985; Cheung and Zienkiewicz, 1965) by simply treating the unknown contact stresses between the structure and the soil as being made up of a series of uniform blocks of pressure. For a raft foundation, the blocks of pressure are assumed to act over a

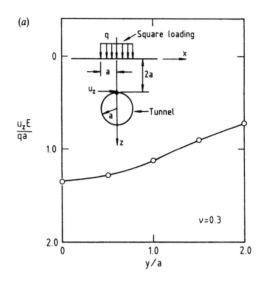

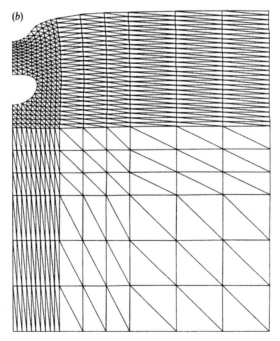

FIG. 10. (a) Surface loading applied to soil/rock mass containing a tunnel. (b)
Deformed finite element mesh at centre of loading $y/a = 0$.

rectangular area surrounding each node in the finite element mesh used in the analysis of the raft. The finite layer method may be used to determine the response of the soil to the applied uniform loads, and it is found that

$$\mathbf{w} = \mathbf{F_f P} \tag{53}$$

where $\mathbf{w}$ is the vector of displacements at the nodal points, $\mathbf{P}$ is the vector of contact stresses for each block and $\mathbf{F_f}$ is the foundation flexibility matrix, the columns of which represent the deflections at all nodal points due to unit contact pressure applied over a single rectangular block.

Equation (53) may be inverted to give the stiffness relation

$$\mathbf{P} = \mathbf{K_f\, w} \tag{54}$$

where

$$\mathbf{K_f} = \mathbf{F_f^{-1}}$$

If a set of nodal forces $\mathbf{f}$ is applied to the structure it is found that

$$\mathbf{K_s w} = \mathbf{f} - \mathbf{P} \tag{55}$$

where $\mathbf{K_s}$ is the stiffness matrix of the structural system. Equations (54), (55) may now be combined to give the complete set of stiffness equations for the soil–structure system

$$(\mathbf{K_f} + \mathbf{K_p})\mathbf{w} = \mathbf{f} \tag{56}$$

Equation (56) may now be directly solved to determine the displacements at nodal points. The magnitudes of each block of pressure can then be determined from eqn (54).

An improvement to this method may be made by assuming that the contact stress may be approximated by a series of terms. This is particularly convenient for circular structures such as anchors, liquid storage tanks and screw plate devices.

Suppose that the contact stress $q_c(r)$ can be written as

$$q_c(r) = \begin{cases} \sum_{n=1}^{N} F_n \varphi_n(r), & 0 < r \leqslant a \\ 0, & a < r \leqslant \infty \end{cases} \tag{57}$$

where $\varphi_n(r)$ are the chosen functions of radius r and F_n are the unknown coefficients of the series.

It is possible to proceed in one of two ways. The first is the approach adopted by Small and Brown (1988) for analysis of a smooth rigid embedded circular plate (shown schematically in Fig. 11). Each of the loading functions is applied to the layered soil in turn and the deflections

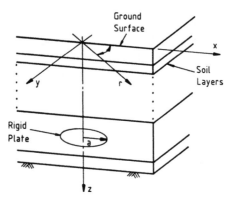

FIG. 11.   Rigid plate within a layered soil.

found at a number of collocation points which may be chosen at any position, but are most conveniently chosen to be evenly spaced across the plate. The unknown multipliers F_n are then found such that (in the case of a rigid plate) all of the deflections at the collocation points are equal. More explicitly this may be written

$$\mathbf{I_s F = \delta}$$

where $\mathbf{I_s}$ is an influence matrix whose nth column consists of the deflections at the N collocation points due to the load $q_c = \varphi_n(r)$, $\boldsymbol{\delta} = [\delta_1 \quad \ldots \quad \delta_N]^T$ is the vector of deflections at the collocation points and $\mathbf{F} = [F_1 \quad \ldots \quad F_N]^T$ are the unknown coefficients in the series.

For a rigid anchor undergoing a deflection Δ

$$\boldsymbol{\delta} = \Delta[1 \quad 1 \quad 1 \quad \ldots \quad 1]^T$$

The influence matrix for the soil $\mathbf{I_s}$ may be obtained by use of the finite layer method provided it is possible to obtain the integral transform of the loading functions. Once this is done, the deflections due to the loading $\varphi_n(r)$ may be computed using a standard finite layer program, and therefore it is possible to treat problems in which there are layered soil profiles or soils which are either isotropic or anisotropic. Vertical equilibrium of force must also hold and so the load applied to the plate (P_A) must be equal to the total contact force, i.e.

$$2\pi \sum_{n=1}^{N} q_n F_n = P_A$$

where (58)

$$q_n = \int_0^a \varphi_n(r) r \, dr$$

We therefore may solve for the unknown coefficients of the series F_n from the following set of equations

$$\begin{bmatrix} \mathbf{I}_s & -\mathbf{a} \\ -\mathbf{q}^T & 0 \end{bmatrix} \begin{bmatrix} \mathbf{F} \\ \Delta \end{bmatrix} = \begin{bmatrix} 0 \\ -P_A/2\pi \end{bmatrix}$$

(59)

where

$$\mathbf{a} = [1 \quad 1 \quad 1 \quad \ldots \quad 1]^T$$

$$\mathbf{q} = [q_1 \quad q_2 \quad \ldots \quad q_N]^T$$

Small and Brown (1988) chose the functions $\varphi_n(r)$ to be uniform blocks of pressure applied over annuli together with a term which varied in a similar way to the stress distribution acting on a perfectly rigid surface loading. Hence

$$\varphi_n(r) = \begin{cases} 1, & r_{n-1} < r < r_n \\ 0, & \text{elsewhere} \end{cases}$$

and

$$r_0 = 0, \quad r_{N-1} = a$$

$$\varphi_N(r) = 1/(r^2 - a^2)^{1/2}$$

The functions $\varphi_n(r)$ are shown in Fig. 12 with results for the rigid deflection of a circular plate of radius a carrying an average applied pressure q in a uniform soil with modulus E and Poisson's ratio v. Results for deflection of the plate are shown in Fig. 13 for both an isotropic material and an anisotropic material with soil properties $E_h/E_v = 4$, $v_h = 0.125$, $v_{vh} = 0.1875$ and $G/E_v = 0.65$ (subscripts v, h refer to vertical and horizontal directions).

A more advanced method is not to match displacements at selected

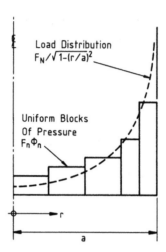

FIG. 12.   Chosen form of loading functions.

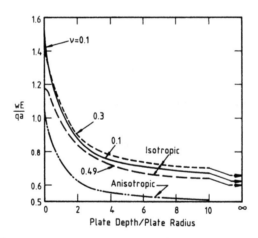

FIG. 13.   Variation of plate displacement with depth.

collocation points, but to use an 'average' deflection for each of the functions $\varphi_n(r)$. This approach has been used by Rowe *et al.* (1982) for the analysis of circular fluid storage tanks. The average deflection δ_n is defined as

$$\delta_n = \int_0^a w(r)\varphi_n(r)r\,\mathrm{d}r \qquad (60)$$

where $w(r)$ is the surface deflection of the soil. By evaluating a similar 'average' deflection for the base of the tank and matching this 'average' deflection with that of the soil, a complete solution to the problem may be obtained.

In the case where the structure has a finite flexibility, the choice of functions

$$\varphi_n = (1 - (r/a)^2)^n, \qquad n = 0, 1, 2, 3, ..., N - 1 \qquad (61)$$
$$\varphi_N = 1/(1 - (r/a)^2)^{1/2},$$

has been found most useful since φ_0 has the form of the contact stress distribution for a very flexible structure (i.e. a uniform loading) and φ_N is the form of the distribution for a rigid loading. The addition of other terms $(\varphi_n, n = 1, ..., N - 1)$ enables solutions to problems where the structural foundation is intermediate of the rigid and perfectly flexible cases.

Once again the finite layer method may be used to obtain the response of the soil to the applied loading functions, provided it is possible to find the integral transform of the loading functions $\varphi_n(r)$. For a circular foundation this involves taking Hankel transforms, and is discussed in detail by Rowe *et al.* (1982).

6.2.5   Horizontal Loading

All of the problems discussed thus far have been concerned with vertical loading, however there are many engineering problems which involve lateral or horizontal loadings applied to foundations. For example, lateral forces may be generated by traffic braking or turning and these forces are transmitted to the layered pavement. Such problems are ideally suited to analysis by the finite layer method.

An example of such a problem is shown schematically in Fig. 14 where a uniform shear t is shown applied to the surface of a layered elastic material over a circular region of radius a.

As an example, the variation of vertical u_z and horizontal u_x displacements with depth is calculated for the three layered system shown in the inset to Fig. 15 on the line $x/a = 0.5$, $y/a = 0$. For this paticular problem the material properties of each of the sublayers A, B, C are described below:

Layer A: $E_h/E_v = 2$, $G_v/E_v = 0.4$, $v_h = 0.3$, $v_{vh} = 0.2$

Layer B: $E_h/E_v = 2$, $G_v/E_v = 0.4$, $v_h = 0.4$, $v_{vh} = 0.2$

Layer C: $E_h/E_v = 1$, $G_v/E_v = 1/3$, $v_h = 0.5$, $v_{vh} = 0.5$

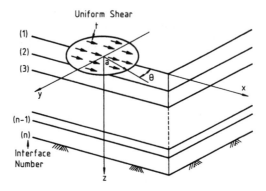

FIG. 14.   Uniform shear loading applied over a circular region.

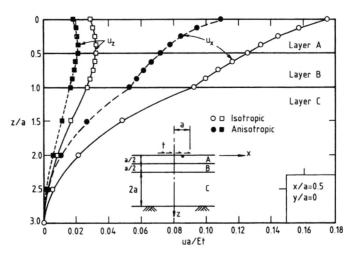

FIG. 15.   Horizontal and vertical displacements in layered material (effect of anisotropy).

The ratio of Young's modulus in the layers is assumed to be

$$(E_v)_A:(E_v)_B:(E_v)_C = 25:5:1$$

Details of the method of taking integral transforms of such shear loadings are given by Booker and Small (1988) and Small and Booker (1986a). For non-symmetric loadings such as shear loadings, two sets of uncoupled finite layer equations result in contrast to symmetric loading problems which result in only one set of finite layer equations. Because the sets of equations are uncoupled, they may be solved independently and only a slight increase in computing time is required in order to obtain solutions for the non-symmetric case as compared to the symmetric case.

6.3 Other Orthogonal Series

In the previous sections, attention has been restricted to Fourier or Fourier-Bessel series or Fourier or Hankel transforms. However, there are other orthogonal functions which may be used. For example Guo *et al.* (1987) in their analysis of piles have used Fourier series to expand field variables in a tangential (θ) direction and a series to expand the field variables in the radial (r) direction. For example the radial displacement around a pile of radius r_0 is given by

$$u_r = \sum_{m=0}^{M} Z^r \cos m\theta \{ \Gamma_{r_0 R} \sum_{n=1}^{N} H_n^1(r)\alpha + \Gamma_{R\infty} \sum_{l=1}^{L} H_l^2(r)\beta \} \tag{62}$$

In the tangential (θ) direction the displacement is expanded as a Fourier series, while in the radial direction the displacements are represented by two series, one valid between $r_0 \leqslant r \leqslant R$ and another valid for the far field $R < r < \infty$. The multipliers Γ reflect this as $\Gamma_{pa} = 1$, $p \leqslant r \leqslant a$ but zero elsewhere. The series which represent the near and far fields are given by the functions $H_n^1(r)$ and $H_l^2(r)$, respectively and α and β are the corresponding interpolating parameters. Z^r is the interpolating function with respect to depth (for the radial displacement) and is determined in the same manner as for finite layer methods which use Fourier series.

The particular functions used by Guo *et al.* (1987) are given below. In the near field ($r_0 \leqslant r \leqslant R$)

$$H_1^1(r) = 1$$

$$H_2^1(r) = \frac{r}{R}$$

$$H_3^1(r) = 3\frac{r}{R} - 4\left(\frac{r}{R}\right)^2$$

$$\vdots$$

$$H_n^1(r) = \sum_{k=1}^{n-1} (-1)^{k-1} \frac{(n+k-1)!}{(n-k-1)!(k+1)!(k-1)!} \left(\frac{r}{R}\right)^k$$

while in the far field $(R < r < \infty)$

$$H_l^2(r) = \left(\frac{R}{r}\right)^l \quad (r > R)$$

The finite layer equations corresponding to each Fourier term are uncoupled as is usual, but are coupled for the radial terms.

7 TIME-DEPENDENT PROBLEMS

The finite layer method is useful not only for obtaining solutions to elastic problems but may also be applied to the whole range of time-dependent problems, such as those involving consolidation, creep, thermo-elasticity and ground water extraction as well as many others.

The analysis of such problems becomes extremely time consuming and expensive to solve when using other numerical techniques such as finite element or finite difference methods, especially when the problem is three-dimensional in nature. However, for the class of three-dimensional problems in which the ground may be assumed to be horizontally layered, the application of the finite layer technique essentially reduces the three-dimensional problem to a sequence of problems which involve only a single spatial dimension and this leads to great savings in computer storage.

7.1 Consolidation

Various techniques have been used to obtain solutions to the consolidation problem. An early approach (Small and Booker, 1979), involved using a Fourier series to represent the displacements and pore pressures in the consolidating soil layer. The coefficients of the Fourier series were approximated by using one-dimensional elements throughout the depth of the layer and assuming that variation within the element could be found by linear interpolation of nodal values. A 'marching' type solution was used to find the solution at any time, each solution being found from the solution at the previous time step. Cheung and Tham (1983) also used Fourier series to obtain solutions to the consolidation of a layered soil. They demonstrated the use of both a quasi-variational and a least square formulation of the problem and examined different forward marching schemes.

The above solutions are based on the theory of Biot (1941) where the soil is treated as a poro-elastic medium. The flow of water out of the pores of the soil results in a compression of the soil skeleton or matrix, and so the fluid flow problem and the stress-deformation problem are coupled together.

Booker and Small (1982a,b) first demonstrated the use of integral transforms in the solution of consolidation problems involving Biot's theory. Instead of approximating the Fourier coefficients by the use of one-dimensional elements and linear interpolation functions, the Fourier transforms of the field quantities were approximated by linear interpolation. Again a 'marching' solution was used to obtain the solution at any particular time during the consolidation process.

Fourier transforms were also used by Vardoulakis and Harnpattanapanich (1986), and Harnpattanapanich and Vardoulakis (1987) for the solution of Biot consolidation problems. They demonstrated the use of direct numerical inversion to obtain the time-dependent solution. This approach involves not only applying the usual Fourier or Hankel transform to the field quantities (i.e. the displacements, pore pressures and loadings), but applying a Laplace transform as well. The effect of this is to effectively eliminate the time dimension. The solution is found by solving the resultant equations and then using numerical methods to invert both the Fourier or Hankel transform and the Laplace transform. The solution may thus be found directly at any time without the need to use a 'marching' scheme.

The approach of Vardoulakis and Harnpattanapanich is based on McNamee and Gibson's (1960) representation of the field quantities and therefore necessitates a 10×8 layer matrix for each layer of material. A similar approach was taken by Booker and Small (1987), whereby a Fourier or Hankel transform was applied to the field quantities together with a Laplace transform. However the method of formulation was different and led to a 6×6 finite layer matrix for each layer of material.

Vardoulakis and Harnpattanapanich (1986) presented results for the problem of a uniform step loading F_z applied to the surface of a layer of soil in which the modulus increases with depth. The shear modulus $\mu(z)$ was assumed to increase with depth according to the equation

$$\mu(z) = \mu_0(1 + z/l_m) \tag{63}$$

where μ_0 is the surface value of μ and l_m represents the depth at which the modulus reaches twice its value at the surface.

The central settlement (u_z) of the strip footing is shown in Fig. 16 plotted against non-dimensional time τ_0 where $\tau_0 = c_0 t/l^2$. The coefficient

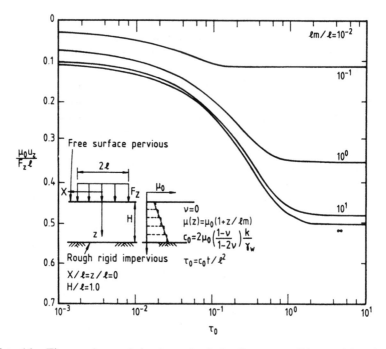

FIG. 16. Time-settlement behaviour of strip loading on a soil layer with a shear modulus which increases with depth (after Vardoulakis and Harnpattanapanich, 1986).

of consolidation c_0 is defined on the figure in terms of the Poisson's ratio of the soil v and its permeability per unit weight of water k/γ_w. It may be observed from the time-settlement plots that as the rate of increase of shear modulus with depth l_m/l becomes larger the final deflections become smaller and occur at a smaller time τ_0.

Because it is possible to directly invert the Laplace transform at any time t, with this type of formulation, problems which involve time-dependent loadings may be easily dealt with, provided that the Laplace transform of the loading function may be taken. An example of this has been given by Small and Booker (1988) for an embankment shaped loading which is increasing with time.

For this problem the loading distribution q may be expressed as

$$q = q_{max}, \qquad |x| \leqslant a$$

$$q = q_{max}\left(2 - \frac{|x|}{a}\right), \qquad a \leqslant |x| \leqslant 2a$$

(64)

This loading was assumed to be applied to a soil layer which had a permeability in the horizontal direction four times that in the vertical direction (i.e. $k_h/k_v = 4$). Two different loading programmes (a,b) were considered. Firstly (a) the loading was assumed to have been applied in a linear fashion starting from zero at $\tau = 0$ and reaching a maximum at $\tau = 1$, where τ is the time factor defined in Fig. 17. In the second case (b)

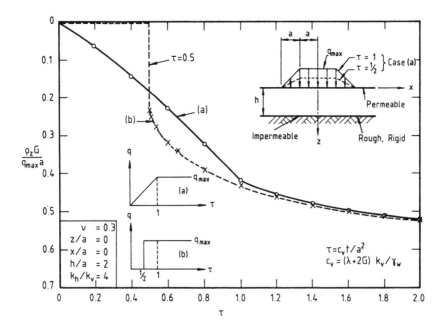

FIG. 17. Effect of loading history on time-settlement behaviour of embankment.

the load was considered to have been applied as a step loading at $\tau = \frac{1}{2}$ and thereafter held constant.

A plot of the central deflection (ρ_z) of the embankment with time is shown in Fig. 17 for the two different loading histories (which are shown schematically in insets (a) and (b) to the figure). It can be seen that for times greater than about $\tau = 1$ the 'step' loading gives quite a good approximation to the deflection calculated for the same loading applied linearly within the time range $0 < \tau \le 1$.

A slightly different approach to the solution of consolidation problems has also been presented by Runesson and Booker (1982a) which involves expressing the field quantities which are functions of position and time in terms of a double Fourier series, e.g. for a two-dimensional problem

$$f_k = \sum_{n=-\infty}^{+\infty} \sum_{m=-\infty}^{+\infty} F_{knm} \exp[i(\alpha_m x + \omega_n t)] \qquad (65)$$

where $f_k(x, t)$ is a field quantity, $F_{knm}(z)$ are the Fourier coefficients and x, z are the horizontal and vertical coordinate directions, t is time and $\alpha_m = 2m\pi/L_x$, $w_n = 2n\pi/T$. L_x and T are the periods in the horizontal direction and time, respectively. Using this approach Runesson and Booker were also able to deal with loadings which were time-dependent.

The same authors (Runesson and Booker, 1982b) have applied discrete Fourier analysis to the consolidation problem. They demonstrated the use of the method for obtaining solutions to strip and square loadings and examined the effects of the artificial boundaries which are introduced by assuming that the loading is periodic.

7.2   Visco-elastic Problems

Problems involving the time-dependent deformation of materials under a constant applied loading may often be treated by assuming that the materials display visco-elastic behaviour, i.e. that the material properties are themselves dependent on time. Creep in concrete under load, or in other materials at elevated temperatures as well as creep in soil or rocks are some of the problems which may be solved by treating the materials involved as being visco-elastic.

It is well known that creep in materials such as soil and rock, which are of primary interest in the field of geomechanics, is a highly complex process (Singh and Mitchell, 1968) and the creep rate may be dependent on many factors such as deviator stress level, mean stress level, tempera-ture, etc. To deal with such complexities, it is still necessary to use a full finite element analysis, however if we can make some simplifying assum-ptions, then the finite layer method may be used for analysis.

For example, we may make the assumption that the behaviour of the material may be separated into a deviatoric and a volumetric behaviour. That is to say there will be a different time-dependent response to a mean stress increase, than to a deviatoric or shear stress increase. We may therefore propose that these responses are caused by a time-dependent change in the bulk modulus and the shear modulus of the material.

Use may then be made of the elastic-visco-elastic analogy or 'corre-spondence principle'. For example, we may write the relationship between the mean or volumetric stress σ_v and the deviatoric σ_d stress and the volumetric ε_v and deviatoric strains ε_d by use of relaxation functions $R_v(t)$

and $R_d(t)$, e.g.

$$\sigma_v(t) = \varepsilon_v(0)R_v(0) + \int_0^t R_v(t-\tau)\frac{\partial \varepsilon_v}{\partial \tau} d\tau \qquad (66)$$

$$\sigma_d(t) = \varepsilon_d(0)R_d(0) + \int_0^t R_d(t-\tau)\frac{\partial \varepsilon_d}{\partial \tau} d\tau \qquad (67)$$

Applying a Laplace transform to the above equations we would obtain

$$\bar{\sigma}_v = s\bar{R}_v\bar{\varepsilon}_v = \bar{K}\bar{\varepsilon}_v$$
$$\bar{\sigma}_d = s\bar{R}_d\bar{\varepsilon}_d = \bar{G}\bar{\varepsilon}_d \qquad (68)$$

where the superior bar denotes the Laplace transform.

Hence by transforming the equations in this manner, the relationship between the transformed stress and strain is analogous to the relationship between stress and strain for an elastic material.

This means that if the associated elastic problem can be solved, then so can the time-dependent problem (in Laplace transform space). All that remains then, is to invert the transformed quantities to obtain the solutions in real time. This approach was taken by Booker and Small (1985). In order to invert the Laplace transforms it was assumed that the relaxation functions could be written in terms of an exponential series, e.g.

$$R_v(t) = A_0 + \sum_{j=1}^n A_j \exp(-\lambda_j t) \qquad (69)$$

and this led to a method of obtaining a 'marching' solution, in which a solution at any time is found from the solution at the previous time. However fitting a relaxation function with an exponential series is a time consuming process, and may not always be possible. A much improved method is to directly invert the Laplace transforms using numerical techniques.

Direct inversion has been demonstrated by Booker and Small (1986). In this paper the equations of visco-elasticity were written in terms of creep functions $J(t)$ rather then relaxation functions

$$\varepsilon_v(t) = J_v(t)\sigma_v(0) + \int_0^t J_v(t-\tau)\frac{\partial \sigma_v}{\partial \tau} d\tau$$

hence

$$\bar{\varepsilon}_v = \bar{C}_v\bar{\sigma}_v \qquad (70)$$

$$\bar{\varepsilon}_d = \bar{C}_d\bar{\sigma}_d \qquad (71)$$

This enabled the corresponding 'elastic' problem to be solved. The solution for the field quantities must then be found by numerical inversion, and the extremely efficient algorithm reported by Talbot (1979) was used in this case.

As an example of the type of problem that may be solved by using direct inversion of the Laplace transform, consider the problem shown in the inset to Fig. 18. It involves a strip or circular loading applied to a soil which consists of an upper visco-elastic layer and a lower elastic layer.

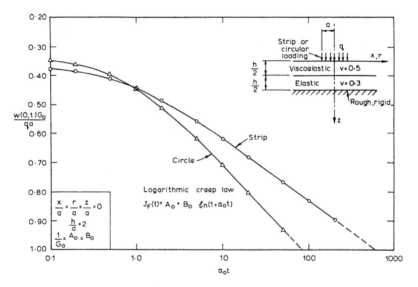

FIG. 18.   Deflection vs time for central point of a strip or circular loading, resting on a layered material with a visco-elastic upper layer.

Both materials were assumed to have a constant Poisson's ratio v, that of the upper layer being $v = 0.5$ (i.e. incompressible) and that of the lower layer $v = 0.3$. It was assumed that the deformation of the material was due to a time-dependent shear modulus. Hence the shear strain γ_{xy} would be linked to a constant shear stress τ_{xy} applied at time $t = 0$ by

$$\gamma_{xy} = J(t)\tau_{xy}(0) = \frac{\tau_{xy}(0)}{G(t)} \tag{72}$$

where $J(t)$ is the creep function. The creep function in this example was chosen to be approximately linear with the logarithm of time as is often observed for soil and rock, so that

$$J(t) = A_0 + B_0 \, ln(1 + \alpha_0 t) \tag{73}$$

where A_0, B_0, α_0 are material constants to be determined from experiment.

Results for the central vertical displacement of the loaded region w are shown in Fig. 18 plotted in non-dimensional form against non-dimensional time $\alpha_0 t$ (see eqn (73)). Because of the form of the chosen creep function, a linear plot is obtained when time is plotted to a logarithmic scale.

For the strip loading, lateral strains ε_{xx} along the line of symmetry have been plotted in Fig. 19 at various times. The strains may be seen to increase with time in the upper visco-elastic layer, but increase only very marginally in the lower elastic layer.

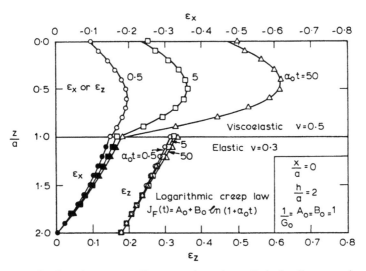

FIG. 19.   Strains along centre line at various times. Strip loading on a layered material with a visco-elastic upper layer.

7.3   Consolidation and Creep

Often geotechnical problems involve not only consolidation or creep but a combination of the two, with both processes occurring simultaneously. As the finite layer method has proved largely successful for the solution of consolidation or creep problems, little further effort is required in applying the method to the combined problem.

For example, if we can write the relaxation functions for the soil in terms of an exponential series (as was previously discussed in Section 7.2) we can use a 'marching' process to obtain the solution at any given time. A complete outline of the solution process is contained in the paper by Small and Booker (1982). In this paper the shear modulus of the soil was assumed to be time-dependent with the shear modulus G being represented by a series of springs and viscous dashpots. In such a case, the transformed shear modulus may be expressed in the form

$$\bar{G} = B_0 + \frac{B_1 s}{s + \lambda_1} + \frac{B_2 s}{s + \lambda_2} + \cdots + \frac{B_n s}{s + \lambda_n} \tag{74}$$

where B_n, λ_n are constants determined for the soil type.

The inset of Fig. 20 shows a uniform loading q applied over a strip, circular or rectangular area on the surface of a visco-elastic layer of soil of thickness h. The dimensions of the loaded region are as shown on the figure. The material is assumed to have a constant Poisson's ratio $v_0 = 0.3$ and a constant permeability per unit weight of water k/γ_w. However the shear modulus of the soil is assumed to change with time in such a way that the shear strain − shear stress relationship can be modelled by the series of springs and dashpots shown in the inset to Fig. 20.

The settlement at the central point of the loaded region w_0 is plotted against time t in Fig. 20 and it may be noticed that as time is plotted on

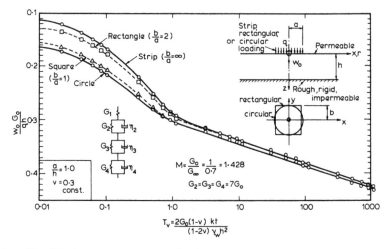

FIG. 20.   Displacement at centre of strip, circular or rectangular load on a soil exhibiting primary and secondary consolidation characteristics.

a logarithmic scale, the settlement-time plot becomes linear once primary consolidation is complete. This behaviour is often observed for soils which tend to display creep behaviour. With the terminating creep model used here, (i.e. the spring and dashpot model) the settlement of the foundation will not continue to creep in this way but will eventually reach an ultimate value at large times. If it is necessary to model a non-terminating creep behaviour, then the use of the direct inversion technique described in Section 7.2 for purely visco-elastic materials could be applied.

7.4 Thermo-elasticity

The flow of heat through an elastic material and the stress and deformation caused by the expansion due to the temperature increase is of interest in a number of fields, especially those which deal with engines and machinery which generate heat. In the field of geomechanics the interest is mainly in the effects caused by cooling tunnels or mine shafts which are present at some depth in hot rock, or in problems associated with atomic waste disposal where heat generated by the waste induces thermal stresses into the surrounding rock.

The heat flow problem is much like that of consolidation, where the flow of water away from regions of high excess water pressure causes a decrease in volume in the soil. For heat flow it is the flow of heat from regions of high temperature which causes a shrinkage or volume reduction. However unlike consolidation problems it is not necessary to couple the flow and stress analysis as this effect is small, that is to say that an increase in temperature will cause an increase in strain, but an increase in stress will only cause a small temperature increase (see Booker and Smith, 1989).

Examples of the use of the finite layer method in the solution of thermo-elastic problems have been given by Small and Booker (1986b). One such problem involves a decaying heat source in a uniform elastic rock matrix. The solution method once again involves applying Laplace transforms to the field quantities, solving the resultant finite layer equations, and then using numerical inversion to obtain the solution in real time.

The point heat source of strength Q is shown in the inset to Fig. 21 at a depth h beneath the surface, and the strength of this source is such that it is decaying exponentially, i.e.

$$Q = Q_0 \, e^{-\lambda t} \tag{75}$$

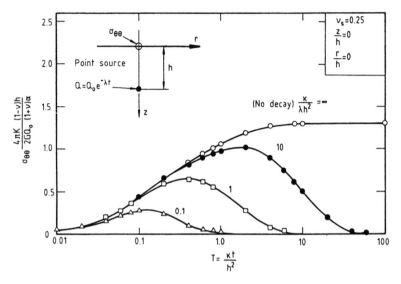

FIG. 21.   Variation of tensile stress at surface of rock mass containing a decaying
heat source.

where Q_0 = initial strength of source, $\lambda = \ln 2/t_1$ with $t = t_1$ the half-life of the source.

The heating will cause the surrounding rock to expand and cause tensile stresses in the rock at various times and positions. The tensile stresses induced into the rock at the surface directly above the source are shown in Fig. 21. The tangential stresses $\sigma_{\theta\theta}$ are plotted in non-dimensional form against non-dimensional time for different decay rates indicated by the parameter $\kappa/\lambda h^2$. Symbols for the elastic and thermal properties of the rock used on this figure are as follows: G is shear modulus, v is Poissons' ratio, K is the coefficient of conductivity, κ is the diffusivity and α is the coefficient of linear expansion of the material.

It may be seen from the figure that as the rate of decay of the source becomes slower (i.e. $\kappa/\lambda h^2$ has a larger value) the tensile stress at the surface becomes larger and peaks at a greater time t after the initial placing of the heat source. Studies such as these have shown that such heat sources, which may be due to high level atomic waste, can induce tensile stresses into the rock and lead to possible cracking of the rock and contamination of the groundwater.

8 OTHER APPLICATIONS

Finite layer methods may be applied to other problems which do not exactly fall into the categories of 'elastic' or 'time-dependent' problems which were discussed in the previous sections.

It is possible, for example, to carry out non-linear analysis of horizontally layered soils by making use of discrete Fourier series.

An example of the use of discrete Fourier series for non-linear problems has been presented by Runesson and Booker (1983) for a uniform strip loading W applied to the surface of a layer of soil with boundaries at $6B$ (side) and $4B$ (base) away from the loading where B is the half-width of the loading. This is shown schematically in Fig. 22.

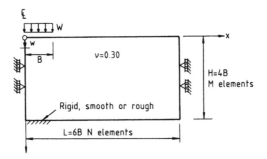

FIG. 22.   Flexible strip footing on a homogeneous cohesive subsoil (after Runesson and Booker, 1983).

The soil was chosen to be elastic-perfectly plastic and obeying Tresca's yield criterion (i.e. $\varphi = 0$), and having an undrained shear strength c. The elastic modulus of the soil is E and its Poisson's ratio $v = 0.3$. The discrete Fourier series approach involves representing the field variables at a number of nodal columns N as discussed by Runesson and Booker, each column containing M nodal points. For this problem M was chosen to be 10 and N was varied in order to assess the effect on the collapse load.

The load-deflection curve for the strip loading (w is the vertical deflection at the centre of the load) is shown in Fig. 23 for each of the N values considered. As may be expected, the theoretical collapse load of $(2 + \pi)c$ is approached more closely as the number of terms (N) increases.

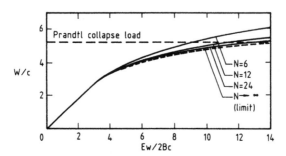

FIG. 23. Influence of number of Fourier terms N (after Runesson and Booker, 1983).

9 APPLICATION TO MICROCOMPUTERS

9.1 FLEA, FLAC

Since the finite layer method is particularly suited to use on microcomputers, most of the results presented in this chapter have been obtained on personal computers. Very little data preparation and computer memory is necessary, and solution times are extremely fast, which means that a program can be run many times in succession in order to assess the sensitivity of solutions to say, the effects of anisotropy or layer thickness. Even for three-dimensional loadings, input is extremely simple and CAD packages or mesh generators are not required as they would be if finite element methods were used.

For example, the progam FLEA (Finite Layer Elastic Analysis) which has been developed at the Centre for Geotechnical Research at the University of Sydney, has a simple screen editor which is used for data preparation. Prompts are issued to the user for details such as the shape of the lcading, whether a vertical and/or shear loading is required, layer thicknesses and material properties of each layer. Data are checked to make sure that they are of the correct type (i.e. real or integer) and that the program limits have not been exceeded (e.g. too many layers). The data may be changed if so desired before the computation is initiated, through use of the editing facility. Stresses, strains and displacements are then computed at the specified locations. Such a program is particularly useful for the design of pavements which are subjected to vertical and horizontal traffic loads applied over circular regions.

As an example of run times, the problem of a vertical circular loading applied to a single material layer was carried out on a machine with an 80286 processor. The time taken to evaluate the solution was 31 s when using 80 Gauss points to carry out the numerical inversion. Good accuracy can be achieved with fewer Gauss points which means run times will be faster, however 80 points are used in the pre-set integration scheme of FLEA and so this number will be used here for comparing run times.

If an 80287 math coprocessor is used, the run time is reduced to 8 s. If the problem involves four material layers, run times (CPU time) are 103 s (80286) and 23 s (80287).

One advantage of the FLEA program is that it is able to deal with incompressible materials such as undrained clays since it is based on a flexibility formulation (i.e. the inverse of the stiffness matrix given in Appendix 1). Hence no solution problems occur if a Poisson's ratio $v = 0.5$ is used.

The results for vertical loadings applied to strip, circular and rectangular regions shown in Fig. 6 were obtained using FLEA as were the horizontal loading results of Fig. 15.

For consolidtion analysis, program FLAC (Finite Layer Analysis of Consolidation) has been developed. The program is based on a more conventional finite layer approach, where discretization within each material layer is required in order to approximate the transformed field quantities with depth. Inversion of the transformed quantities is carried out by using numerical integration (Gaussian quadrature). For each Gauss point taken the solution is 'marched' forward in time, with each solution being found from the solution at the previous timestep.

The approach results in lower computational times than if direct numerical inversion is used. Program CONTAL (CONsolidation using TALbot inversion) is based on numerical inversion of the Laplace transform of the field quantities; however, because an inverse Fourier transform and an inverse Laplace transform must be carried out to obtain the solution at any particular time, computation times are high. Approximately the same amount of computer time is taken to evaluate 20 solutions at various times using the 'marching' technique as is taken to obtain a single solution at a specified time using direct inversion. One advantage of direct inversion is, however, that the solutions are generally more accurate as they are not dependent on previous solutions as are 'marching' schemes.

Computational times taken by FLAC to solve the problem of a strip loading applied to a layer of soil, where the layer was divided into 11 linear elements and solutions are evaluated at 15 time steps were 446 s on an 80286 based machine and 108 s on a machine with an 80287 math coprocessor.

REFERENCES

BIOT, M.A. (1941). General theory of three dimensional consolidation, *J. Appl. Phys.*, **12**, 155–67.

BOOKER, J.R. and SMALL, J.C. (1982a). Finite layer analysis of consolidation. I *Int. J. Numer. Anal. Methods Geomech.*, **6**, 151–71.

BOOKER, J.R. and SMALL, J.C. (1982b). Finite layer analysis of consolidation. II *Int. J. Numer. Anal. Methods Geomech.*, **6**, 173–94.

BOOKER, J.R. and SMALL, J.C. (1985). Finite layer analysis of layered viscoelastic materials under three-dimensional loading conditions, *Int. J. Numer. Methods Eng.*, **21**, 1709–27.

BOOKER, J.R. and SMALL, J.C. (1986). Finite layer analysis of viscoelastic layered materials, *Int. J. Numer. Anal. Methods Geomech.*, **10**, 415–30.

BOOKER, J.R. and SMALL, J.C. (1987). A method of computing the consolidation behaviour of layered soils using direct numerical inversion of Laplace transforms, *Int. J. Numer. Anal. Methods Geomech.*, **11**, 363–80.

BOOKER, J.R. and SMALL, J.C. (1988). Finite layer analysis of layered pavements subjected to horizontal loading, *Proc. 6th Int. Conf. Numer. Methods Eng.*, Innsbruck, Vol. 3, pp. 2109–113.

BOOKER, J.R. and SMITH, D.W. (1989). Behaviour of a heat source in a fully coupled saturated thermoelastic soil. *Numer. Models in Geomech., NUMDG III*, ed. S. Pietruszczak & G.N. Pande. Elsevier Applied Science, London, pp. 399–406.

CHEUNG, Y.K. (1976). *Finite Strip Method in Structural Mechanics*, Pergamon Press, Oxford.

CHEUNG, Y.K. and FAN, S.C. (1979). Analysis of pavements and layered foundations by finite layer method, *Proc. 3rd Int. Conf. Numer. Methods Geomech.*, Aachen, Vol. 3, pp. 1129–35.

CHEUNG, Y.K. and THAM, L.G. (1983). Numerical solutions for Biot's consolidation of layered soil, *J. Eng. Mech. ASCE*, **109**(3), 669–79.

CHEUNG, Y.K. and ZIENKIEWICZ, O.C. (1965). Plates and tanks on elastic foundations, an application of the finite element method, *Int. J. Solids Struct.* **1**, 451–61.

CHEUNG, Y.K., THAM, L.G. and GUO, D.J. (1985). Applications of finite strip and layer methods in micro-computers, *Proc. 5th Int. Conf. Numer. Methods Geomechs.*, Nagoya, Vol. 4, pp. 1755–62.

CHEUNG, Y.K., YEO, M.F. and CUMMING, D.A. (1976). Three-dimensional analysis of flexible pavements with special reference to edge loads, *1st Conf. Road Eng. Assoc. of Asia and Australasia*, Bangkok.

GUO, D.J., THAM, L.G. and CHEUNG, Y.K. (1987). Infinite layer for the analysis of a single pile, *Comput. Geotech.*, **3**, 229–49.

HARNPATTANAPANICH, T. and VARDOULAKIS, I. (1987). Numerical Laplace–Fourier transform inversion technique for layered-soil consolidation problems: II. Gibson soil layer, *Int. J. Numer. Anal. Methods Geomech.*, **11**, 103–12.

MAIER, G. and NOVATI, G. (1988). Elastic analysis of layered soils by boundary elements: comparative remarks or various approaches, *Proc. 6th Int. Conf. Numer. Methods Geomech.*, Innsbruck, Vol. 2, pp. 925–33.

McNAMEE, J. and GIBSON, R.E. (1960). Plane strain and axially symmetric problems of the consolidation of a semi-infinite clay stratum, *Q. J. Mech. Appl. Math.*, **XIII**(2), 210–27.

ROWE, R.K. and BOOKER, J.R. (1981a). The behaviour of footings resting on a non-homogeneous soil mass with a crust. Part I. Strip footings., *Can. Geotech. J.*, **18**, 250–64.

ROWE, R.K. and BOOKER, J.R. (1981b). The behaviour of footings resting on a non-homogeneous soil mass with a crust. Part II. Circular footings, *Can. Geotech. J.*, **18**, 265–79.

ROWE, R.K. and BOOKER, J.R. and SMALL, J.C. (1982). The influence of soil nonhomogeneity upon the performance of liquid storage tanks, *Proc. 4th Int. Conf. Numer. Methods Geomech.*, **2**, 757–66.

RUNESSON, K.R. and BOOKER, J.R. (1982a). Exact finite layer method for the plane strain consolidation of isotropic elastic layered soil, *Proc. Int. Conf. Finite Element Methods.*, Peking, pp. 781–5.

RUNESSON, K.R. and BOOKER, J.R. (1982b). Efficient finite element analysis of 3D consolidation, *Proc. 4th Int. Conf. on Numer. Methods Geomech.*, Edmonton, pp. 365–71.

RUNESSON, K.R. and BOOKER, J.R. (1983). Finite element analysis of elastic-plastic layered soil using discrete Fourier series expansion, *Int. J. Numer. Methods Eng.*, **99**(12), 473–78A.

SINGH, A. and MITCHELL, K.M. (1968). General stress-strain–time function for soils, *J. Soil Mech. Found. Div. ASCE.*, **94**(SM1), 21–46.

SMALL, J.C. and BOOKER, J.R. (1979). Analysis of the consolidation of layered soils using the method of lines, *Proc. 3rd Int. Conf. on Numer. Methods in Geomechs.*, Aachen, Vol. 1, pp. 201–11.

SMALL, J.C. and BOOKER, J.R. (1982). Finite layer analysis of primary and secondary consolidation, *Proc. 4th Int. Conf. Numer. Methods in Geomech.*, Edmonton, Vol. 1, pp. 365–71.

SMALL, J.C. and BOOKER, J.R. (1984). Finite layer analysis of layered elastic materials using a flexibility approach. Part I–Strip loadings, *Int. J. Numer. Methods Eng.*, **20**, 1025–37.

SMALL, J.C. and BOOKER, J.R. (1986a). Finite layer analysis of layered elastic materials using a flexibility approach. Part 2 – Circular and rectangular loadings, *Int. J. Numer. Methods Eng.*, **23**, 959–78.

SMALL, J.C. and BOOKER, J.R. (1986b). The behaviour of layered soil or rock containing a decaying heat source, *Int. J. Numer. Anal. Methods Geomech.*, **10**, 501–19.

SMALL, J.C. and BOOKER, J.R. (1987). A method of computing the consolidation behaviour of layered soils using direct numerical inversion of Laplace transforms, *Int. J. Numer. Anal. Methods Geomech.*, **11**, 363–80.

SMALL, J.C. and BOOKER, J.R. (1988). Consolidation of layered soils under time-dependent loading, *Proc. 6th Int. Conf. Numer. Methods Geomech.*, Innsbruck, Vol. 1, pp. 593–7.

SMALL, J.C. and BROWN, P.T. (1988). Finite layer analysis of the effects of a sub-surface load, *Proc. 5th Aust-N.Z. Conf. Geomech.*, Sydney, pp. 123–7.

SMALL, J.C. and WONG, H.K.W. (1988). The use of integral transforms in solving three dimensional problems in geomechanics, *Comput. Geotech.*, **6**, 199–216.

SNEDDON, I.N. (1951). *Fourier Transforms*, McGraw-Hill, New York.

TALBOT, A. (1979). The accurate numerical inversion of Laplace transforms, *J. Int. Math. Appl.*, **23**, 97–120.

VARDOULAKIS, I. and HARNPATTANAPANICH, T. (1986). Numerical Laplace–Fourier transform inversion technique for layered-soil consolidation problems: I. Fundamental solutions and validation, *Int. J. Numer. Anal. Methods Geomech.*, **10**(4), 347–365.

APPENDIX 1

The layer stiffness matrix for a layer of depth $2h$ of a transversly isotropic material having the stress strain relation

$$\begin{bmatrix} \sigma_{xx} \\ \sigma_{zz} \\ \tau_{xz} \end{bmatrix} = \begin{bmatrix} A & -B & 0 \\ -B & C & 0 \\ 0 & 0 & F \end{bmatrix} \begin{bmatrix} \varepsilon_{xx} \\ \varepsilon_{zz} \\ \gamma_{xz} \end{bmatrix}$$

can be expressed in the form

$$\mathbf{F}_l = \mathbf{K}_l \mathbf{A}_l$$

where

$$\mathbf{F}_l = \begin{bmatrix} -T_l & -N_l & T_m & N_m \end{bmatrix}^\mathrm{T}$$
$$\mathbf{A}_l = \begin{bmatrix} U_l & W_l & U_m & W_m \end{bmatrix}^\mathrm{T}$$

and

$$\mathbf{K}_l = \frac{\alpha A}{2} \mathbf{K}_\mathrm{A} + \frac{\alpha B}{2} \mathbf{K}_\mathrm{B}$$

where

$$\mathbf{K}_\mathrm{A} = \begin{bmatrix} \Omega_s & \chi_s & \Omega_d & -\chi_d \\ \chi_s & \psi_s & \chi_d & -\psi_d \\ \Omega_d & \chi_d & \Omega_s & -\chi_s \\ -\chi_d & -\psi_d & -\chi_s & \psi_s \end{bmatrix}$$

$$\mathbf{K}_\mathrm{B} = \begin{bmatrix} 0 & -\Delta_s & 0 & \Delta_d \\ -\Delta_s & 0 & -\Delta_d & 0 \\ 0 & -\Delta_d & 0 & \Delta_s \\ \Delta_d & 0 & \Delta_s & 0 \end{bmatrix}$$

and

$$\psi_s = \frac{\psi_a}{\Delta_a} + \frac{\psi_b}{\Delta_b}, \quad \psi_d = \frac{\psi_a}{\Delta_a} - \frac{\psi_b}{\Delta_b}$$

$$\chi_s = \frac{\chi_a}{\Delta_a} + \frac{\chi_b}{\Delta_b}, \quad \chi_d = \frac{\chi_a}{\Delta_a} - \frac{\chi_b}{\Delta_b}$$

$$\Omega_s = \frac{\Omega_a}{\Delta_a} + \frac{\Omega_b}{\Delta_b}, \quad \Omega_d = \frac{\Omega_a}{\Delta_a} - \frac{\Omega_b}{\Delta_b}$$

$$\Delta_s = \frac{1}{\Delta_a} + \frac{1}{\Delta_b}, \quad \Delta_d = \frac{1}{\Delta_a} - \frac{1}{\Delta_b}$$

and

$$\Delta_a = A^2 \Omega_a \psi_a - (A\chi_a - B)^2$$
$$\Delta_b = A^2 \Omega_b \psi_b - (A\chi_b - B)^2$$
$$\Omega_a = pq(p^2 - q^2) S_p S_q / D_a$$
$$\Omega_b = pq(p^2 - q^2) C_p C_q / D_b$$

$$\chi_a = \frac{pq}{\alpha^2} \frac{D_b}{D_a}$$

$$\chi_b = \frac{pq}{\alpha^2} \frac{D_a}{D_b}$$

$$\psi_a = \alpha^2 (p^2 - q^2) C_p C_q / D_a$$
$$\psi_b = \alpha^2 (p^2 - q^2) S_p S_q / D_b$$
$$D_a = \alpha^3 (p S_p C_q - q S_q C_p)$$
$$D_b = \alpha^3 (p C_p S_q - q C_q S_p)$$
$$C_p = \cosh(ph), \quad C_q = \cosh(qh)$$
$$S_p = \sinh(ph) \quad S_q = \sinh(qh)$$

The values of p,q are given by

$$\left(\frac{p}{\alpha}\right)^2 = \{-(2B - F) + ((2B - F)^2 - 4AC)^{\frac{1}{2}}\}/2C$$

$$\left(\frac{q}{\alpha}\right)^2 = \{-(2B - F) - ((2B - F)^2 - 4AC)^{\frac{1}{2}}\}/2C$$

Chapter 8

THE EXPLICIT FINITE DIFFERENCE TECHNIQUE
APPLIED TO GEOMECHANICS.
PART I: CONTINUA

N.C. LAST[a] and R.M. HARKNESS[b]

[a]*British Petroleum Research, Sunbury-on-Thames, UK*
[b]*Department of Civil Engineering, University of Southampton, UK*

ABSTRACT

The purpose of the next two chapters is to provide an introduction to the explicit finite-difference technique as applied to geomechanics problems. The more traditional finite-difference techniques for solving partial differential equations have been well documented (for example, Hildebrand, 1968). However, neither the explicit finite-difference scheme which is used to solve the governing solid mechanics equations nor the technique in its various forms (for continua and discontinua) have received wide coverage in the geomechanics context. Furthermore, there has been a general trend towards using the more recently developed finite element and boundary element techniques over the last twenty years or so. There are justifiable reasons for this trend, but the intention of this discussion is to alert the interested reader to the explicit finite-difference technique as a viable alternative which, in most instances will match, and in many may exceed the capabilities and flexibility offered by the aforementioned methods.

Whilst every attempt is made to cover the essentials of the method, it is not possible to describe comprehensively all facets of the various models; wherever possible reference is made to more detailed descriptions of specific issues. The method has now been developed to the stage at which both continua and discontinua can be modelled, an important consideration in geomechanics applications. The basic concept is the same in both cases, however. This common basis is described, and then specific examples of application of the models are given. In this chapter the continuum formulation is presented.

1 INTRODUCTION

The term geomechanics is assumed to include problems involving the behaviour of geomaterials in the widest sense, from the design of foundations on soils, to the assessment of the stability of underground excavations in rock, to the interpretation of large scale tectonic deformation in structural geology, and many other related topics such as fluid flow in rock masses. The assertion is that the explicit finite-difference models described here and in the next chapter can be applied usefully across this wide range of disciplines. However, the problems are often complex, and a straightforward deterministic approach may not always be appropriate. In particular, in the general area of rock mechanics, there is usually a shortage of data (unavailable or difficult to obtain) so that the problem is not fully defined. Indeed, most problems in the broader geomechanics context fall into the data-limited category as defined by Starfield and Cundall (1988). These authors promulgate the idea that a shortage of data does not preclude the use of models to acquire a better understanding of the problem at hand; the simplest possible model should be chosen to include the available information and any perceived mechanisms which are thought to be important. The model should then be used as an experiment whose response to probing can be investigated. The results can then be used to falsify or confirm the anticipated response and may be instructive in identifying new mechanisms. Even if a unique answer remains intangible, the model will often provide bounds on the behaviour of the real system.

Ideally a simpler model can be deduced from the observed behaviour, particularly if some effects are found to be insignificant. Alternatively, a more complex model will have to be sought if the original model is found to be deficient in some way. In either case, the modelling is likely to be at least instructive in assessing the performance of the system. The models and example described here follow this philosophy closely and are used to support the assertion that the explicit finite-difference technique as described provides a flexible and adaptable method for tackling a wide range of complex problems in geomechanics.

2 THE COMMON BASIS:
THE EXPLICIT FINITE-DIFFERENCE TECHNIQUE

The models described here are characterised by the solution method adopted – the explicit finite-difference technique. Any local disturbance of equilibrium is propagated at a materially dependent rate consistent with Newton's Laws of Motion. Thus the modelling follows a sequence of locally determined dynamic (d'Alembert) equilibrium states rather than a series of globally determined static equilibrium states. In order for the equilibrium to be assessed locally in this way, the incremental time-step between successive sets of calculations (each representing a state in the evolution process) must be small enough to prevent propagation of information beyond neighbouring calculation points within such a step (the scheme is conditionally stable). Numerically, therefore, many relatively simple calculations are substituted for the fewer, but invariably complex, inversions or multiple iterations that would be required for the solution of the equivalent series of global equilibrium states. It is the ability to perform all calculations at the local level which renders the scheme explicit. In other words, at the local level for each time-step, unknown quantities, such as displacements, can be calculated by establishing simple equations which relate the unknown quantity to currently known variables at, or immediately surrounding, the local point. In an equivalent implicit scheme, calculation of specific unknown quantities at a given point will involve similar unknowns from the surrounding points. Hence a set of simultaneous equations must be established (usually through specifying overall static equilibrium) and solved globally.

To demonstrate in a simple form the conventional explicit finite-difference technique, consider the non-dimensional diffusion equation

$$\frac{\partial U}{\partial T} = \frac{\partial^2 U}{\partial X^2} \tag{1}$$

which can be used to represent, for example, one-dimensional heat conduction along a bar, or one-dimensional consolidation of a soil layer. In the case of heat conduction, U represents the temperature at time T at a position X along the bar. If the bar is discretised into equal sections of length h, and increments of time k are used, an appropriate finite-difference representation of the governing equation at $X = ih$ and $T = nk$ is

$$\frac{U_{i,n+1} - U_{i,n}}{k} = \frac{U_{i+1,n} - 2U_{i,n} + U_{i-1,n}}{h^2} \tag{2}$$

which can be rearranged to give

$$U_{i,n+1} = U_{i,n} + r(U_{i-1,n} - 2U_{i,n} + U_{i+1,n}) \tag{3}$$

in which $r = k/h^2$. Therefore, the unknown temperature $U_{i,n+1}$ at the position $X = ih$ can be determined at time $T = (n+1)k$ from known temperatures at the previous time-step $T = nk$, and for this reason the calculation is termed *explicit*. Hence the spatial and temporal variation of temperature along the bar can be calculated by applying eqn (3) to all calculation points (individually and in any order) for a series of time-steps. Although this scheme is computationally simple, it has, as previously noted, the drawback that the time-step, k, is limited (the scheme is conditionally stable). In fact, it can be demonstrated that $0 < 2k < h^2$ is the critical condition. Outside of this range, the simple calculations become unstable.

The calculations can be rendered unconditionally stable (no restriction on r) by using an implicit formulation. For example, Crank and Nicolson (1947) replaced the second derivative on the right-hand side of eqn (2) by the mean of its finite-difference representation at time $(n+1)k$ and nk. Hence eqn (2) becomes

$$\frac{U_{i,n+1} - U_{i,n}}{k} = \frac{1}{2}\left[\frac{(U_{i+1,n+1} - 2U_{i,n+1} + U_{i-1,n+1})}{h^2}\right.$$

$$\left. + \frac{(U_{i+1,n} - 2U_{i,n} + U_{i-1,n})}{h^2}\right] \tag{4}$$

which can be rearranged to give (compare eqn (3))

$$-rU_{i-1,n+1} + 2(1+r)U_{i,n+1} - rU_{i+1,n+1}$$

$$= rU_{i-1,n} + 2(1-r)U_{i,n} + rU_{i+1,n} \tag{5}$$

In general, the left hand side of eqn (5) contains three unknown values of U. Hence if there are N points of calculation for U along the bar, application of eqn (5) to each point yields N simultaneous equations which must be solved (usually by iteration) for the N unknown values of U at the next time-step. Thus it is impossible to calculate the new value of U in isolation. This scheme is therefore termed *implicit*.

In the implicit scheme, the calculations per time-step are necessarily more complex than those of the explicit scheme. However, fewer steps are usually required in the implicit case. In both schemes, due regard must be given to the value of h if sufficient accuracy is to be achieved. One of the main advantages, however, of the explicit scheme is that if the problem is highly non-linear, for example the material coefficients change as a function of temperature, the evolution of the spatial distribution of temperature can be followed almost as easily as with the linear case. With the implicit technique, either much more computation per time-step would be required, or more probably, depending on the severity of the non-linearity, the time-step would have to be reduced.

Application of the explicit finite-difference scheme to mechanics problems can be illustrated by considering a simple system, a one dimensional array of springs and lumped masses (Fig. 1). In this case, the 'calculation

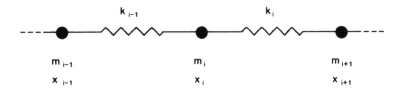

FIG. 1. Simple mechanical system: one-dimensional array of masses and springs.

points' are the positions x_i of the masses (m_i) and the 'material behaviour' is represented by the stiffness k_i of the interconnecting springs. The damped motion of the lumped masses can be represented by the momentum equation in the form

$$m\ddot{x} = F - m\alpha\dot{x} \tag{6}$$

in which F is the out-of-balance force acting on the mass m, whose position is x, and the term $m\alpha\dot{x}$ provides velocity proportional viscous damping. A superposed dot denotes time differentiation. The momentum equation (eqn (6)) can be solved incrementally through time by applying a central finite-difference equation in the form

$$(\dot{x}_i^{t+\Delta t/2} - \dot{x}_i^{t-\Delta t/2})/\Delta t = F_i/m_i - \alpha(\dot{x}_i^{t+\Delta t/2} + \dot{x}_i^{t-\Delta t/2}) \tag{7}$$

which can be rearranged to give

$$\dot{x}_i^{t+\Delta t/2}(1 + \alpha\Delta t/2) = F_i/m_i - \dot{x}_i^{t-\Delta t/2}(1 - \alpha\Delta t/2) \tag{8}$$

By using the velocities at the mid-points of the series of time-steps a central difference scheme is preserved, and the mean velocity is used in the damping term.

The basic calculation cycle can be summarised in a simple flow diagram (Fig. 2). In the first stage, the momentum equation (eqn (6)) is

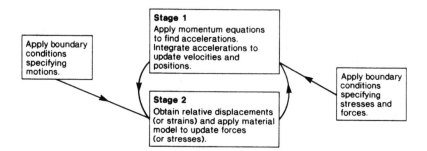

FIG. 2.   Basic calculations cycle.

solved for each mass in turn. The out-of-balance forces F_i are calculated as

$$F_i = f_{i+1} - f_i \qquad (9)$$

and are assumed to remain constant during the (small) calculation time-step. The new velocities can then be calculated from eqn (8). A simple integration with respect to time (assuming that the velocity remains constant during the step) then yields the displacement increments

$$\Delta x_i^{t+\Delta t} = \dot{x}_i^{t+\Delta t/2} \Delta t \qquad (10)$$

and the position of each lumped mass can be updated from

$$x_i^{t+\Delta t} = x_i^t + \Delta x_i^{t+\Delta t} \qquad (11)$$

Thus the accelerations (forces), velocities and displacements are calculated in a staggered manner using a Euler-type integration scheme as indicated in Fig. 3.

In the second stage, the new forces f_i in each spring are calculated. In the simple mechanical system the change in force is proportional to the change in length of the spring so that

$$f_i^{t+\Delta t} = f_i^t + k_i(\dot{x}_{i+1}^{t+\Delta t} - \dot{x}_i^{t+\Delta t})\Delta t \qquad (12)$$

The completion of these two stages constitutes one calculation cycle and represents an increment of time equal to the time-step. When all of the new forces have been determined, stage 2 of the calculation cycle of Fig. 2 is complete. To advance further in time, a new time-step is started by re-evaluating the motion of the lumped masses, beginning with eqn (9). The complete calculation cycle is then repeated until the required model time is achieved. Although the scheme appears very simple, the important

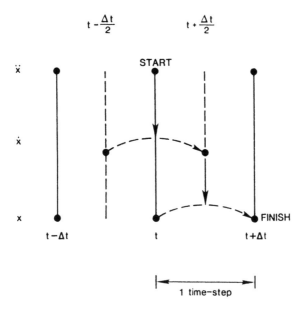

FIG. 3. Temporal integration scheme.

point is that the calculations can only be performed in this way if the difference equations are formulated explicitly. The application of the explicit finite-difference technique to continua and discontinua follows the same basic calculation cycle.

3 CONTINUUM FORMULATION

In a solid mechanics application the aim is to solve the static or dynamic equations of equilibrium over a region subjected to specified boundary conditions so that a complete description of the deformation is provided. If a large displacement/large strain (i.e. geometrically non-linear) analysis is required, it is important to choose the kinematic and kinetic variables carefully so that the deformations are properly represented. The choice is strongly influenced by the method of mapping adopted in the computational procedure used to solve the equations. In the scheme described here an updated Lagrangian formulation is used to specify the configuration of the deforming region. Hence, as will be demonstrated later, it is sufficient to consider the governing equations appropriate to geometrically linear problems. A consideration of the conservation of linear momentum leads to the equations of motion in tensor form as

$$\text{div } \boldsymbol{\sigma} + \mathbf{b} = \rho \ddot{\mathbf{u}} \qquad (13)$$

in which $\boldsymbol{\sigma}$ is the Cauchy stress, $\mathbf{b}$ the body force (per unit volume), ρ the density and $\mathbf{u}$ the displacement. The relationship between deformation and strain is specified in rate form so that

$$\mathbf{D} = (\mathbf{L} + \mathbf{L}^{\mathrm{T}})/2 \qquad (14)$$

in which $\mathbf{D}$ is the velocity strain (or rate of deformation) and $\mathbf{L}$ is the velocity gradient. Finally, a constitutive law relates the stress increments $\Delta\boldsymbol{\sigma}$ to the strain increments $\Delta\boldsymbol{\varepsilon}$ in the general form

$$\Delta\boldsymbol{\sigma} = \mathbf{C}\Delta\boldsymbol{\varepsilon} \qquad (15)$$

in which $\mathbf{C}$ (a fourth-order tensor) represents the constitutive properties.

3.1 Numerical Implementation

The computational scheme is similar to that used in other explicit calculations (Wilkins, 1964; Cundall, 1976). The region of interest is subdivided into quadrilateral zones which are connected to their nearest neighbours by grid-points placed at their corners (Fig. 4). The deforma-

FIG. 4.   Finite-difference zone showing local grid-point numbering. (From Last and Harkness, 1989.)

tion of the region is traced by following the motion of the grid-points through a sequence of small time-steps Δt. The configuration (at time t) is described by the material co-ordinates $\mathbf{x}(\mathbf{X}, t)$ of the grid-points and prior to each successive time-step (representing a small displacement increment) the reference state $\mathbf{X}$ is replaced by the instantaneous current position **x**. In this way the trajectory of material points marked by the grid-points is traced. This updated Lagrangian formulation enables the kinematic variables to be measured in terms of infinitesimal strain increments $\Delta\varepsilon$ because, providing that the displacements in any one step are small the difference between the material co-ordinates **x** and the spatial co-ordinates **X** can be ignored and the velocity gradient can be assessed by differentiation with respect to the current configuration. However, care must still be exercised in specifying the kinetic variables so that the stresses remain conjugate with the strains and the stress rate is objective; appropriate measures are the Cauchy stress together with the Jaumann rate of Cauchy stress.

Within the spatial discretisation, a linear variation of velocity is assumed between the grid-points. The velocity field defined by the motion of the four grid-points surrounding each zone must be used to estimate the strain rate. This is achieved by using Gauss's divergence theorem to express the derivative in a zone in terms of an integral around its boundary. From Gauss's theorem, if a vector field **a** has continuous first-order partial derivatives in a region Ω bounded by a surface Γ, then

$$\int\int\int_\Omega \nabla a \, dV = \int\int_\Gamma \mathbf{a} \cdot \mathbf{n} \, dS \tag{16}$$

in which **n** is the outward normal to the surface and dV and dS are elements of volume and surface area, respectively. If a two-dimensional region of area A is considered and if the derivative is assumed to be constant over the area, then in component form

$$\mathbf{a}_{i,j} = A^{-1} \int_\Gamma a_i n_j \, dL \tag{17}$$

in which dL is a line segment. This allows the evaluation in a region of the spatial derivative of **a** based on the distribution of **a** around the closed boundary. In the numerical scheme, the integration involved in eqn (17) can be represented by a summation around the four sides of the quadrilateral. For a linear distribution along the sides, an exact evaluation of the integral is given if the mid-side value of a is used in the expression

$$a_{i,j} = A^{-1} \sum_{N=1}^{4} (\bar{a}_i^{(N)} n_j^{(N)} L^{(N)}) \tag{18}$$

in which L is the length of the Nth side.

Equation (18) is the basic 'contour integral' operator that is used to determine the spatial derivatives in the finite-difference scheme. It is worth noting that, unlike traditional finite-difference schemes, the use of this type of operator does not restrict the mesh to being rectangular – in fact the enclosed region can be of arbitrary shape and have any number of sides. This means that complex geometries, large deformations and boundary shapes can be handled in a straightforward way.

Consider the velocity strain given in eqn (14). For a typical element of the discretised solution domain represented by a single quadrilateral, the components of the velocity gradient can be represented through expression (18) as

$$\dot{u}_{i,j} = A^{-1} \sum_{N=1}^{4} \bar{u}_i^{(N)} e_{ij} \Delta x_j^{(N)} \tag{19}$$

in which $\Delta \mathbf{x}^{(N)}$ is the vector representing the Nth side and $\mathbf{e}$ is the permutation matrix.

Equation (18) is also used to determine the divergence of the stress tensor as required in the momentum equation (eqn (13)). The equation is solved at the grid-points which represent the connection points between neighbouring zones (Fig. 5). Hence the components of the stress gradient can be represented by

$$\sigma_{ij,j} = A_g^{-1} \sum_{M=1}^{4} \bar{\sigma}_{ij}^{(M)} e_{jk} \Delta x_k^{(M)} \tag{20}$$

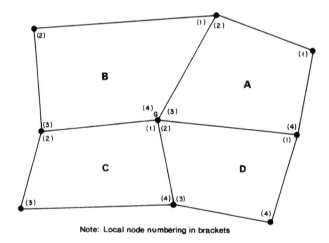

Note: Local node numbering in brackets

FIG. 5.   Patch of zones showing topological connectivity and local zone lettering around a grid-point G. (From Last and Harkness, 1989.)

in which $\bar{\sigma}$ is the mid-side stress on the Mth side and A_g the area of the quadrilateral over which the gradients are to be evaluated. However, the stress σ does not have continuous first-order partial derivatives within the enclosed path (stresses and strains are assumed to be uniform in each individual zone) and eqn (20) must therefore be considered approximate since it does not comply with the conditions of Gauss's divergence theorem. Furthermore, the area A_g representing the closed path around which the integral is applied must be selected – the choice of path is not immediately obvious. If the gradients (20) are considered in the context of the momentum equation (13), the following equation is realised:

$$\ddot{u}_i = (\rho A_g)^{-1} \sum_{M=1}^{4} \bar{\sigma}_{ij}^{(M)} e_{jk} \Delta x_k^{(M)} + b_i \rho^{-1} \tag{21}$$

which can be restated in the equivalent form as

$$\ddot{u}_i = m_g^{-1} f_i + g_i \tag{22}$$

so that m_g represents the grid-mass point, f the out-of-balance force at the grid-point and g the acceleration due to gravity (assuming only gravitational body forces exist). Hence the summation term in eqn (21) can be viewed as a direct application of the equilibrium condition (excluding body forces) to the boundaries of the region of area A_g (mass m_g). To be consistent, the contour integral in eqn (20) should be taken around the boundary of the region which encloses the inertial mass that the force f is to accelerate. In other words, the area A_g should represent the volume of material from which the grid-point mass m_g is derived. The continuous mass of the physical system must be distributed to the grid-points in such a way that the whole domain is covered. The grid-point mass is interpreted as a lumped mass with contributions derived from the surrounding zones. Hence, once an appropriate mass distribution has been specified, the area A_g and the summation path implied in eqn (21) are effectively fixed.

3.2 Mass Distribution

The basic consideration is that the centre of gravity of the continuously distributed mass of a zone should not be altered by lumping the mass at the four corners. For rectangles, this can be achieved by simply placing one-quarter of the mass of the zone at each corner (Hancock, 1973). However, this is clearly in error if irregular quadrilaterals are used. The following approach is proposed. If a quadrilateral is subdivided into two triangles by one of the diagonals (Fig. 6(a)), the centre of mass is maintained if one-third of the mass of each triangle is allocated to the appro-

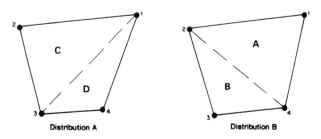

FIG. 6.   Mass distribution. (From Last and Harkness, 1989.)

priate grid-point. Hence

$$m_1 = (A_C + A_D)\rho/3 = A_q\rho/3$$
$$m_2 = A_C\rho/3$$
$$m_3 = m_1$$
$$m_4 = A_D\rho/3 \tag{23}$$

in which A_C and A_D are the areas of the two triangles, A_q is the area of the quadrilateral, ρ is the material density and m_1 to m_4 are the contributions to the four surrounding grid-points (see Fig. 5). However, there is equal justification for dividing the quadrilateral along the other diagonal (Fig. 6(b)). In this case the contributions to the grid-points become

$$m_1 = A_A\rho/3$$
$$m_2 = (A_A + A_B)\rho/3 = A_q\rho/3$$
$$m_3 = A_B\rho/3$$
$$m_4 = m_2 \tag{24}$$

in which A_A and A_B are the areas of the two triangles. Since each subdivision is equally valid, the average of the two is adopted so that

$$m_1 = (A_q + A_A)\rho/6$$
$$m_2 = (A_q + A_C)\rho/6$$
$$m_3 = (A_q + A_B)\rho/6$$
$$m_4 = (A_q + A_D)\rho/6 \tag{25}$$

This procedure correctly distributes the mass for an arbitrarily shaped quadrilateral and even allocates the mass correctly if a quadrilateral degenerates to a triangle. The requirement that the mass allocated to a grid-point be enclosed by the path of the contour integral used to evaluate the corresponding grid-point force is also met and can be proved (see Last and Harkness, 1989). Furthermore, the computational effort is simpler than with other schemes (for example, Cundall *et al.*, 1980).

The mass of a grid-point is calculated by summing the contributions from the surrounding zones. Hence, for a typical patch of zones (Fig. 5), the mass m_g lumped at the common grid-point C is given by

$$m_g = m_3^{(A)} + m_4^{(B)} + m_1^{(C)} + m_2^{(D)} \tag{26}$$

in which the bracketed, superscripted letters refer to the adjoining zones. The establishment of the mass distribution scheme fixes the summation path in the momentum equation (eqn (21)). Conceptually, this path passes through the mid-points of the adjacent zone edges and the centres of mass of the surrounding zones (Fig. 7). In fact, because the stresses in each zone are assumed to be uniform, the actual path followed through

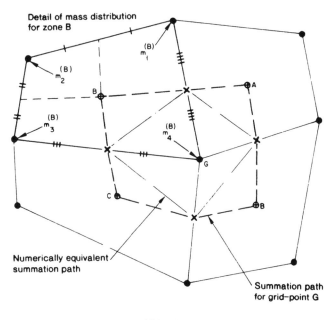

FIG. 7.   Derivation of grid-point mass and summation path for force balance.
(From Last and Harkness, 1989.)

each individual zone is numerically unimportant – the main consideration is the points at which the path crosses into the next zone. To be consistent with the mass distribution, the mid-side points must be used, giving the numerically equivalent summation path indicated in Fig. 7. In applying eqn (21), the stress acting on each segment of the path is assumed to be equal to the uniform stress in the corresponding zone.

3.3 Hourglass (or Kinematic) Modes

Unfortunately there is a drawback to using the simple, constant strain quadrilateral formulation described here. This takes the form of spurious kinematic modes, often referred to as hourglassing because of their shape in plane stress or plane strain configurations (Fig. 8). These occur in *both*

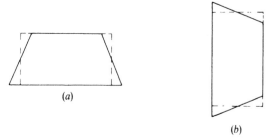

FIG. 8.   Hourglass modes for individual zones. (a) X-mode; (b) Y-mode. (From Last and Harkness, 1989.)

finite element and finite difference schemes if similar constant strain quadrilateral elements are used (see, for example, Cook, 1977; Belytschko *et al.*, 1983; Maenchen and Sack, 1964). The problem arises because the hourglass deformations do not generate any nodal forces to resist this particular pattern of mesh displacements. In other words, two unresisted degrees of freedom exist at the zonal level. The problem can be apparently overcome by using constant strain triangles or linear strain quadrilaterals. Whilst this will remove the hourglass modes, another problem termed locking will be introduced which will lead to incorrect predictions of displacements and collapse loads to over-stiffening at the zonal level. Nagtegaal, *et al.* (1974) offered the first lucid account of this effect. They explained that, in essence, accurate collapse loads will not be achieved using elements which cannot represent a pointwise incompressibility condition. Many elements, including the linear strain quadrilateral, fail to meet this condition. The constant strain quadrilateral does meet this condition, but at the expense of introducing, in some instances, the unwanted hourglass distortions. This dilemma can also be resolved by introducing higher order elements as suggested by Nagtegaal *et al.* (1974). However, there is strong evidence that the computational efficiency of the simpler elements outweighs the apparent advantages of the more complex elements, particularly when solving transient or highly non-linear problems. Thus use of the simpler element remains attractive, but if they are to remain competitive, the unwanted distortions must be removed or controlled in a computationally efficient way.

To demonstrate the source of the spurious modes in the finite-difference formulation the derivation of the velocity strains (eqn (14)) from the contour integral equation (eqn (19)) needs to be closely examined. Consider the normal component of strain in the x_1 direction. For the quadrilateral zone shown in Fig. 4, application of eqn (19) leads to

$$\dot{u}_{1,1} = (1/2A)\{\dot{u}_1^{(4)} + \dot{u}_1^{(1)})(x_2^{(1)} - x_2^{(4)}) + (\dot{u}_1^{(1)} + \dot{u}_1^{(2)})(x_2^{(2)} - x_2^{(1)})$$
$$+ (\dot{u}_1^{(2)} + \dot{u}_1^{(3)})(x_2^{(3)} - x_2^{(2)}) + (\dot{u}_1^{(3)} + \dot{u}_1^{(4)})(x_2^{(4)} - x_2^{(3)})\} \qquad (27)$$

in which the bracketed superscript numbers refer to the local grid-point numbering (Fig. 4). On multiplying out and rearranging, this leads to

$$D_{11} = (1/2A)\{\dot{u}_1^{(1)} - \dot{u}_1^{(3)})(x_2^{(2)} - x_2^{(4)}) - (\dot{u}_1^{(2)} + \dot{u}_1^{(4)})(x_2^{(1)} - x_2^{(3)})\} \qquad (28)$$

The other components of the velocity strain can be determined in the same way.

Equation (28) reveals that whilst the uniform strain rate component

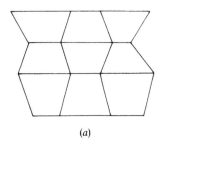
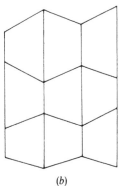

(a)

(b)

FIG. 9.   Mesh hourglass instabilities. (a) X-pattern; (b) Y-pattern. (From Last and Harkness, 1989.)

D_{11} is uniquely determined by the velocity components around the quadrilateral the reverse is not true. In other words, there is a particular velocity field (independent of the rigid-body motions) that does not activate the strain tensor.

These modes are simply a reflection of the fact that the difference

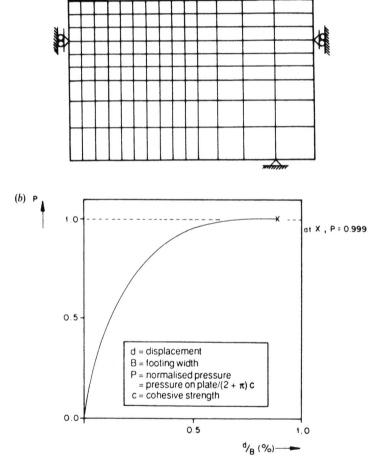

FIG. 10. Validation involving plastic flow: indentation of a rigid punch. (a) Initial mesh and boundary conditions; (b) collapse load. (From Last and Harkness, 1989.)

equations that are used to decompose the velocity field into the orthogonal deformational modes of each zone are over-determined. There are eight independent components of velocity per zone (two at each grid-point) while only six modes are used to define the deformation of the zone; three rigid-body motions plus three components of uniform strain. This over-determinacy is a direct result of the assumption that uniform strains exist in the quadrilateral zone; the spurious modes correspond to modes that comprise linear variations of strain. Hence, under certain boundary conditions (which do not locally preclude this pattern of displacement), this system can lead to either of the two global kinematic mesh instabilities shown in Fig. 9 (or any linear combination of them).

In the finite-difference context, Hancock (1973) and Marti and Cundall (1982) proposed cures. The former was based on a velocity redistribution, the latter on so-called mixed discretisation. However, both have some

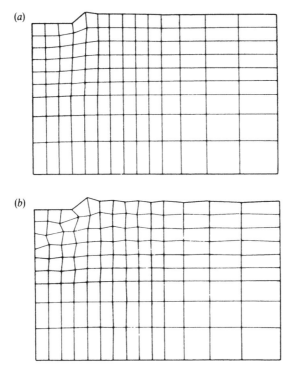

FIG. 11. Deformations produced during punch indentation (a) with and (b) without hourglass control. (From Last and Harkness, 1989.)

inherent difficulties. The authors have successfully used a new scheme (Last and Harkness, 1989) which overcomes many of the deficiences in two-dimensional calculations and has the advantage of being extendable to hourglass control in three dimensions. This scheme is based on a zone-by-zone evaluation of the hourglass components of the deformation followed by a momentum redistribution of the unwanted displacements. Although somewhat similar to Hancock's scheme, the new technique offers better control for irregular geometries and large displacements, is easier to implement and is theoretically more rigorous.

3.4 Validation

This formulation has been tested and validated extensively (see, for example, Last and Harkness, 1989). In particular, as noted earlier, formulations which do not permit constant volume flow during plastic deformations are likely to lock and cause over-estimation of collapse loads. A classic problem for testing this is the plane strain indentation of a punch (Prandtl, 1921). The numerical results for the indentation of a smooth, rigid plate into a cohesive material represented by a relatively coarse mesh of rectangular zones are shown in Fig. 10. The collapse load is predicted accurately, and the efficacy of the hourglass control scheme is confirmed by the comparison given in Fig. 11. Indeed, the spurious modes are progressively amplified and the solution generally degrades with increased indentation in the uncontrolled mesh.

4   AN EXAMPLE APPLICATION: DEFORMATION OF A GEOLOGICAL STRUCTURE

Creep flow in mobile salt or shale substrata has long been recognised as an important factor in the tectonics of salt basins (Trusheim, 1960) and major deltas (Merki, 1972; Evamy *et al.*, 1978). The available evidence suggests that salt and shale layers may flow in the direction of decreasing overburden load to produce features such as shale waves, salt pillows or ridges, and overburden faulting, which are of particular interest, since any mechanical model capable of explaining their occurrence in a coherent manner will be helpful, for example in quantitative studies of hydrocarbon migration and accumulation patterns.

A simplified theoretical approach was used earlier by Lehner (1977) to predict the evolution of the shape of the interface in an idealised, two-layer salt (or shale)/overburden sequence. The theory provides an estimate of the flow rate within the mobile (viscous) layer and predicts the occurrence of travelling waves (or ridges) and associated trailing depressions (or basins) due to lateral movement within the mobile layer and depressions in the basement. Support for the occurrence of these features can be drawn from field observations and experiments. Moreover, if a numerical model is used, assumptions concerning the velocity profile in the substratum and the behaviour of the overburden can be less restrictive. Indeed, comparisons between the numerical simulations and the predictions of Lehner's model should provide useful quantitative and qualitative data to assess the range of applicability of the simplified analysis. The objective was to develop a numerical model for investigating this type of geological setting (Last, 1988).

4.1 Additional Features

To enable the modelling of the slowly evolving, time-dependent processes associated with sedimentation and with creep of the mobile layer, additional features were developed. These are a direct consequence of the attempt to simulate the deformations that occur in a geological setting.

4.1.1 Geological Time Scales

The limiting time-step Δt associated with the explicit finite-difference scheme is related to the local rate of propagation of compressional waves so that

$$\Delta t \leqslant l\sqrt{(\rho/k)} \tag{29}$$

in which l is a characteristic length of the spatial discretisation (usually the minimum diagonal length of a zone), ρ is the material density and k the appropriate elastic modulus. Using typical numerical meshes and material properties for geomaterials, Δt is usually in the range from 0·001 to 0·1 s. Clearly the modelling of events which take place over geological time is not feasible unless a suitable scaling rule can be adopted. Previously an 'adaptive density scaling' technique (Cundall, 1982) was used: essentially the inertial density in eqn (29) is scaled to artificially increase Δt provided the inertial force term remains small in comparison to a reference value, for example the gravitational body forces. In a problem where deformation rates are constrained by slow viscous flow, an alternative approach is possible: the viscosity can be scaled to speed up the process, provided that the inertial forces remain small in comparison to the viscous forces. The Reynolds' number (=kinetic energy per unit volume/viscous stress) is a suitable measure of this ratio,

$$Re = \rho v d / \eta \tag{30}$$

in which η and ρ are the viscosity coefficient and the density respectively, v is an average velocity and d a characteristic length of the viscous flow regime. For the scaled system, the coefficient of viscosity becomes

$$\eta^s = \eta / \lambda \tag{31}$$

Under steady quasi-static conditions, this scaling will simply increase the real velocities v by the same factor so that the computed velocities become

$$v^s = \lambda v \tag{32}$$

and the Reynolds number for the computations becomes

$$Re^s = \lambda^2 Re \tag{33}$$

Hence the scaling factor λ should be selected such that $Re^s \ll 1$. It is immediately clear that a scaling by λ^2 of the material density in eqn (30) would have the same effect on the Reynolds number and would increase the time-step of eqn (29) by λ; this is in fact the basis of density scaling. Both approaches have been implemented and successfully used. While adaptive density scaling remains the more generally applicable scheme, viscosity scaling appears to be particularly appropriate for the class of problems studied here. Furthermore, under near steady state conditions, numerical results indicate that the schemes are indeed equivalent.

4.1.2   Sedimentation and Erosion

To model sedimentation, the top zone (surface zones) of the mesh grow vertically to reflect the specified rate of growth and are treated differently from the underlying regular zones (Fig. 12). Within each time-step, the change of stress in the growing zone due to sedimentation corresponds to the addition of a thin slice of frictional material whose stresses are consistent with the stress state in an infinite Rankine slope. The stress state for the whole zone is then evaluated in the usual way by reference to a constitutive model. The only difference is that the surface zone is treated in a step-by-step, semi-Eulerian fashion so that the vertical sides of the zone remain vertical. The surface grid points (Fig. 12) are not material points (they have no inertial mass) and their velocity reflects the movement of the sedimentation boundary rather than the local material velocity. When a surface zone reaches a specified size, it is switched to

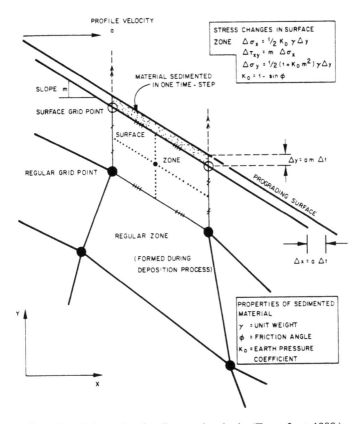

FIG. 12. Schematic of sedimentation logic. (From Last 1988.)

being treated as a regular zone, and a new surface zone is formed.

The scheme ensures that the growing surface always conforms to a prescribed, time-dependent profile, which implies that locally the effective rate of deposition (or erosion, simply 'negative deposition') varies accordingly.

4.1.3 Material Behaviour

The material behaviour assumed for the frictional overburden is linearly elastic with a limiting Mohr–Coulomb plastic yield surface and a non-associated flow rule (Davies *et al.*, 1974). To model the mobile (salt) substratum, a model of viscous behaviour has been implemented. The model has an elastic volumetric response and portrays Maxwellian behaviour in shear so that the behaviour becomes time-dependent. The viscosity coefficient is prescribed initially (with a spatial variation, if appropriate) and is assumed to remain constant thereafter.

4.2 Validation for Viscous Flow

For an ideal incompressible Newtonian fluid, the Navier–Stokes equations can be solved for some simple boundary value problems involving plane viscous flow. The analytic solutions apply to an ideal incompressible Newtonian fluid but in the steady state, numerical solutions obtained with the compressible Maxwellian material should agree. The examples of Plane Couette flow, Plane Poiseuille flow and Jaeger's approaching parallel plates (Jaeger, 1956) have been used. In the latter case an excellent test is realised by checking the total force P exerted on the plates (separation $2d$, width $2b$, velocity of approach $2v_0$),

$$P = 2 \int_0^b \sigma_y(x)\, dx = 2\eta v_0 (b/d)^3 \tag{34}$$

For the mesh shown in Fig. 13, the computed load is within 1·5% of the

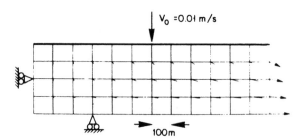

FIG. 13.   Squeezing of viscous material between parallel plates (steady state velocity vectors superimposed). (From Last 1988.)

analytic solution. In fact, for slow rates of flow and steady or near steady state conditions the analytical predictions can be reproduced for all three classes of flow. Furthermore, the viscosity scaling option was used successfully to obtain these results. These validations encompass the main classes of plane viscous flow that are likely to be encountered in the geological models.

4.3 Geological Setting

The geological setting consists of a gently sloping overburden on a thin mobile substratum of varying thickness. The assumed sequence of events is that an initially uniform overburden (Fig. 14(a)) is subsequently built up by sedimentation to form a wedge-like differential load (Fig. 14(b)) that will cause flow of the substratum in the direction of decreasing load. The time varying deformations in the two-layer sequence and in particular, of the interface, should lead to the features that are of geological interest.

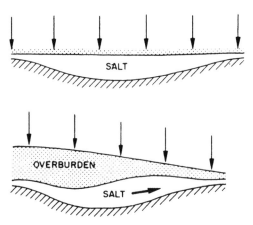

FIG. 14. Geological setting. (From Last 1988.)

4.4 Review of the Analytical Model Based on Lubrication Theory

Lehner (1977) examined the deformation of the mobile layer by treating the material as an incompressible, Newtonian viscous fluid. Along the upper and lower surfaces the substratum adheres to the adjacent layers and the upper surface is loaded vertically to reflect the presence of the sloping overburden. Using these basic assumptions, 'Reynolds equation' can be shown to govern the time-dependent thickness h of the mobile layer (Fig. 15) in the form

$$\frac{\partial}{\partial x}\left(h^3\frac{\partial p}{\partial x}\right)6\eta\frac{\partial}{\partial x}[(V_1+V_2)h]-12\eta\frac{\partial h}{\partial t}=0 \qquad (35)$$

This equation arises in the theory of hydrodynamic lubrication (for example, Langlois, 1964) where it describes the lubricating effect of a thin viscous layer. In the geological context, to obtain the time-varying shape of the mobile substratum, eqn (35) must be integrated. This requires a knowledge of the pressure gradient $\partial p/\partial x$ and the bounding velocities V_1 and V_2. Lehner considered situations in which the mobile layer adheres to a laterally immobile overburden and basement ($V_1=V_2=0$). This

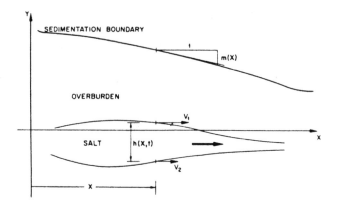

FIG. 15.   Idealised two-layer sequence used by Lehner (1977).

corresponds, for example, to a rigid basement and an overburden which
suffers negligible horizontal displacements, but is allowed to deform in
inhomogeneous simple shear along the vertical. In other words, the
second component of flow associated with eqn (35) is removed and only
the Poiseuille flow remains. This flow is driven by the pressure gradient
$\partial p/\partial x$ which, on neglecting the overburden's resistance to shear and
assuming a constant slope (m) cn be expressed by

$$\partial p/\partial x \simeq \gamma m \tag{36}$$

in which γ is the unit weight of the overburden. Equation (35) then readily
yields (cf. Lehner, 1977).

$$\left.\frac{\mathrm{d}x}{\mathrm{d}t}\right|_h = \frac{\gamma m h^2}{4\eta} \tag{37}$$

which represents the speed at which a vertical section of thickness h
propagates horizontally in the downslope direction. Hence, for a given
initial geometry, the evolution of the deforming layer can be predicted.

4.5   Numerical Simulation

The initial geometry and dimensions of the idealised two-layer sequence
are outlined in Fig. 16 and the principal material properties are given in
Table 1.

   The overburden was modelled in three ways:

— Case A: A 'thick' frictional layer consisting of elasto-plastic (Mohr–
   Coulomb) material elements.

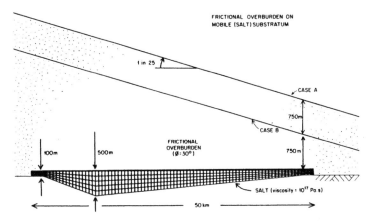

FIG. 16.   Detail of numerical models. (From Last 1988.)

TABLE 1
PRINCIPAL MATERIAL PROPERTIES

Property	Over-burden	Salt	Units
Unit weight	11·0	11·0	kN/m^3
Bulk modulus	150·00	1500·0	MN/m^2
Shear modulus	60·0	600·0	MN/m^2
Viscosity	–	10^{17}	Ns/m^2
		$(=10^{18}$	poise)
Friction angle	30·0	–	degrees
Cohesion	0·0	–	MN/m^2
Dilation angle	0·0	–	degrees

— Case B: A 'thin' frictional layer (material elements modelled as in Case A).
— Case C: corresponds most closely to the lubrication theory model. However, it is not exactly the same because in the numerical simulation no a-priori assumption is made concerning the velocity profile through the viscous layer.

The overburden was deposited during a short period of geological time, with erosion and/or deposition maintaining a fixed sedimentation boundary during subsequent deformation. Figure 17 indicates the deformed numerical grids after about 0·5 million years.

4.6   Interpretation of Computer Simulations

The 'thick' overburden (Case A, Fig. 17(a)) has remained almost intact but flow has occurred in the substratum in the direction of decreasing

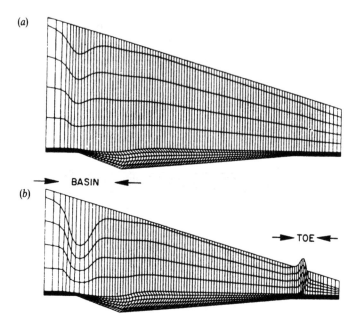

FIG. 17.   Deformed numerical grids (a) Case A and (b) Case B (not to scale).
(From Last 1988.)

overburden load. This flow, combined with some lateral movement of the
overburden, has created a small depression or basin above the upslope
edge of the substratum and a region of compression (with some uplift)
above the downslope edge (the 'toe'). The 'thin' overburden (Case B, Fig.
17 (b)) undergoes more severe deformations and a deeper basin (ap-
roximately 1 km, Fig. 18(a)) and an anticlinal structure (associated with
thrusting, Fig. 18(b)) are produced in the corresponding locations. The
horizontal motion of the thick overburden is prevented while the bulk of
the thin overburden rides out on the creeping substratum (travelling
approximately 2 km after 0·5 million years). The horizontal velocity
profiles in the substratum exhibit predominantly Poiseuille flow under
the thick overburden and Couette flow under the thin overburden. There
is no plastic flow in the thick overburden, but in the thin overburden
yielding occurs in the regions of concentrated extension (the basin) and
compression (the toe) and indicate the areas where faulting might be
anticipated. In other words the stresses caused by the tendency for the
overburden to move downslope on the lubricating substratum locally
exceed the frictional strength of the material.

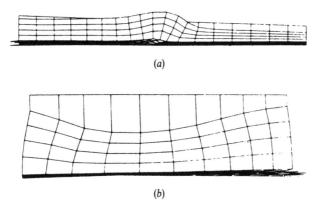

(a)

(b)

FIG. 18. Detail of toe and basin region, Case B, to scale. (a) Toe; (b) basin. (From Last 1988.)

This result is clearly significant, because two modes of overall response, yielding and non-yielding, are found to be associated with different overburden thicknesses.

In the simpler case in which the (frictionless) overburden is represented by a distributed load (Case C), points on the interface are constrained to move vertically and the solution for the interface shape depends only on the overburden gradient (and not on thickness). The velocity profile in the substratum indicates Poiseuille type flow, as assumed in the lubrication theory model.

4.7 Simple Classification Based on the Computed Results

Slope instability is defined as the conditions which allow significant horizontal motion of the bulk of the overburden (the 'wedge') and will be assumed to occur when the toe and basin regions yield in idealised passive and active Rankine states, respectively. Furthermore, for stable slopes the substratum will be treated as a lubricating layer in the manner postulated by Lehner (1977). These assumptions are supported by the results of the computer simulations. Incipiently unstable slopes can then be assessed by examining the overall static equilibrium of the overburden wedge (Fig. 19).

In particular, horizontal equilibrium requires

$$F_a + T - F_p = 0 \tag{38}$$

The active and passive resistances are given through the Rankine earth

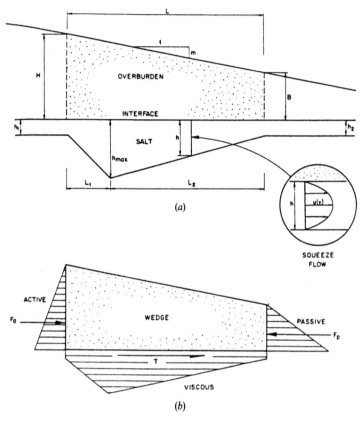

FIG. 19. Horizontal equilibrium of the overburden wedge. (a) Geometry; (b) forces. (From Last 1988.)

pressure coefficients (k_a and k_p) by

$$F_a = \tfrac{1}{2}\gamma H^2 k_a \tag{39}$$

$$F_p = \tfrac{1}{2}\gamma B^2 k_p = \tfrac{1}{2}\gamma B^2 (k_a)^{-1} \tag{40}$$

Note that the assumption that k_p is simply equal to the inverse of k_a is only strictly true for a horizontal, cohesionless layer. In the numerical model there was no cohesion and the overburden slope was very gentle so that use of this relationship is justified. The shear force T acting along the base of the wedge can be obtained by integrating the shear traction τ at the interface which results from plane Poiseuille flow in a substratum of thickness h. Expressed in terms of the pressure gradient in the substratum, this traction is

$$\tau = -\frac{h}{2}\frac{\partial p}{\partial x} \tag{41}$$

and by assuming a constant slope,

$$\tau = \frac{h}{2}\gamma m \tag{42}$$

The thickness h varies gently through the lateral extent of the substratum and can be described by

$$h(x)h_1 + (h_{max} - h_1)x/L_1, \quad \text{for } x = 0, \dots, L_1$$

$$h(x) = h_{max} - (h_{max} - h_2)(x - L_1)/(L - L_1), \quad \text{for } x = L, \dots, L_1$$

Hence the shear force can be evaluated as

$$T = \int_0^L \tau(x)\,dx$$

$$= \tfrac{1}{2}\gamma m \int_0^{L_1} h(x)\,dx + \tfrac{1}{2}\gamma m \int_{L_1}^L h(x)\,dx$$

$$= \tfrac{1}{4}\gamma m(Lh_{max} + L_1 h_1 + L_2 h_2) = \tfrac{1}{2}\gamma m A \tag{43}$$

in which A is the cross-sectional area of the substratum. This simple result holds true only when the overburden slope (and therefore the pressure gradient) is constant. Furthermore, an averaged value h for the substratum depth is introduced and defined by

$$A = Lh \tag{44}$$

so that substitution of eqns (39), (40), (43) and (44) into eqn (38) leads to

$$(H/L)^2(1 - k_a^2) - 2m(H/L) + [m^2 - k_a(mh/L)] = 0,$$
$$\text{for } B/L = H/L - m \geqslant 0 \tag{45}$$

in which h/L represents a normalised measure of the average depth of the substratum which should be small if the conditions assumed in the analysis are to remain applicable. This result (45) is interesting since it involves neither the coefficient of viscosity nor the position of maximum depth of the substratum. Solutions to this equation are shown in Fig. 20 for a range of appropriate values of the geometric and material parameters.

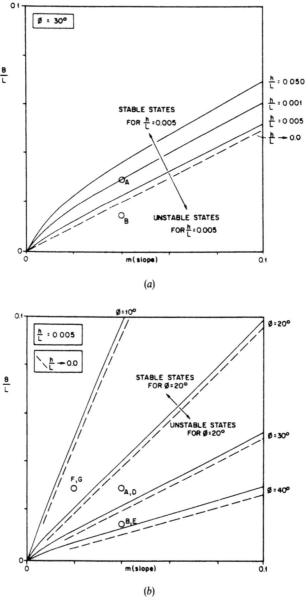

(a)

(b)

FIG. 20. Stable and unstable states for the idealised overburden wedge. (a) Effect of substratum depth; (b) effect of overburden strength (ϕ=friction angle). (From Last 1988.)

Notice that for relatively shallow substrata ($h/L \leqslant 0 \cdot 01$ say) the effect of the shear force (43) is quite small. Indeed, if this force is ignored, solutions to eqn (45) simplify to

$$H/L = m/(l - k_a), \quad \text{for } B/L = H/L - m \geqslant 0 \tag{46}$$

and these are shown as dashed lines in Fig. 20.

The curves in Fig. 20 separate in a broad sense into unstable (yielding) and stable (non-yielding) overburden states. The conditions pertaining to the computer simulations are marked as points A and B. Furthermore, additional simulations have been performed (Table 2) and are included in Fig. 20.

Clearly the additional computer simulations (Cases D to G) support the general trends indicated by eqn (45).

TABLE 2
SUMMARY OF PARAMETRIC STUDY

Case	m	B/L	ϕ°	Stable?
A	0·04	0·030	30	Yes
B	0·04	0·015	30	No
C	0·04	0·015	—	—
D	0·04	0·030	20	No
E	0·04	0·015	40	Yes
F	0·02	0·030	20	Yes
G	0·02	0·030	10	No

5 CONCLUDING REMARKS

The aim of this chapter was to introduce the explicit finite-difference technique and its application to geomechanics problems in which the material can be treated as a continuum.

The explicit finite-difference technique has been described by demonstrating its application to the solution of two simple problems, namely heat conduction along a bar and the motion of a one-dimensional array of springs and masses. This provided the framework in which to proceed with the description of the continuum formulation. This included a novel mass distribution scheme and a new hourglass control procedure.

The specific example selected to illustrate the application of the continuum formulation involved the deformation of a large scale geological structure. Several additions and modifications were needed to make the modelling possible, including sedimentation logic, viscous material behaviour and large time scales. The successful implementation of these features demonstrates the adaptability of the technique. Furthermore, subsequent simulations of the evolving geological structure were used to characterise the behaviour of the two-layer sequence and to test the applicability of a simplified analytical solution to the problem. In essence, a competent overburden produces deformations in the underlying viscous layer which are similar to the analytical predictions, but if the frictional overburden yields, the deformation of the viscous layer is modified by the translation and deformation of the overburden, and large extensional and compressional regions are formed in the overburden. The modelling has therefore prompted a better qualitative and quantitative understanding of the geological setting.

Some further general comments are reserved for the concluding section of the next chapter.

REFERENCES

BELYTSCHKO, T., ONG, S.J., LIU, W.K. and KENNEDY, J. (1983). Hourglass control in linear and non-linear problems, *Comput. Methods. Appl. Mech. Eng.*, **43**, 251–76.

COOK, R.D. (1977). Ways to improve the bending response of finite elements, *Int. J. Numer. Methods Eng.*, **11**, 1029–39.

CRANK, J. and NICOLSON, P. (1947). A practical method for numerical evaluation of solutions of partial difference equations of the heat conduction type, *Proc. Camb. Philos. Soc.*, **43**, 50–67.

CUNDALL, P.A. (1976). Explicit finite difference methods in geomechanics, *Int. Conf. Numer. Methods Geomech.*, Blacksburg, Virginia, Vol. 1, pp. 132–50.

CUNDALL, P.A. (1982). Adaptive density scaling for time-explicit calculations, *4th Int. Conf. Numer. Methods in Geomech.*, Edmonton, Canada, Vol. 1, pp. 23–6.

CUNDALL, P.A., HANSTEEN, H., LACASSE, S. and SELNES, P. (1980). NESSI – Soil structure interaction program for static and dynamic problems, Norwegian Geotechnical Institute Report 51509-8.

DAVIS, E.H., RING, G.J. and BOOKER, J.R. (1974). The significance of the rate of plastic work in elasto-plastic analysis, Univ. of Sydney, School of Civ. Eng., report 242.

EVAMY, B., HAREMBOURE, J., KAMERLING, P., KNAAP, W., MALLOY, F. and ROWLANDS, P. (1978). Hydrocarbon habitat of tertiary Niger Delta, *AAPG Bulletin*, Vol. 62, pp. 1–39.

HANCOCK, S. (1973). An hourglass subtraction procedure, Physics International Company, Tech. Memo TCAM 73-6, San Leandro, California.

HILDEBRAND, F.B. (1968). *Finite-Difference Equations and Simulations*, Prentice Hall, Englewood Cliffs, NJ, 338 pp.

JAEGER, J.C. (1956). *Elasticity, Fracture and Flow with Engineering and Geological Applications*, Methuen, London, Chap. 3.

LANGLOIS, W.E. (1964). *Slow Viscous Flow*, Macmillan, London, Chap. 9.

LAST, N.C. (1988). Deformation of a sedimentary overburden on a slowly creeping substratum, *Proc. 6th Int. Conf. Numer. Methods Geomech.*, Innsbruck, Balkema, Rotterdam, pp. 577–85.

LAST, N.C. and HARKNESS, R.M. (1989). Kinematic (or hour-glass) mode control for a uniform strain quadrilateral by an assumed strain technique, *Int. J. Numer. Anal. Methods Geomech.*, **13**, 381–410.

LEHNER, F.K. (1977). A theory of substratal creep under varying overburden with applications to tectonics, AGU Spring Meeting, Washington DC, EOS Abstr, Vol. 58, p. 508.

MAENCHEN, G. and SACK, S. (1964). The TENSOR code, *Methods Comput. Phys.*, **4**, 181–210.

MARTI, J. and CUNDALL, P.A. (1982). Mixed discretisation procedure for accurate modelling of plastic collapse, *Int. J. Numer. Anal. Methods Geomech.*, **6**, 129–39.

MERKI, P. (1972). Structural geology of the Cenozoic Niger Delta, *Proc. 1st Conf. African Geomech.*, Ibadan Univ., Nigeria, pp. 635–46.

NAGTEGAAL, J.C., PARKS, D.M. and RICE, J.R. (1974). On numerically accurate finite element solutions in the fully plastic range, *Comput. Methods Appl. Mech. Eng.*, **4**, 153–77.

PRANDTL, L. (1921). On the penetrating strengths (hardness) of plastic construction materials and the strength of cutting edges, *Zeit. Angnew. Math. Mech.*, **1**(1), 15–20.

STARFIELD, A.M. and CUNDALL, P.A. (1988). Towards a methodology for rock mechanics modelling, *Int. J. Rock Mech. Min. Sci. Geomech. Abstr.*, **25**, 99–106.

TRUSHEIM, F. (1960). Mechanism of salt migration in Northern Germany, *AAPG Bulletin*, Vol. 44, No. 9, pp. 1519–40.

WILKINS, M.L. (1964). Calculation of elasto-plastic flow, *Methods Comput. Phys.*, **3**, 211–63.

Chapter 9

THE EXPLICIT FINITE DIFFERENCE TECHNIQUE APPLIED TO GEOMECHANICS. PART II: DISCONTINUA – THE DISTINCT ELEMENT METHOD

N.C. LAST

British Petroleum Research, Sunbury-on-Thames, UK

ABSTRACT

In this chapter the application of the explicit finite-difference technique to discontinuous materials is described. This implementation has become known as the distinct element method, but has not been widely covered in the open literature. For this reason a fairly detailed account of the technique is presented which should allow the interested reader to gain a full appreciation of its capabilities. The response of a joined rock mass to fluid injection is used as an example application. This required the addition of several features to the basic model, including fluid flow along the joints and into the blocks. The results show that it is essential in certain circumstances to explicitly model the discontinuities in a discontinuous material, and demonstrate the flexibility and applicability of the technique. Finally, some overall conclusions are drawn concerning the utility of the explicit finite-difference technique in the broader geomechanics context.

1 INTRODUCTION

Problems in geomechanics are often characterised by the discontinuous nature of the material being modelled. This is true across a wide range of scales, from the individual grains which exist at the micro scale, through joints at an intermediate scale to major faults at the large scale. These discontinuities will usually dominate or at least have a significant effect on the mechanical behaviour of the structure. In these situations the discontinuities must be represented at the appropriate level of detail.

As an example, consider the behaviour of granular soils. A considerable effort has been put into quantifying the mechanical behaviour of these materials by formulating constitutive laws which are characterised through continuum quantities such as stress and strain. However, in reality, the medium is discrete, and the continuum quantities do not exist (or cannot be defined) at every point. In fact, to investigate these materials through a continuum approach, the assumption that a large number of discrete particles behaves in an essentially homogeneous way, with uniform applied boundary conditions being representative of the internal state, is required. The very nature of granular materials usually renders this assumption invalid (see, for example, the experimental work of de Josselin de Jong and Verruijt, 1969), particularly if anything other than infinitesimal strains are encountered. Investigators have therefore been prompted to examine the micromechanics of these materials. This requires discrete quantities such as forces between particles and displacements and rotations of individual particles to be measured. This is extremely difficult in physical experiments. However, numerical models which can simulate the interaction of an assemblage of discrete particles, in two and three dimensions, can be usefully employed in this context. In particular, the distinct element (DE) method, pioneered by Cundall (1971), has been used to model interacting discs or spheres; see, for example, Cundall and Strack (1979) and Cundall (1988). The aim of this work has been to gain an understanding of the micromechanics so that better constitutive models can be developed.

2   DISCONTINUUM FORMULATION: THE DISTINCT ELEMENT METHOD

In general the DE method can be used to simulate the interaction of arbitrarily shaped particles. In the basic form as first proposed by Cundall (1971), the block medium is represented in two dimensions by a set of polygons which are assumed to be rigid and interact with their immediate neighbours through corner to edge contacts.

The equations of rigid body mechanics are then solved so that the motions of the individual blocks can be traced. The rigid body displacement of each block is referred to the block centroid in a fixed, cartesian framework (Fig. 1) so that the equations of conservation of linear and angular momentum take the form

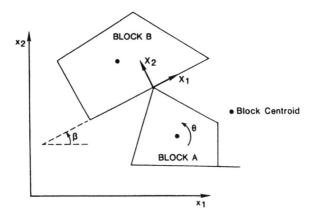

FIG. 1.   Block to block interaction: definition of co-ordinate systems. (From Last and Harper, 1990.)

$$\mathbf{M\ddot{x}}=\mathbf{f}, \quad I\ddot{\theta}=m \tag{1}$$

in which, for translation (2 modes), $\mathbf{M}$ is the mass matrix, $\mathbf{x}$ the position vector and $\mathbf{f}$ the vector of resultant forces. A superposed dot represents time differentiation. In rotation, I is the moment of inertia, θ is the angular displacement and m the resultant moment. Contributions to the force sum $\mathbf{f}$ (and moment m) arise from the forces produced through contacts with adjacent blocks, body forces (for example gravitational) and applied loading. Damping forces are also included to enable energy dissipation (in eqn (1)). Solution of eqn (1) provides the translational and rotational velocities $\dot{\mathbf{x}}$ and $\dot{\theta}$ of the block centroid which, for a rigid material, also determines the motion of any other point within the body. This means that the velocity at any point of contact with adjacent blocks can readily be obtained.

Interaction between neighbouring blocks occurs by the transfer of forces at points of contact (Fig. 2). Once a contact is established, the force of interaction $\mathbf{c}$ is governed by contact laws of the form

$$\mathbf{c}=\mathbf{Ku} \tag{2}$$

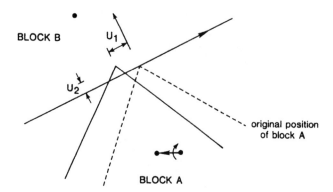

FIG. 2.   Edge to corner contact: displacements during one time-step. (From Last and Harper, 1990.)

in which **u** is the relative displacement (which includes normal and tangential components) at the point of contact between the two blocks and **K** represents the constitutive properties of the contact. In the simplest case, the forces **c** and displacements **u** are linearly related and **K** then consists of elastic stiffness coefficients.

Typically the governing equations of motion (1) and interaction (2) must be solved for a prescribed geometry of blocks subject to initial and boundary conditions. These equations are closely coupled and the most suitable way of solving for a general non-linear problem is to use an incremental procedure. An explicit finite-difference scheme similar to that described in the previous chapter is used in the DE method.

2.1   Numerical Implementation

The basic calculation scheme is similar to that described in the previous chapter and is outlined in Fig. 3. In the first stage, the momentum equations (1) are solved by using the appropriate finite-difference representation. For each block in turn, the force (and moment) sums at the centroid (including boundary forces) are calculated and the momentum equations are used to obtain the instantaneous components of acceleration. These acceleration components are assumed to be constant within the (small) calculation time-step so that a simple integration with respect to time yields the velocity increments. These increments are added to the last known velocity components to obtain the new current values. These velocities are then integrated with respect to time to obtain the increments of displacement (assuming that the velocities remain constant

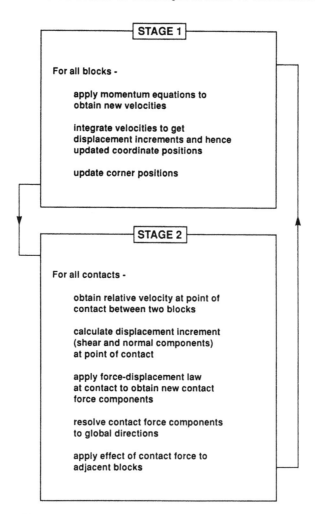

FIG. 3. Basic calculation scheme for the DE method.

over the time-step). Finally, the displacement increments are added to the last known co-ordinates of the block centroid to obtain the new position vector. An analogous set of calculations is performed for the rotational degree of freedom of each block to give the angular velocity, incremental rotation and accumulated rotation. In the second stage, at each contact the incremental relative displacement between the two juxtaposed blocks is calculated and used to update the forces of interaction by application of the force-displacement law (eqn (2)).

The completion of these two stages constitutes one calculation cycle and represents an increment of time equal to the time-step.

2.1.1 Stage One Calculations

The block accelerations can be evaluated directly from eqn (1) if the mass matrix **M** takes a diagonal form. This simplification is achieved by assuming a lumped mass representation whereby the whole mass M of the block acts at its centroid, corresponding to the point at which the rigid body velocity components are defined. The block motion is referred to a cartesian framework (Fig. 1) so that, in component form the momentum equations become

$$M\ddot{x}_i = f_i + Mg_i + h_i$$
$$I\ddot{\theta} = m \tag{3}$$

in which Mg_i represents the components of the gravitational body force and h_i are the components of force arising from any externally applied loads (and f_i are the components of force resulting from inter-block contacts). For a small increment of time Δt for which the forces (and moments) are assumed to remain constant, the time-centred finite-difference representation of these equations gives

$$M(\dot{x}_i^{t+\Delta t/2} - \dot{x}_i^{t-\Delta t/2})/\Delta t = f_i^t + Mg_i + h_i^t$$
$$I(\dot{\theta}^{t+\Delta t/2} - \dot{\theta}^{t-\Delta t/2})/\Delta t = m^t \tag{4}$$

in which the superscript t denotes the current time (measured from zero), equivalent to total time of $n\Delta t$ if the time-step remains constant and n calculation steps have been completed. Hence, if the forces **f** (and moment m) are known together with the previous velocity components (corresponding to a time of $t - \Delta t/2$), the new velocity components at time $t + \Delta t/2$ can be obtained from

$$\dot{x}_i^{t+\Delta t/2} = \dot{x}_i^{t-\Delta t/2} + (f_i^t + h_i^t)\Delta t/M + g_i\Delta t$$
$$\dot{\theta}_i^{t+\Delta t/2} = \dot{\theta}_i^{t-\Delta t/2} + m^t\Delta t/I \tag{5}$$

In other words, the accelerations at time t are being integrated over a time interval Δt to obtain the new velocity components. A second integration gives the incremental displacements and thus the updated coordinate positions and rotation of the block at time $t + \Delta t$:

$$x_i^{t+\Delta t} = x_i^t + \dot{x}_i^{t+\Delta t/2}\Delta t$$
$$\theta_i^{t+\Delta t} = \theta_i^t + \dot{\theta}_i^{t+\Delta t/2}\Delta t \tag{6}$$

If the velocity of the block centroid is prescribed as a boundary condition eqns (5) are unnecessary and the updated co-ordinates can be evaluated directly through eqns (6).

The velocity of the pth corner of the block can then be calculated from the centroidal velocity and rate of rotation of the host block

$$\overset{p}{\dot{x}}{}_i^{t+\Delta t/2} = \dot{x}_i^{t+\Delta t/2} - \dot{R}_{ij}^{t+\Delta t/2} \overset{p}{(x_j^t} - x_j^t)$$ (7)

in which $\mathbf{R}$ is the skew symmetric rotation matrix so that

$$\dot{R}_{ij}^{t+\Delta t/2} = \begin{bmatrix} 0 & \dot{\theta}^{t+\Delta t/2} \\ -\dot{\theta}^{t+\Delta t/2} & 0 \end{bmatrix}$$

The position of the corner then becomes

$$\overset{p}{x}{}_i^{t+\Delta t} = \overset{p}{x}{}_i^t + \overset{p}{\dot{x}}{}_i^{t+\Delta t/2} \Delta t$$ (8)

Thus the accelerations, velocities and displacements are calculated in a staggered manner by using a Euler-type integration scheme as described in the previous chapter. Stage one of the scheme is then complete. Stage two involves the calculation of the forces of interaction between blocks.

2.1.2 Stage Two Calculations

For each contact such as that illustrated in Fig. 2, the relative velocity (superposed r) at the point of contact $\overset{p}{x}$ between block A (superposed a) and block B (superposed b) at time $t + \Delta t/2$ is calculated from

$$\overset{r}{\dot{x}}{}_i^{t+\Delta t/2} = \overset{a}{\dot{x}}{}_i^{t+\Delta t/2} - \overset{a}{\dot{R}}{}_{ij}^{t+\Delta t/2} \overset{c}{(x_j^t} - \overset{a}{x_j^t})$$

$$- \overset{b}{\dot{x}}{}_i^{t+\Delta t/2} + \overset{b}{\dot{R}}{}_{ij}^{t+\Delta t/2} \overset{c}{(x_j^t} - \overset{b}{x_j^t})$$ (9)

In the basic model it is assumed that the contact co-ordinates $\overset{c}{x}$ can be represented by the co-ordinates $\overset{c}{x}$ of the corner involved in the corner-to-edge contact. If the angle between the global co-ordinate axes x and the local axes u (aligned with the edge of block B) is β, then the incremental relative normal (local direction 2) and shear (local direction 1) displacement components for the contact are calculated from

$$\Delta u_i^{t+\Delta t} = S_{ij}^t \overset{r}{\dot{x}}{}_i^{t+\Delta t/2} \Delta t$$ (10)

in which

$$S_{ij}^t = \begin{bmatrix} \cos \beta^t & \sin \beta^t \\ -\sin \beta^t & \cos \beta^t \end{bmatrix}$$

Note that the component of relative normal displacement at the contact represents a numerical overlap of the two interacting blocks. If the overlap at the numerical contact is to remain small so that the mechanical response is correctly modelled, the stiffness in the normal direction must obviously be large enough to prevent excessive overlaps.

The incremental displacements are then used to evaluate changes in force at the contact. This requires a contact law which, due to the incremental nature of the calculations, can take a very general form. In the simplest form, the normal and shear forces are linearly related to the displacements through elastic stiffness moduli (units: force per unit length) with a frictional strength providing a limit on the allowable shear force. Hence the increments of contact force, Δc_i, in the local framework are given by

$$\Delta c_i^{t+\Delta t} = K_{ij} \Delta u_j^{t+\Delta t} \tag{11}$$

in which $K_{11} = K_s =$ unit shear stiffness, $K_{22} = K_n =$ unit normal stiffness, $K_{12} = K_{21} = 0$ and the new contact force is given by

$$c_i^{t+\Delta t} = c_i^t + \Delta c_i^{t+\Delta t} \tag{12}$$

subject either to the overriding condition of no tension

$$c_1^{t+\Delta t} = c_2^{t+\Delta t} = 0, \quad \text{if } c_2^{t+\Delta t} > 0 \text{ (tension positive)} \tag{13}$$

or otherwise to the shear strength limit given by

$$c_1^{t+\Delta t} = \pm \tau_{max}, \quad \text{if } |c_1^t + \Delta c_1^{t+\Delta t}| > \tau_{max} \tag{14}$$

in which

$$\tau_{max} = \mu |c_2^{t+\Delta t}|$$

and μ is the coefficient of friction at the contact. The sign of the limiting shear stress (eqn 14) depends on the local direction of shearing.

The contact force c must then be resolved back into the global framework x to give components b_i that can be appropriately added to the centroidal force sums of the surrounding blocks. Hence

$$b_i^{t+\Delta t} = T_{ij}^{t+\Delta t} c_j^{t+\Delta t} \tag{15}$$

in which

$$\mathbf{T} = \mathbf{S}^{\mathrm{T}}$$

Finally, the contribution of the contact forces to the resultant force acting at the centroids of the two blocks must be evaluated. Hence, for the contact between blocks A and B in Fig. 2, the contributions to block A are

$$\Delta \overset{a}{f}_i^{t+\Delta t} = -b_i^{t+\Delta t}$$

$$\Delta \overset{a}{m}^{t+\Delta t} = -b_j^{t+\Delta t}\,(\overset{c}{x}_i^{t+\Delta t} - \overset{a}{x}_i^{t+\Delta t})e_{ij} \tag{16}$$

and to block **B**

$$\Delta \overset{b}{f}_i^{t+\Delta t} = b_i^{t+\Delta t}$$

$$\Delta \overset{b}{m}^{t+\Delta t} = b_j^{t+\Delta t}\,(\overset{c}{x}_i^{t+\Delta t} - \overset{b}{x}_i^{t+\Delta t})e_{ij} \tag{17}$$

in which e is the two-dimensional permutation tensor. These calculations (eqns 9–17) are performed for all the contacts in the problem to obtain the net contribution of the interaction forces to the out-of-balance force f at each block centroid. When all contacts have been processed, stage two of the calculation cycle is finished. To advance further in time, a new time-step is started by re-evaluating the block motions, beginning with eqn (3). The complete calculation loop is then repeated until the required model time is achieved.

2.2 Stability of Finite-Difference Scheme

As with the continuum formulation, the explicit finite-difference scheme used to represent the governing equations of the discontinuum is conditionally stable and a suitable, bounded value of Δt must be calculated. However, there is no method for determining an exact value of Δt for an arbitrary assemblage of blocks. Instead, an estimate must be made based on the known solution of a simplified system.

For a single degree of freedom system, the limiting time-step is given by (see, for example, Last and Harper, 1990)

$$\Delta t < 2\sqrt{(M/K)} \tag{18}$$

where M is the mass and K the spring stiffness. Numerically, the distinct element consists of a lumped mass surrounded by an arbitrary number of springs (contacts). The larger the number of springs, the greater the apparent stiffness and therefore the smaller the time-step given by eqn (18). In practice, a factor f is introduced to allow for the possibility of multiple contacts so that

$$\Delta t < f\sqrt{(M/K)} \tag{19}$$

For an arbitrary array of blocks, the minimum value of Δt is used as the calculation time-step. The value of f is typically 0·1–0·5.

During the course of a simulation the critical time-step may vary, particularly if the apparent stiffness of individual blocks changes. Thus the calculation time-step may be increased or decreased to reflect the changing situation, but care must be taken not to make large changes during a single step because the centering of the finite-difference equations must be maintained.

2.3 Damping

If natural energy dissipation such as inter-block sliding accompanies a static or quasi-static simulation, unwanted vibrations due to initial or transient force imbalance will be absorbed. However, if a predominantly elastic analysis is required, it is necessary to provide some artificial damping when a static solution is expected. This is applied to the block centroids and to the contacts between blocks to give an overall effect analogous to Rayleigh damping (Seed and Idriss, 1970), a type of modal damping used in finite element frequency domain analyses. At the block centroids, velocity-proportional viscous damping is applied to the rigid body motions so that damping force terms $\mathbf{d}$ for the translational degrees of freedom in the momentum equation (eqn (1)) take the form

$$\mathbf{d} = -\alpha \mathbf{M}\dot{\mathbf{x}} \tag{20}$$

and in component form this becomes

$$d_i = (-\alpha M/2)(\dot{x}_i^{t+\Delta t/2} + \dot{x}_i^{t-\Delta t/2}) \tag{21}$$

in which the damping force is assumed to respond to the average velocity during the time-step Δt. Inclusion of these terms into eqns (3) and rearranging gives

$$\dot{x}_i^{t+\Delta t/2}(1 + \alpha\Delta t/2) = (1 - \alpha\Delta t/2)\,\dot{x}_i^{t-\Delta t/2} + (f_i^t + h_i^t)\Delta t/M + g_i\Delta t \tag{22}$$

A similar expression applies for the rotational degree of freedom.

A vibrational energy generated at contacts between blocks can be damped by applying relative-velocity proportional viscous damping at the points of contact. Hence the damping force $\mathbf{s}$ is given by

$$\mathbf{s} = \beta \mathbf{K}\overset{r}{\dot{\mathbf{u}}}$$

and in component form

$$s_i^{t+\Delta t} = \beta K_{ij}\overset{r}{\dot{u}}_j^{t+\Delta t/2} \tag{23}$$

and these forces must be added to the forces of interaction (eqn 11). The damping force $\mathbf{s}$ is omitted if the contact is sliding because frictional dissipation then provides natural damping. Qualitatively the velocity-proportional part tends to act on the lower frequency modes which are usually associated with the movement in unison of several blocks ('sloshing') while the stiffness proportional part damps higher frequency inter-block vibrations ('rattling'). However, the level of damping is frequency-dependent and the values of α and β must be chosen to provide a suitable fraction of critical damping. The frequency of the dominant mode(s) is found either by using an analogy (for example the vibration of an equivalent elastic half-space) or by monitoring a short undamped run so that the important mode(s) can be identified and appropriately damped in subsequent simulations.

2.4 Some Examples of the Use of the Basic DE Method

The basic model (rigid blocks, point contacts) has been used mainly in the area of civil engineering where typical problems have been slope stability and underground excavation in jointed rock. These are near surface problems in which relatively low stresses prevail and consequently deformation of the rock mass is likely to be dominant by block movements resulting from joint slip rather than by deformation of the intact rock.

Many simple examples have been used to demonstrate specific features of the basic model. However, there are very few solutions to boundary value problems which can be used to validate the model. One documented example (Fig. 4, taken from Cundall *et al.*, 1978) is a comparison between the DE model prediction and the rock slope first analysed by Goodman and Bray (1976). Based on limit equilibrium techniques, the analytical solution predicts the toe force T that is required to maintain stability of the slope. For the given slope (Fig. 4(a)), the DE method gives a limiting toe force that is 1% lower than the value calculated from the analytical solution. Close inspection of the DE method result indicated that a slightly different mechanism had been realised. Although in this case the differences between the two predictions are small, the usefulness of the DE method is illustrated. Firstly, the two methods give very similar quantitative results. Secondly, the DE simulation reveals a slightly different mechanism which would probably not have otherwise been perceived and leads to a closer understanding of what possibly occurs in reality. This is often the course of events with DE modelling of this kind; the computed results will lead to a better understanding of the qualitative physical response of a given system and may promote new ideas concerning possible mechanisms of deformation and

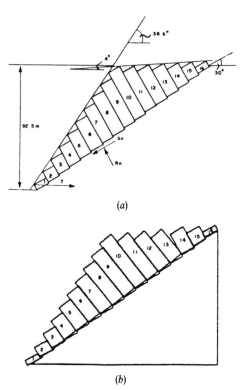

(a)

(b)

FIG. 4. Validation of DE method against the rock slope theory of Goodman and Bray (1976). (a) Slope configuration from Goodman and Bray's paper (1976). (b) Snapshot plot from DEM showing failure mode for a friction coefficient of 0·65.

collapse. In addition, of course, quantitative information on forces and displacements (or velocities or accelerations) is provided.

The full potential of the DE method is realised when applied to problems for which no feasible alternative analytical technique exists. For example, Figs 5 and 6 show the predicted mechanisms of failure in two complex rock slopes (Cundall *et al.*, 1976). Physical model studies would probably be the only plausible alternative for examining these slopes. Indeed, the first slope (Fig. 5) was tested in a base friction model and the mode of collapse for physical test and the numerical simulation was remarkably similar. For the second slope, which is geologically very complicated, even physical model studies would be extremely difficult to conduct and interpret. The numerical simulation (Fig. 6) gives a clear

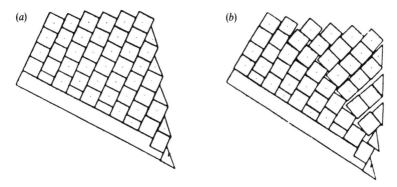

FIG. 5.   Predicted rock slop failure, example 1 (from Cundall *et al.*, 1976.)

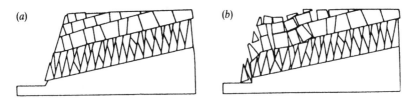

FIG. 6.   Predicted rock slop failure, example 2 (from Cundall *et al.*, 1976.)

picture of the failure mode and furthermore, the mode of collapse was found to be strongly dependent on the inter-block friction coefficient; this would have been very difficult to test using physical models.

These examples illustrate the use of the basic DE method but it is also clear that additional features are needed if a more complete description of a general blocky rock mass is to be realised. Many such features were investigated individually by Cundall *et al.* (1978) and later (Cundall, 1980) several of these were implemented into a single code UDEC (for Universal Distinct Element Code). This code has provided the starting point for most subsequent development and analysis (for example, Cundall and Hart, 1983; Lemos *et al.*, 1985; Lorig *et al.*, 1986; Last and Harper, 1990).

2.5   Key Developments

The motivation to develop UDEC (Cundall, 1980) was provided by the need to combine several previous developments into a single code and to establish a framework into which new features could easily be incorporated. In particular, revised point contact behaviour, block deformability and cracking, and edge-to-edge contacts (joints) have been implemented, together with many features to make the code more flexible and adaptable.

2.6   Contact Behaviour

In the early model (Cundall *et al.*, 1978), an unrealistic response was sometimes observed when interaction occurred close to or at two opposing corners (Fig. 7(a)). Blocks were sometimes observed to become hung-up or locked. This results from the modelling assumption that corners are sharp and of infinite rock strength. In reality a stress concentration would occur that would cause crushing or cracking. Explicit modelling of these effects was considered to be impracticable. However, a more realistic representation can be achieved by treating the corners of blocks as circular arcs (Fig. 7(b)) so that blocks may smoothly slide past one another when two opposing corners interact (Cundall, 1980). The circular arcs are defined by specifying the distance from the true apex to the point of tangency with the adjoining edges. By specifying this distance, rather than a constant radius, sharp acute angled vertices are not severely truncated (Fig. 7(c)). The point of contact between an edge and a corner is then located at the intersection between the edge and the normal taken from the centre of radius of the circular arc to the edge (Fig. 8(a)). The normal force of interaction occurs along the same normal. If two corners are in contact, then the normal force acts along the line joining the two opposing centres of radii, and the point of action occurs where the circular arcs cross the line of action(Fig. 8(b)). These two types of contact are sufficient to cover all possibilities within the blocky system. The important class of planar, edge-to-edge contacts is represented by a domain (which forms the void between adjacent blocks) that contains two such point contacts. This redefinition of contacts requires that when referring to the position of a contact, the true contact coordinates must be used in the appropriate finite-difference equations, instead of those of the corresponding corner (as assumed in the basic model). Note that it is only the contact mechanics which is based on the rounded corner logic; all other properties such as block mass and moment of inertia are based on the complete block.

2.7   Block Behaviour

The rigid block idealisation allows the modelling of low stress situations in which sliding along joints is dominant and block deformation can be neglected. However, at higher stress levels the deformation of the intact

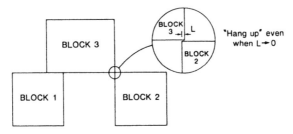

(a) Hang up caused by assumption of sharp, infinitely strong corners.

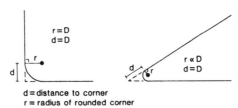

d = distance to corner
r = radius of rounded corner

(b)   Rounding of corners using constant rounding length D.

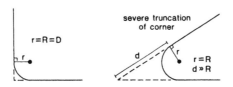

(c)   Rounding of corners using constant radius R demonstrating
unacceptable truncation of acute angled corner.

FIG. 7.   Hang-up of sharp corners and definition of rounded corners. (From
Last and Harper, 1990.)

material is likely to become important. This prompted the development
of 'simply deformable' distinct elements (Cundall et al., 1978). The
concept was to introduce three additional degrees of freedom to each
block to enable a first-order representation of uniform extensional and
shear deformations. To maintain the characteristics of the model, equations governing the dynamics of interacting deformable bodies are
utilised (Last and Harper, 1990). For the simply deformable distinct
elements, the three components of rigid body motion are combined with
three additional modes that correspond to the three components of
uniform strain in two dimensions (Fig. 9). The choice of these strain

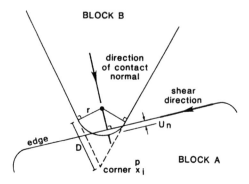

Detail of rounded corner to edge contact (rounding length exaggerated).

(a)

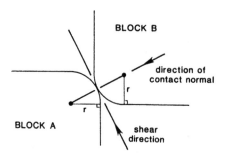

Smooth interaction of corner to corner contact

(b)

FIG. 8. Redefinition of contacts for rounded corners. (From Last and Harper, 1990.)

modes is consistent with the assumption that the velocity gradients are linear within an element and that a uniform internal stress state exists.

In addition to block deformability, blocks may also subdivide (Cundall *et al.*, 1978). If this option is used, the internal stress state of each block is checked against a failure criterion and the block is split if the allowable stress state is exceeded. For example, a simple tensile failure limit requires that, when the minimum principal stress reaches a predefined tensile strength, the block will be split along a line running through the block centroid and parallel to the maximum principal stress.

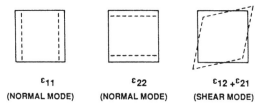

ε_{11} ε_{22} $\varepsilon_{12} + \varepsilon_{21}$

(NORMAL MODE)   (NORMAL MODE)   (SHEAR MODE)

FIG. 9. Modes of deformation for simply deformable distinct elements.

2.8 Joint Behaviour

The physical testing of rock joints (for example, Snow, 1968; Barton, 1976; Yoshinaka and Yamabe, 1986) has shown that in general they exhibit extremely complex behaviour. The approach to quantify the response has been to assign joint unit stiffnesses (for compression and shear) and to limit the shear strength by a failure criterion. However, this usually represents an over-simplification. In reality, the deformation of rock joints is a highly non-linear process: the stiffness and strength are history dependent, dilation or contraction accompanies shearing and the overall behaviour is sensitive to the geometry and roughness, and to any infilling (or cementing) which may have occurred. A joint model that captures all of these features has not yet been realised, but the model described here portrays those features which are thought to be dominant.

The stress–displacement relationship in the normal direction is assumed to be linear and governed by the simple unit stiffness modulus k_n so that

$$\sigma_n = k_n u_n \tag{24}$$

in which σ_n is the normal effective stress and u_n is a measure of the normal displacement (requiring that the unit stiffness has units of stress per unit displacement). Similarly, in shear the response is controlled by a constant shear stiffness k_s but the shear stress τ_s is limited by a combination of cohesive (c) and frictional (ϕ) strength so that

$$\tau_s = k_s u_s^e, \quad |\tau_s| \leqslant c + \sigma_n \tan(\phi) = \tau_{max}$$

or

$$\tau_s = \text{sign}(u_s)\,\tau_{max}, \quad |\tau_s| \geqslant \tau_{max} \tag{25}$$

in which superscript e denotes the elastic component of displacement. This behaviour (Fig. 10) is analogous to Mohr–Coulomb plasticity and

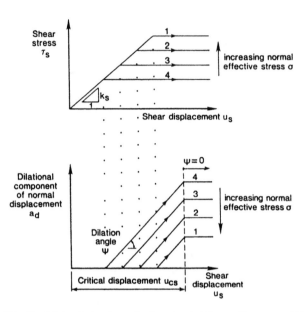

FIG. 10.   Idealised joint behaviour during shear. (From Harper and Last, 1989.)

embodies the main features observed in experiments (for example, Schneider 1974; Yoshinaka and Yambe, 1986). However, in principal, any quantifiable behaviour can be incorporated. For example, in the model described here, the onset of sliding is accompanied by joint dilation a_d which is governed by a specified dilation angle ψ (Fig. 10). The accumulated dilation is limited by either a high normal stress level, or by a large accumulated sliding displacement which exceeds the critical shear displacement u_{cs}. This restriction reflects the observation (for example, Schneider, 1974) that expansion due to shearing is bounded because shearing and/or high normal stresses tend to crush and degrade the joint asperities which would otherwise cause further dilation. Furthermore, the sliding and dilational displacements are irreversible so that a natural hysteresis is included. Hence,

$$\psi = 0, \quad \text{if } |\tau_s| \leqslant \tau_{max}$$

and

$$\psi = 0, \quad \text{if } |\tau_s| = \tau_{max} \quad \text{and} \quad u_s \geqslant u_{cs} \qquad (26)$$

in which u_s is the total shear displacement (magnitude of elastic displacement plus accumulated plastic displacement) and u_{cs} is the critical shear displacement beyond which shear dilation ceases. This is equivalent to assigning a non-associated flow rule to the Mohr–Coulomb strength envelope (Fig. 10) such that the dilation angle depends in a simple way on the level of joint damage.

Numerically a joint is represented as the contact surface formed between two subparallel block edges (Fig. 11). The joint is assumed to

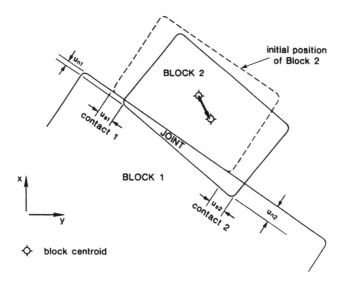

FIG. 11.   Mechanical interaction of two distinct elements along a joint (block overlap is exaggerated). (From Harper and Last, 1989.)

extend between the two point contacts (one at each end) and to be divided in half with each half-length supporting its own contact stress so that stress gradients within a single joint can be modelled. The point contact associated with each half-length is used to calculate the incremental normal and shear displacements (eqns (9) and(10)). These displacements are used to calculate the contact stresses for each half-length. In the normal direction

$$\Delta \sigma_n^{t+\Delta t} = k_n \Delta u_n^{t+\Delta t}$$

$$\sigma_n^{t+\Delta t} = \sigma_n^t + \Delta \sigma_n^{t+\Delta t} \tag{27}$$

subject to the condition of no tension. In shear,

$$\Delta\tau_s^{t+\Delta t} = k_s \Delta u_s^{t+\Delta t}$$

$$\tau_s^{t+\Delta t} = \tau_s^t + \Delta\tau_s^{t+\Delta t}, \quad \text{if } |\tau_s^{t+\Delta t}| < c + \sigma_n^{t+\Delta t} \tan(\phi) = \tau_{max}$$

$$\text{otherwise } \tau_s^{t+\Delta t} = \tau_{max}, \tag{28}$$

If the half-joint is sliding the dilation angle is selected from conditions (26) and used to calculate the dilational component of normal displacements from

$$\Delta u_d^{t+\Delta t} = \Delta u_s^{t+\Delta t} \tan(\psi) \tag{29}$$

and is implemented as an additional component to the normal contact stress

$$\sigma_n^{t+\Delta t} = \sigma_n^t + \Delta\sigma_n^{t+\Delta t} + k_n \Delta u_d^{t+\Delta t}$$

The magnitude of the force of interaction between the two adjacent blocks is obtained by integration of the stresses over the appropriate half-length and it is assumed to act through the associated point contact. Hence, for each half of the joint (total length L),

$$c_1^{t+\Delta t} = \tau_s^{t+\Delta t} L^{t+\Delta t}/2$$

$$c_2^{t+\Delta t} = \sigma_n^{t+\Delta t} L^{t+\Delta t}/2 \tag{30}$$

and the contact forces c can then be resolved and applied to the two blocks (eqns (15), (16) and (17)).

3 AN EXAMPLE APPLICATION: RESPONSE OF JOINTED ROCK TO FLUID INJECTION

Fluid is injected into geological formations primarily for the purpose of resource extraction or storage (usually hydrocarbon, water, or heat), or disposal of waste. These formations are often naturally fractured. If this is the case, both the intrinsic hydraulic conductivity of the fractures and the changes in hydraulic conductivity resulting from the injection/extraction process (due to change of aperture or development of new flow paths) will impact the performance of the scheme. In particular, the coupling between the mechanical and hydraulic behaviour of the fractures may be significant. The UDEC program has been utilised to investigate this effect (Harper and Last, 1989; Last and Harper, 1990; Harper and Last, 1990a, b). Specifically, the intention was to examine the characteristic behaviour resulting from injection into a continuous or a discontinuous (fractured) rock mass in the context of well stimulation, an oilfield process used to enhance productivity. However, the work has lead to conclusions of a more generic nature. Selected results are used here for demonstrating the overall capabilities of the DE method.

The most significant addition to UDEC that was required for this investigation was the incorporation of fluid flow through the joints and into the blocks.

3.1 Fluid Flow Modelling

Fluid flow through the blocky system was modelled as the diffusion of a single, saturated compressible phase through the linked network of porous blocks and conducting inter-block joints and voids which together span the whole solution domain. The individual, fluid-filled elements (blocks, joints and voids) are assumed to form uniformly pressured reservoirs between which the transfer of fluid occurs according to simple one-dimensional flow laws. The dominant path for fluid flow is along the joints but flow into blocks is also permitted.

3.1.1 Flow Through Joints

The flow of viscous fluid in a joint has been the subject of many investigations (for example, Bear, 1972; Witherspoon *et al.*, 1980). The general approach to modelling has been to assume a flow law of the Darcy type (flux proportional to gradient of hydraulic head) and to calculate the hydraulic conductivity k_f by adopting the analogy of planar parallel plates to represent the walls of the joint. Hence, the steady laminar flow through an idealised joint of height w and aperture a, the local flux is given by

$$q = wak_f \, dp/dl \tag{31}$$

in which dp/dl is the hydraulic gradient referred to the direction along the joint, and the hydraulic conductivity is given by

$$k_f = a^2 \rho g/12\eta \tag{32}$$

in which ρ and η are the fluid density and viscosity, respectively.

For ideal uniform conditions over the whole joint length L, eqns (31) and (32) can be combined to give

$$q = w(a^3/12\eta)(P/L) \tag{33}$$

in which (P/L) is the fluid pressure gradient along the joint. This equation forms the basis of what is usually called the cubic law for flow in a fracture. However, the equation has been derived for an 'open' fracture in which the planar surfaces remain parallel and do not contact each other at any point. Several investigators have introduced an empirical factor f in an attempt to allow for departures from the idealised conditions:

$$q = (w/f)(a^3/12\eta)(P/L) \tag{34}$$

Witherspoon et al. (1980) investigated this relationship for a wide range of conditions and demonstrated clearly the general applicability of the cubic law. The factor f was found to vary between 1·04 and 1·65 showing that, as might be anticipated, the flow rate through real fractures is always somewhat less than that for the idealised situation.

In UDEC the flow law has been implemented in the form

$$q = Ca^3(P/L), \quad C = w/(12\eta f) \tag{35}$$

in which C is the fluid flow joint properly that is assumed to remain constant (the value of w is unity). Evidently the rate of fluid flow is critically dependent on the third power of the aperture a. Measurements have shown that the conductivity of fractures is a function of the confining pressure (for example, Iwai, 1976; Brace, 1978) and this has been attributed to the change of conducting aperture caused by a change in the normal effective stress. Various relationships have been suggested (e.g. Barton et al., 1985). For the examples reported here a simple, but nevertheless realistic relationship has been implemented (Fig. 12) giving a conducting aperture

$$a = a_{res} + r\sigma_n k_a + a_d \tag{36}$$

for which

$$r = \begin{cases} 1, & \text{for } 0 < \sigma_n < \sigma_c \\ 0, & \text{otherwise} \end{cases}$$

This shows that the conducting aperture depends linearly on the confining stress (though the compliance k_a) until the closure stress σ_c is reached. For higher confining stress, the conducting aperature remains constant at the residual value, a_{res}. Hence, in keeping with experimental observation, some fluid conductivity is always maintained.

Any dilational component a_d which arises is assumed to be irrecoverable and modifies the basic relationship as illustrated by the dashed line in Fig. 12. If a joint loses all compressive effective stress, lift-off occurs

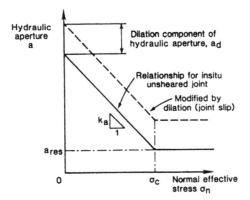

FIG. 12.   Joint hydraulic aperture model: effect of dilation on effective stress–
hydraulic aperture relationship. (From Harper and Last, 1990.)

and the aperture is then controlled by movement and deformation of the
adjacent blocks. If the same joint is remade, the history is remembered
and any previously accumulated dilation is used in the aperture calculation (36), thereby modelling mismatch of the block faces.

The pressure change in the joint is calculated by applying the continuity equation in the form

$$\dot{p}=(Q/V+\dot{V})C_\mathrm{s} \qquad (37)$$

in which Q is the net flow rate into the joint, V is the joint volume, $\dot{V}$ the
volumetric strain rate of the joint and C_s is the fluid bulk stiffness. Note
that an idealised Newtonian fluid is incompressible. However, if a
realistic value of the fluid bulk stiffness is used, the fluid is only slightly
compressible and the departure from idealised Newtonian behaviour is
negligible.

3.1.2   Flow into Blocks

During injection of fluid into a jointed rock mass, the loss of fluid into
the intact material may be significant. In oilfield hydraulic fracturing this
is referred to as 'leak-off'. The loss of fluid will affect the pressure
distribution in the joints and will lead to changes of the pore pressure in
the intact rock. Hence the effective stresses in the joints and blocks will
be influenced by fluid loss and therefore the mechanical response may be
altered. In the extreme case, the reduction in effective stress in the intact
material might promote cracking and the formation of new fractures. The

influence of fluid loss depends to a large extent on the relative time scales associated with joint flow and flow into the blocks. A simple scheme to represent fluid loss and its effects was implemented into the current model.

Fluid loss is assumed to be governed by a flow law of the Darcy type so that for a single block at a given instant in time the rate of fluid flow into a block can be expressed by a line integral of the form

$$Q = \int_0^s c(\partial p/\partial n) \, ds \tag{38}$$

in which c is a fluid loss coefficient, $(\partial p/\partial n)$ is the fluid pressure gradient normal to the block edge, and the integral is taken over s, the block perimeter. For a single segment, defined as the distance around the block between two successive contacts, the rate of flow is

$$q_1 = \int_0^{s_1} c_1 (\partial u/\partial n)_1 \, ds_1 \tag{39}$$

The instantaneous net flow rate into the block Q, can then be calculated by summation of all similar contributions q_1 around the block. Note that the individual flow components q_1 also contribute to the net flow into the joint from which they originate.

Once the net flow rates have been determined, the pressure change in each block can be calculated from the continuity equation in the form (compare eqn (37))

$$\dot{p} = (Q/V + \dot{V}) C_s/n \tag{40}$$

in which V is the block volume, $\dot{V}$ is the block volumetric strain and n the block porosity.

3.1.3   Implementation of Fluid Flow Model

Numerically the connected system of joints is treated as a flow network in which uniform fluid pressures p are defined at the centre of each joint and at the centre of each block. These pressures are used to evaluate the pressure gradients which drive the fluid flow either along the joint or into the block. For the joints, the conducting aperture and pressure gradient are calculated for each half-length (consistent with the area over which the local effective stress is applied) so that, for a typical half joint, the aperture is calculated from (eqn (36))

$$a^{t+\Delta t} = a_{res} + r k_a \sigma_n^t + a_d^t \tag{41}$$

in which a is the accumulated dilation, and the flow rate (eqn (35)) becomes

$$q^{t+\Delta t} = C(a^{t+\Delta t})^3(2P^t/L^{t+\Delta t}) \tag{42}$$

in which P is the fluid pressure drop over the half-length.

The flow rate into a block through each block face (eqn (39)) is calculated from

$$q_1^{t+\Delta t} = C_1 s_1^t p_1^t/n_1^t \tag{43}$$

in which (P/n) is the pressure gradient (assumed linear) for segment 1 along the line drawn perpendicularly from the block face to the block centroid.

Hence, once all the flow rate calculations have been performed throughout the block assemblage, the net flow rates into each joint and block can be evaluated by simple summation of the relevant contributions.

The pressure changes in the blocks and in the joints are then evaluated by applying eqns (37) and (40). For example, for each joint,

$$p^{t+\Delta t} - p^t = (Q^{t+\Delta t}/V^{t+\Delta t} + \dot{a}^{t+\Delta t})C_s\Delta t \tag{44}$$

in which $\dot{a}$ is the rate of closure of the joint.

In preparation for the start of the next calculation step the effects of the fluid flow on the equilibrium of each block must be evaluated and applied through the force term f_i (and m) in eqns (4). In addition to the fluid pressure p, which acts normal to the block face, a shear force will be generated in the tangential direction during fluid flow. For laminar flow between parallel plates, the shear stress acting on each wall is

$$\tau = \frac{a}{2}\frac{\partial p}{\partial l}$$

in which a is the aperture. Hence for each half of a joint the additional forces arising from fluid flow are evaluated from

$$c_1^{t+\Delta t} = p^{t+\Delta t}a^{t+\Delta t}/2$$
$$c_2^{t+\Delta t} = p^{t+\Delta t}L^{t+\Delta t}/2 \tag{45}$$

These forces are then resolved into the global framework and applied to the two juxtaposed blocks in the usual way (eqns (15), (16) and (17)). While the shear stress resulting from the flow of water (Newtonian viscosity) may be negligibly small, the shear stress resulting from a highly viscous, non-Newtonian fluid (as used in some fracturing treatments) may be significant.

3.2 The Model Reservoirs

3.2.1 Geometry

Continuous and discontinuous joint systems were realised by using square blocks $(10 \times 10\,\text{m})$ in one of two different arrangements as illustrated in Figs 13 and 14. With the regular array of blocks in Fig. 13

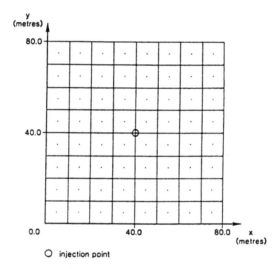

FIG. 13.   Continuously jointed reservoir. (From Harper and Last, 1990.)

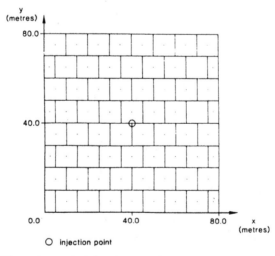

FIG. 14.   Discontinuously jointed reservoir. (From Last and Harper, 1990.)

continuous joints are formed. If the blocks are offset as in Fig. 14 the joints in the x direction remain continuous while those in the y direction become discontinuous.

3.2.2 In-Situ Stress State

The magnitudes of the principal stresses were chosen to reflect conditions at approximately 3 km (10 000 ft) depth. Both sets of joints are assumed to be vertical, implying that the geometries shown in Figs. 13 and 14 are plan views. At the depth concerned, the overburden stress is approximately 70 MPa (10 000 psi assuming a gradient of 1 psi per foot). If the ratio between the overburden stress and the minimum horizontal stress is assumed to be 0·5, the minimum horizontal total stress is 35 MPa (4500 psi). Similarly, if the ratio between the overburden stress and the maximum horizontal stress is assumed to be 0·65, the maximum horizontal total stress is 45·5 MPa (6500 psi). Hereafter the two stresses in the horizontal plane will be referred to as the maximum principal stress (45·5 MPa) and the minimum principal stress (35 MPa). The reservoir is assumed to be slightly overpressured at a fluid pressure of 31·5 MPa ($= 1·5$ MPa overpressure). This gives a maximum effective principal stress of 14 MPa and a minimum effective principal stress of 3·5 MPa, with a ratio of 4 between the maximum and minimum effective principal stresses. Two alignments of the principal stresses were considered. In the first, the principal stress axes are aligned with the joint directions (Fig. 15(a)) and in the second the axes are rotated through 30° from the joints running in the y direction (Fig. 15(b)). These two stress states will be referred to as the aligned and the rotated stresses, respectively.

These stresses and the initial reservoir pressure are maintained at the boundaries throughout the simulations.

3.2.3 Material Parameters

A summary of the material parameters is given in Tables 1, 2 and 3. The rock parameters were selected to represent a tight sandstone and the fluid parameters were selected to represent water containing some gas. By far the greatest uncertainty lies in the value of the joint properties. General guidance for the selection of parameters was taken from a number of sources (for example, Barton et al., 1985; Pine and Cundall, 1985; Yoshinaka and Yamabe, 1986; Rosso, 1976). There has also been a lack of modelling of the type reported here and consequently there is little previous experience on which to call. The only other work which is comparable has been performed to model the stimulation of a geothermal reservoir in jointed granite (Pine and Cundall, 1985).

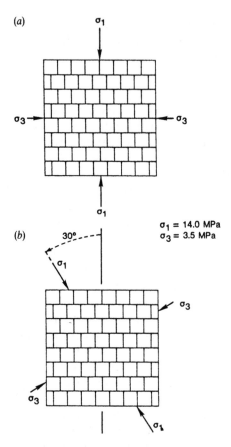

FIG. 15. Reservoir stress states. (From Last and Harper, 1990.)

TABLE 1
BLOCK PROPERTIES

Property	Symbol	Value	Units
Density	ρ	2500·0	kg/m^3
Porosity	n	0·1	–
Shear modulus	G	14·0 (2×10^6)	GPa (psi)
Bulk modulus	K	20·0 ($2·9 \times 10^6$)	GPa (psi)
Permeability	k	$1·0 \times 10^{-15}$ (1·0)	m^2 (md)

TABLE 2
JOINT PROPERTIES

Property	Symbol	Value	Units
Tensile strength		0	MPa
Cohesion	c	3·5	MPa
Friction angle	ϕ	37·5	degrees
Shear dilation angle	ψ	15	degrees
Critical shear displacement	u_{cs}	40×10^{-3}	m
Joint normal stiffness	k_n	2·0	GPa/m
Joint shear stiffness	k_s	0·2	GPa/m
Residual aperture	a_{res}	$0·1 \times 10^{-3}$	m
Aperture-stress compliance	k_a	$0·5 \times 10^{-3}$	m/MPa
Closure stress	σ_c	3·8	MPa

TABLE 3
FLUID PROPERTIES

Property	Symbol	Value	Units
Bulk stiffness	C_s	3·0	GPa
Density	ρ	1000·0	kg/m^3
Viscosity	η	$0·35 \times 10^{-3}$ (0·35)	Ns/m^2 (cp)
Injection rate (per metre thickness)		0·004 (4·0)	m^3/s (l/s)

3.3 Test Procedure

The combination of two geometries and two stress states gives a total of four basic models. Initially each model was consolidated under the prescribed stresses to obtain the four initial states. In all models the initial fluid pressure (31·5 MPa) was uniform.

For demonstration purposes, the case of principal stresses aligned with continuous joints (Model 1) will be described. In this case the bounding principal stresses are transferred uniformly through the block assemblage so that the stresses in the blocks are identical to the prescribed boundary stresses, and the normal stresses in the joints are either 3·5 MPa (for joints parallel to the y axis) or 14·0 MPa (for joints parallel to the x axis). This means that there is an anisotropy of hydraulic conductivity under the initial conditions of effective stress because the hydraulic apertures are different in the two sets of joints (Fig. 16). This result is true only if at least one of the two joint sets is above the minimum aperture (the joint

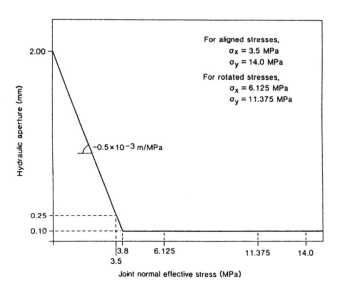

FIG. 16. Pre-injection hydraulic apertures for continously jointed reservoirs. (From Last and Harper, 1990.)

properties were considered to be the same in both joint sets). This in turn requires a difference of normal effective stress between the two joint sets.

The model reservoir was then tested for its initial productivity by lowering the well pressure to 28·0 MPa (representing a drawdown of 3·5 MPa or 500 psi) until steady state conditions were achieved. The flow rates into the model well and into the boundary of the model reservoir are shown in Fig. 17, demonstrating that a mass balance is attained (neglecting the slight compressibility of the fluid). Furthermore, the increases in effective stress in the reservoir caused by the drawdown has closed all of the joints in the vicinity of the wellbore and the initial anisotropy is removed. The flow pattern has become symmetric about the well. The closure of one of the initially open joints is shown in a history plot (Fig. 18). These results show that when dealing with fractures whose conductivity is stress level dependent, the flow pattern (and therefore the swept volume) and the rate of flow into the well depend on the magnitude of the drawdown. Indeed, in some circumstances a reduced drawdown may optimise the productivity.

The original model (prior to drawdown) was then stimulated by

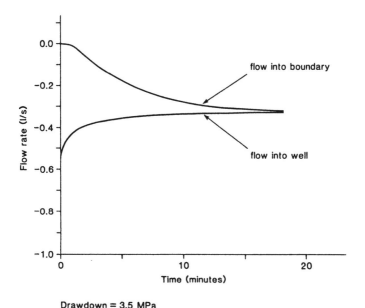

Drawdown = 3.5 MPa

FIG. 17. Initial productivity for continuously jointed reservoir, aligned stresses
(Model 1). (From Last and Harper, 1990.)

injection of fluid at a constant rate of 4 litres per second per metre
thickness of the reservoir. The major effect is to open the continuous joint
which intercepts the well and lies parallel to the major principal stress.
Figure 19 illustrates the fluid pressures and hydraulic apertures produced.
Clearly in this case the boundary restricts the growth of the major
fracture. Notice that there are pressure changes in the orthogonal and
parallel joints, indicating that fluid is being lost from the main fracture.

Following the period of injection, the well was shut-in and the reservoir
was allowed to equilibrate. After some time, the original reservoir
conditions were re-established, indicating that there had been no irreversible changes during the stimulation period. Indeed, a subsequent period
of drawdown (by 3·5 MPa) gave exactly the same productivity as that
obtained prior to stimulation. This is consistent with what would be
expected from this model reservoir.

This basic test sequence (initial productivity, injection, shut-in and final
productivity) was performed on the four reservoir models.

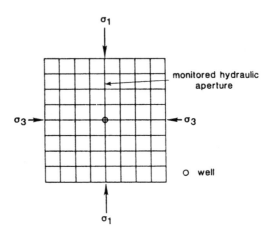

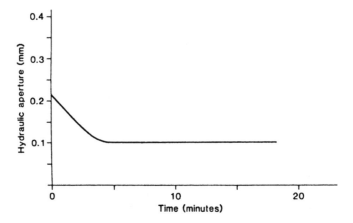

FIG. 18.   Closure of joint during drawdown, Model 1. (From Last and Harper, 1990.)

3.4   General Observations

The following general observations have been made. Further supporting evidence and other effects have been discussed by Harper and Last (1989, 1990a, b). Selected results are presented in Figs 19–23.

The majority of the fluid injected into fracture networks flows in a limited number of distinct pathways (Figs 21(c) and (d)). It may be inferred that, in general, the higher the rate of injection, the greater the number of dominant pathways accepting the majority of the injected·

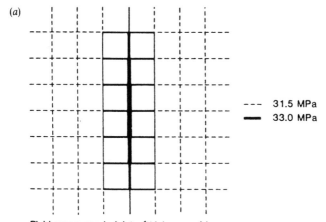

Fluid pressures in joints (thickness of bar
proportional to pressure)

--- 31.5 MPa
— 33.0 MPa

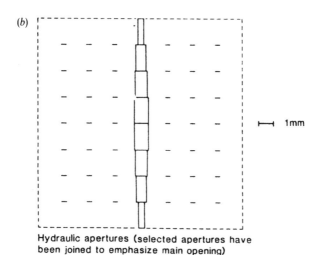

⊢—⊣ 1mm

Hydraulic apertures (selected apertures have
been joined to emphasize main opening)

FIG. 19. Selected results, Model 1. (From Last and Harper, 1990.)

fluid. The hydraulic conductivity of these channels is typically spatially
variable. Additionally, it is probable that the lower the stress difference,
the more diffuse the flow.

Modelling of a discontinuum which has incorporated a dynamic
coupling of stress and fluid flow has demonstrated the very high rates of

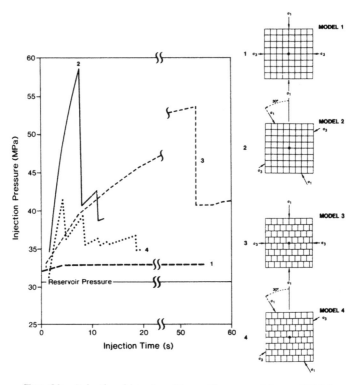

FIG. 20.   Injection histories. (From Harper and Last, 1990*b*.)

pressure transmission which occur in fractured media. Fractures have
been observed to dilate abruptly, the dilation in these UDEC modelling
sequences being essentially instantaneous. During fluid injection, fractures dilate incrementally, joint-by-joint, progressively away from the
point of injection, opening either by shear-induced dilation, dilation
alone, or a combination of the two modes.

There is a clear coupling between fluid pressure and fracture dilation,
and the abrupt fracture dilation may be reflected in abrupt pressure
changes at the point of injection (Fig. 20). Moreover, the dilational
behaviour of fractures is coupled one to another by means of fluid
pressure changes in the fracture network. For example, the abrupt
opening of a fracture may be accompanied by a corresponding abrupt
closure of another fracture in the network. Changes or even reversals of
the direction of fluid flow may occur as a result of such changes (Figs

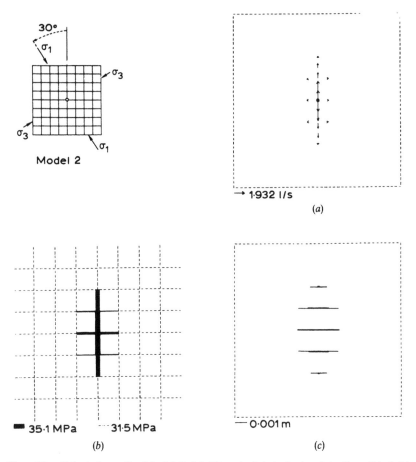

FIG. 21. Selected results, Model 2. (a). Flow in joints during injection; (b). fluid pressures during injection; (c). permanent post-injection shear dilation. (From Harper and Last, 1989.)

22(b) and (c)). Steady state solutions for high rate injection or withdrawal applied to a blocky medium are unlikely to be accurate because of the history dependent nature of fracture conductivity.

The occurrence of a pressure peak shortly after the start of injection appears to be a common occurrence (Fig. 20). This pressure behaviour is qualitatively similar to that observed at 'breakdown' during hydraulic fracturing of a continuum. Furthermore, the magnitude of the pressure peaks is very dependent on the initial geometry and stress state.

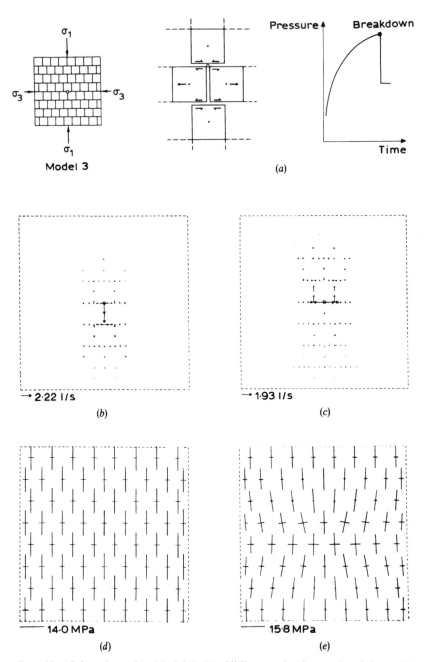

FIG. 22. Selected results, Model 3. (a). Sliding mechanism at breakdown; (b). flow in joints prior to breakdown; (c). flow in joints after breakdown; (d). pre-injection principal stresses; (e). principal stresses after breakdown. (From Harper and Last 1990a.)

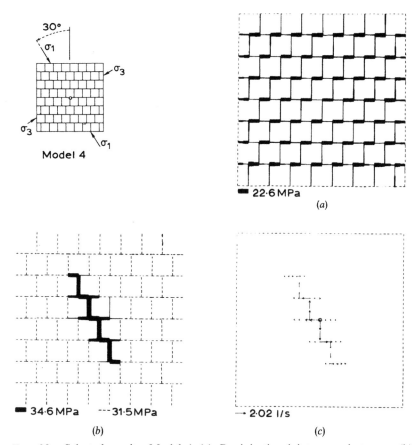

FIG. 23.   Selected results, Model 4. (a). Pre-injection joint normal stresses; (b) fluid pressures during injection; (c). flow in joints during injection. (From Harper and Last, 1989.)

Crosscutting or intersecting fractures provide a discontinuity of stiffness in what would be termed the fracture wall in a continuum. Step changes of fracture aperture at these intersections are common. This stiffness contrast and the non-linear effective stress–aperture relationships for fractures appear to have dominated the model behaviour.

Inelastic deformation manifest by sliding on joints appears to be a common feature of the response of blocky media to fluid injection (Fig. 21(c)). Moreover, the stress distribution around the point of injection may be substantially changed (Figs 22(d) and (e)). These stress changes may remain after the end of injection, as a manifestation of the inelastic behaviour.

If the matrix rock material behaves as an elastic material, the modelling demonstrated that gradients of normal stress along the fracture can occur (Fig. 23(a)). These arise because of the stiffness contrast at fracture intersections. Such gradients develop when fractures intersect but do not crosscut. The result is a range of normal stresses which can extend to substantially greater magnitudes than the range of the magnitudes of applied boundary stresses.

Anisotropy of hydraulic conductivity in fractured rock subject to a non-hydrostatic stress state must be assumed to be commonplace. Moreover, a uniform rise or fall of pore pressure is likely to cause a rotation of the overall directions of maximum and minimum hydraulic conductivity in a discontinuous system. This rotation reflects the non-hydrostatic nature of the boundary stresses typical of many geological environments, in conjunction with non-linear fracture aperture–stress relationships. Where fractures are irregularly distributed, the evolving distribution of hydraulic conductivity induced by a uniformly changing pore pressure is likely to be complex (in both time and space).

The existence of a shear stress in the plane of a fracture prior to fluid injection (including those fractures remote from the wellbore) is not a necessary condition for sliding on that fracture during fluid injection (compare Models 2 and 3, Figs 21 and 22). This is because the process of injection may induce shear stresses on fractures (Fig. 22(a)). Indeed, even reversals of the sense of shear stress may occur in response to injection.

Opening of fractures either side of a point of injection may be asymmetric, with extension of the path of opened fractures alternating from one side of the injection point to another. This would be the equivalent in a continuum of fracture extension occurring by increments alternating regularly between the two fracture wings.

Fluid stiffness influences the response of a discontinuum to constant rate injection. Higher stiffness fluids have been observed to lead to a more dynamic mechanical response (greater rates of change), higher initial peak pressures and larger fracture apertures. However, the result of these changes was that the post-peak pressure at and in the vicinity of the point of injection was lower for the higher stiffness fluid.

It is perhaps not surprising that lowered injection rates gave rise to lower initial peak pressures, lower rates of pressure and fracture aperture change and lower values of aperture.

4 CONCLUSIONS

The overall aim of these two companion chapters was to introduce the explicit finite-difference technique as applied to problems in geomechanics and to demonstrate its use.

The common basis to the continuum and discontinuum models has been described, followed by specific details of the two formulations. Some new features have been included. In particular, extensions that were necessary for the selected examples have been described. This overall account should enable the reader to gain an appreciation of the differences between the explicit finite-difference approach and other methods.

Although both models (continua and discontinua) can be used to solve more conventional problems in engineering (for example, bearing capacity of foundations, or stability of slopes or excavations in rocks), the examples used for demonstration purposes were selected to illustrate the flexibility and applicability of the technique to a wider class of problems in geomechanics. It is perhaps in this context that the technique is most useful, in both a research and a practical engineering environment.

In the continuum example (previous chapter), the modelling of a large scale geological setting required the addition of sedimentation logic and viscous material behaviour. Furthermore, a technique (viscosity scaling) for simulating large timescales was devised. Thus, in a computational scheme which is often viewed as being primarily suitable to dynamic problems only, an effective method for modelling slowly evolving, time-dependent processes has been realised. The simulations indicated that a simplified analytical solution to the problem has restricted applicability. In particular, if a more representative (frictional) overburden is added to the viscous substratum, significant horizontal movement of the interface between the two layers is observed due to translation of the overburden if yielding occurs. Furthermore, large extensional and compressional features, corresponding to a basin and an anticline associated with overthrusting are produced in the overburden. A simple quantitative characterisation of settings leading to this overall behaviour was deduced from the improved understanding gained from the simulations.

In the discontinuum example, the addition of fluid flow and stress-dependent hydraulic conductivity in the joints was required. The primary objective was to investigate the characteristic response of a jointed rock mass subject to fluid injection. The results indicate a general sensitivity of the response to the geometry and initial conditions of the model reservoirs. Even with only two joint patterns and two different regional stress states, a number of mechanisms and corresponding flow regimes have been revealed. Typically, fluctuating injection pressures, anisotropy of hydraulic conductivities (both initial and induced), and irreversible changes of stress and joint dilation are predicted. In addition, the magnitude of the injection pressures and the net effect of the injection period on the final distribution of fracture conductivity are quite different in each case. This response is qualitatively similar to some field observations (for example, Shuck and Komar, 1979). Thus a very idealised model has prompted a better understanding of mechanisms that may explain the observed characteristics of fluid injection into jointed rock.

Both examples fall into the data-limited category as defined by Starfield and Cundall (1988). It would be only too easy to dismiss the modelling attempts on the basis of the lack of information available; there is uncertainty concerning geometry and material behaviour in both cases. However, the implementation of an idealised numerical model which can then be used as an experiment to probe the behaviour of the system has, in both cases, led to a better understanding and a characterisation of the system response. The numerical models described here, based on the explicit finite-difference technique, have proved to be extremely valuable tools in this context.

REFERENCES

BARTON, N.R. (1976). The shear strength of rock and rock joints, *Int. J. Rock Mech. Sci. Geomech. Abstr.*, **13**, 255–79.

BARTON, N.R., BANDIS, S. and BAKHTAR, K. (1985). Strength, deformation and conductivity coupling of rock joints, *Int. J. Rock Mech. Sci. Geomech. Abstr.*, **22**, 121–40.

BEAR, J. (1972). *Dynamics of Fluids in Porous Media*, Elsevier, New York, 387 pp.

BRACE, W.F. (1978). Note on permeability changes in geologic material due to stress, *PAGEOPH*, **116**, 627–33.

CUNDALL, P.A. (1971). A computer model for simulating progressive, large scale movements in blocky rock systems, *Proc. Symp. Int. Soc. Rock Mech.*, Nancy, Paper II-8.

CUNDALL, P.A. (1980). UDEC–A generalised distinct element program for modelling jointed rock, US Army European Research Office and Defense Nuclear Agency, Contract report DAJA 37-39-C-0548.

CUNDALL, P.A. (1988). Computer simulations of dense sphere assemblies, *Micromechanics of Granular Materials*, Elsevier, Amsterdam, pp. 113–23.

CUNDALL, P.A. and HART, R.D. (1983). Development of generalised 2nd and 3rd distinct element programs for modelling jointed rock, US Army Eng., WES Tech. Report.

CUNDALL, P.A. and STRACK, O.D.L. (1979). A discrete numerical model for granular assemblies, *Geotechnique*, **29**, 47–65.

CUNDALL, P.A., VOEGELE, M. and FAIRHURST, C. (1976). Computerised design of rock slopes using interactive graphics, Technical Note, University of Minnesota.

CUNDALL, P.A., MARTI, J., BERESFORD, P., LAST, N.C. and ASGIAN, M. (1978). Computer modelling of jointed rock masses, US Army Eng., WES Tech. Report N-78-4.

GOODMAN, R.E. and BRAY, J.W. (1976). Toppling of rock slopes. In *Rock Engineering for Foundations and Slopes, Proc. ASCE Speciality Conference*, Boulder, Colorado, Vol. 2, p. 141.

HARPER, T.R. and LAST, N.C. (1989). Interpretation by numerical modelling of changes of fracture system hydraulic conductivity during fluid injection, *Geotechnique*, **39**(1), 1–11.

HARPER, T.R. and LAST, N.C. (1990*a*). Response of fractured rock subject to fluid injection. Part II. Characteristic behaviour, *Tectonophysics*, **172**, 33–51.

HARPER, T.R. and LAST, N.C. (1990*b*). Response of fractured rock subject to fluid injection. Part III. Practical application, *Tectonophysics* **172**, 53–65.

IWAI, K. (1976). Fundamental studies of fluid flow through a single fracture, PhD thesis, University of California, Berkeley.

DE JOSSELIN DE JONG, G. and VERRUIJT, A. (1969). Etude photo-elastique d'un epilement de disques, *Cah. Grpe. Fr. Etud. Rheol.*, **2**, 73–86.

LAST, N.C. and HARPER, T.R. (1990). Response of fractured rock subject to fluid injection. Part I. Development of a numerical model, *Tectonophysics*, **172**, 1–31.

LEMOS, J.V., HART, R.D. and CUNDALL, P.A. (1985). A generalised distinct element program for modelling jointed rock, *Proc. Int. Symp. on Fundamentals of Rock Joints*, Bjorkliden, Sweden.

LORIG, L.J., BRADY, B.H.G. and CUNDALL, P.A. (1986). Hybrid distinct element-boundary analysis of jointed rock, *Int. J. Rock Mech. Min. Sci. Geomech. Abstr.*, **23**, 303–12.

PINE, R.J. and CUNDALL, P.A. (1985). Applications of the fluid-rock interaction program (FRIP). to the modelling of hot dry rock Geothermal Energy Systems, *Proc. Int. Symp. on Fundamentals of Rock Joints*, Bjorkliden, Sweden.

ROSSO, R.S. (1976). A comparison of joint stiffness measurements in direct shear, triaxial compression and *in situ*, *Int. J. Rock Mech. Min. Sci. Geomech. Abstr.*, **13**, 167–72.

SCHNEIDER, M.J. (1974). Rock friction, a laboratory investigation, *Proc. 3rd Cong. Int. Soc. Rock Mech.*, Denver, Vol. 2A, p. 311.

SHUCK, L.Z. and KOMAR, A. (1979). The dynamic pressure response of a petroleum reservoir, SPE 8349.

SEED, H.B. and IDRISS, J.M. (1970). Soil moduli and damping factors for dynamic response analysis, Report No. EERC 70-10, Earthquake Engineering Research Centre, Univ. of California, Berkeley.

SNOW, D.T. (1968). Fracture deformation and changes of permeability and storage upon changes of fluid pressure, *Quarterly Report, Colorado School of Mines*, **63**, 201.

STARFIELD, A.M. and CUNDALL, P.A. (1988). Towards a methodology for rock
    mechanics modelling, *Int. J. Rock Mech. Min. Sci. Geomech. Abstr.*, **25**,
    99–106.
WITHERSPOON, P.A., WANG, J.S.Y., IWAI, K. and GAILE, J.E. (1980). Validity of
    cubic law for fluid flow in a deformable rock fracture, *Water Res.*, **16**,
    1016–24.
YOSHINAKA, R. and YAMABE, T. (1986). Joint stiffness and the deformation
    behaviour of discontinuous rock, *Int. J. Rock Mech. Min. Sci. Geomech.
    Abstr.*, **23**, 19–28.

INDEX

Milton Keynes UK
Ingram Content Group UK Ltd.
UKHW021830071024
449327UK00021B/1474